The Clinical Research Process in the Pharmaceutical Industry

DRUGS AND THE PHARMACEUTICAL SCIENCES

A Series of Textbooks and Monographs

Edited by

James Swarbrick
School of Pharmacy
University of North Carolina
Chapel Hill, North Carolina

Volume 1. PHARMACOKINETICS, *Milo Gibaldi and Donald Perrier* (out of print)

Volume 2. GOOD MANUFACTURING PRACTICES FOR PHARMACEUTICALS: A PLAN FOR TOTAL QUALITY CONTROL, *Sidney H. Willig, Murray M. Tuckerman, and William S. Hitchings IV* (out of print)

Volume 3. MICROENCAPSULATION, *edited by J. R. Nixon*

Volume 4. DRUG METABOLISM: CHEMICAL AND BIOCHEMICAL ASPECTS, *Bernard Testa and Peter Jenner*

Volume 5. NEW DRUGS: DISCOVERY AND DEVELOPMENT, *edited by Alan A. Rubin*

Volume 6. SUSTAINED AND CONTROLLED RELEASE DRUG DELIVERY SYSTEMS, *edited by Joseph R. Robinson*

Volume 7. MODERN PHARMACEUTICS, *edited by Gilbert S. Banker and Christopher T. Rhodes*

Volume 8. PRESCRIPTION DRUGS IN SHORT SUPPLY: CASE HISTORIES, *Michael A. Schwartz*

Volume 9. ACTIVATED CHARCOAL: ANTIDOTAL AND OTHER MEDICAL USES, *David O. Cooney*

Volume 10. CONCEPTS IN DRUG METABOLISM (in two parts), *edited by Peter Jenner and Bernard Testa*

Volume 11. PHARMACEUTICAL ANALYSIS: MODERN METHODS (in two parts), *edited by James W. Munson*

Volume 12. TECHNIQUES OF SOLUBILIZATION OF DRUGS, *edited by Samuel H. Yalkowsky*

Volume 13. ORPHAN DRUGS, *edited by Fred E. Karch*

Volume 14. NOVEL DRUG DELIVERY SYSTEMS: FUNDAMENTALS, DEVELOPMENTAL CONCEPTS, BIOMEDICAL ASSESSMENTS, *edited by Yie W. Chien*

Volume 15. PHARMACOKINETICS, Second Edition, Revised and Expanded, *Milo Gibaldi and Donald Perrier*

Volume 16. GOOD MANUFACTURING PRACTICES FOR PHARMACEUTICALS: A PLAN FOR TOTAL QUALITY CONTROL, Second Edition, Revised and Expanded, *Sidney H. Willig, Murray M. Tuckerman, and William S. Hitchings IV*

Volume 17. FORMULATION OF VETERINARY DOSAGE FORMS, *edited by Jack Blodinger*

Volume 18. DERMATOLOGICAL FORMULATIONS: PERCUTANEOUS ABSORPTION, *Brian W. Barry*

Volume 19. THE CLINICAL RESEARCH PROCESS IN THE PHARMACEUTICAL INDUSTRY, *edited by Gary M. Matoren*

Other Volumes in Preparation

The Clinical Research Process in the Pharmaceutical Industry

edited by

Gary M. Matoren
Editor-In-Chief
Clinical Research Practices
and Drug Regulatory Affairs
Middletown, New York

MARCEL DEKKER, INC. New York and Basel

Library of Congress Cataloging in Publication Data
Main entry under title:

The Clinical research process in the pharmaceutical industry.

(Drugs and the pharmaceutical sciences ; v.19)
Includes index.
1. Drugs--Testing. 2. Medicine, Clinical--Research.
I. Matoren, Gary M., [date]. II. Series.
[DNLM: 1. Pharmacology, Clinical. 2. Clinical trials--Standards. 3. Clinical trials--Methods. W1 DR893B v.19 / QV 771 C6408]
RS189.C579 1984 615.1'072 84-18914
ISBN 0-8247-1914-X

MARCEL DEKKER, INC.
270 Madison Avenue, New York, New York 10016

Current printing (last digit):
10 9 8 7 6 5 4 3 2

PRINTED IN THE UNITED STATES OF AMERICA

This book is dedicated to the memory of

Steven J. Eisner
1904–1980

whose interest in the clinical research process provided the inspiration and encouragement for me to undertake the development of this book

and to

my loving wife Susan and darling children Bonnie Lisa and Debbie Lynn

for their patience, support, and understanding while this book was being developed

FOREWORD

Physicians and scientists participating in drug development are in a unique situation because they have the potential for improving the health of millions of individuals worldwide. In contrast, physicians in private practice or in an academic milieu can be directly responsible for helping limited numbers of patients in their immediate geographic locations. It is profoundly satisfying to be involved in developing new, clinically important therapeutic agents. Furthermore, it is essential for industrial, academic, and government researchers to share the responsibility for drug development in order to achieve success.

Most drugs introduced over the past 25 years have been developed primarily by the pharmaceutical industry in cooperation with academia and government. Twenty years ago the cost in dollars and time to develop a new chemical entity was relatively low, compared to recent estimates of approximately $70 million and 10 years to develop a drug from synthesis to final approval of a New Drug Application. Clinical research represents a significant portion of these expenditures. Since clinical research is an important part of this overall process, a reference work on the clinical research process in the pharmaceutical industry is necessary to promote mutual understanding and appreciation among clinical investigators in industry and academia and those individuals involved with the regulatory process (both in industry and government). In general, this book should be useful to anyone involved in the drug development process.

Major changes have occurred in the pharmaceutical industry as a result of the 1962 Kefauver-Harris Amendments to the Food and Drug Administration Act, when efficacy (as well as safety) became a requirement for drug approval. Moreover, other requirements and guidelines have surfaced to reflect advances in the medical and pharmaceutical sciences, including pharmacokinetic profiling, bioavailability, drug interactions, differences in metabolic fate of drugs in healthy vs. ill subjects, and in the old and the young. In many instances, this information is beneficial and useful; at times acquiring it is not justified and at times may border on being unethical.

The authors of this book have attempted to address contemporary issues such as ethics in clinical research, institutional review boards, medical devices, postmarketing surveillance, contract research organizations, and "orphan" drugs, as well as the monitoring and investigation of clinical researchers themselves.

Because of the changes in the clinical research process that have occurred in the past two decades, potential investigators should be aware of the joys and advantages of conducting clinical studies of potential new drugs or marketed drugs, as well as the disadvantages that may be encountered. Such an awareness will aid investigators to avoid the pitfalls and frustrations that result from inadequate attention to details. A reference work should be available to point out the necessity for careful preparation of protocols, subsequent adherence to the protocols, and accurate record-keeping.

In addition, the recognition that various personnel collaborate in team-oriented research is of paramount importance for the overall success of drug development. These individuals include scientists involved with clinical pharmacology, clinical trial monitoring and execution, pharmaceutical formulation, statistics, computer analysis, drug disposition, and those who have not only an in-depth knowledge of the regulatory process, but, more important, know how to deal with it effectively to achieve the desired goal.

Because individuals from multiple disciplines are involved in the development of new drugs for the prevention and cure of disease, as well as for the palliation of pain and suffering, this text is most welcome. It should be enlightening to individuals already committed to the field as well as those planning to enter this exciting and socially important discipline.

Louis Lemberger
Director of Clinical Pharmacology
Lilly Laboratory for Clinical Research
Eli Lilly and Company
and
Professor of Pharmacology, Medicine, and Psychiatry
Indiana University School of Medicine
Indianapolis, Indiana

PREFACE

My twenty years of experience in administration and planning in the pharmaceutical industry, academia, government, and the health care field provided the foundation and impetus for undertaking the development of this comprehensive book on the clinical research process in the pharmaceutical industry.

This book examines the sequence of events and methodology in the clinical research process. This book is devoted to the industrial clinical research process as it relates to the organization and administration of clinical research programs; clinical project coordination; research quality assurance; project management; monitoring process; drug regulatory affairs; legal, moral and ethical problems; clinical information systems including computer applications; biometric and study design; education and career development; market research and the clinical research process; health economics; postmarketing surveillance; drug disposition; drug safety evaluation; the role of the Food and Drug Administration; the history of clinical research; the future; recent trends, and many other exciting topics. This list serves to provide you with a sampling of the scope of this book.

In documenting the industrial clinical research process, the authors look at the sequence of events leading to the development of new therapeutic agents. Intertwined with the methodology and sequence of events is the conceptual framework involving the philosophical, economic, political, historical, regulatory, planning, and marketing aspects of the clinical research process. Support functions involved in the industrial clinical research process, including preclinical activities, are presented to show their relationship to the whole.

This book will serve as a reference source for the multidisciplinary personnel engaged in the industrial clinical research process. Readers will become familiar with the internal and external forces surrounding the flow of events in the clinical research milieu. The rapid development of information in this field makes keeping abreast of developments at once obligatory and difficult. Since the methodology of the industrial clinical research process is constantly evolving, this book is designed to provide readers with a ready source of background information as well as a preview of things to come.

This book will serve to provide personnel involved in the industrial clinical research process with information on the latest state of the art. Clinical monitors, planners, clinical information scientists, clinical study coordinators, drug regulatory personnel, biometricians, computer scientists, medical writers, clinical research associates, clinical investigators, institutional review board members, research administrators, project management personnel, quality assurance staff, pharmaceutical scientists, market research staff, personnel and financial administrators, educators, governmental regulators, clinical scientists, and students of the health professions will find this book to be a valuable reference tool.

I believe this book will serve as a catalyst for colleges of pharmacy, medical schools, undergraduate and graduate programs in the medical sciences, and nursing and allied health programs to develop courses in the clinical aspects of drug development. I hope that this book will serve as a learning tool for the many students in the health professions embarking on a career in the pharmaceutical industry, in particular, in clinical research. Also, the book is ideal for in-service education programs conducted by pharmaceutical corporations, professional organizations, and government agencies. I intend to utilize this book for a new course I have developed, entitled: The Clinical Research Process in the Pharmaceutical Industry.

I believe this book will provide the pharmaceutical industry, academia, government, clinical investigators, physicians, pharmacists, and all personnel mentioned earlier with a long-overdue comprehensive source of information on the clinical research process in the pharmaceutical industry.

Gary M. Matoren
Editor-in-Chief
Clinical Research Practices and Drug Regulatory Affairs
Middletown, New York

ACKNOWLEDGMENTS

I would like to acknowledge the inspiration provided by the late Steven J. Eisner, whose interest in the clinical research process motivated me to undertake the development of this comprehensive book. The renaissance in the drug development process in my professional milieu, coupled with my background in academia, government, health care, and the pharmaceutical industry, added to the impetus for creating this book. My role in clinical as well as in research and development project coordination gave me an appreciation of the need for such a book. Coupled with this was my desire to bring to fruition a book to document the clinical research process; provide a reference source on methodology and sequence of events; and stimulate the development of courses and curricula in clinical aspects of drug development in pharmacy colleges, medical schools, and graduate programs in the basic medical sciences, nursing, and allied health care fields.

The outstanding response from my colleagues in academia, government, and the pharmaceutical industry in accepting my invitation to participate is evidenced by this outstanding book on the clinical research process. I want to express my sincere appreciation to these authors, who took time from pressing professional responsibilities to contribute to this work. They were recruited because of their international reputation in the drug development process. To them, I owe an eternal debt of gratitude.

I would like to express my appreciation to Professor Janet Landau and to Marion Weinreb for proofreading and for developing the index entries for several chapters.

I am deeply indebted to Linda Da Silva and Sharon A. Duritzo-Spocinski for the significant amount of time they devoted to proofreading the entire manuscript and developing and compiling the subject index. Finally, Linda and Sharon spent many hours integrating and typing the completed index. Their attention to detail and perfection are sincerely appreciated.

This book would have not been possible without the encouragement of the staff of Marcel Dekker, Incorporated. In particular, I want to express my sincere appreciation to Dr. Maurits Dekker and Mr. Marcel Dekker.

Finally, I would like to express my eternal gratitude to my loving wife Susan and two darling children, Bonnie Lisa and Debbie Lynn, for their patience, understanding, and support while this book was being developed.

CONTRIBUTORS

William B. Abrams, M.D., Merck Sharp & Dohme Research Laboratories, Division of Merck & Co., Inc., West Point, Pennsylvania

John D. Arnold, M.D., Quincy Research Center, Kansas City, Missouri

Joseph R. Assenzo, Ph.D., The Upjohn Company, Kalamazoo, Michigan

C. L. Bendush, M.D.,* Eli Lilly and Company, Indianapolis, Indiana

E. Keith Borden, M.D., M.S., The Upjohn Company, Kalamazoo, Michigan

T. A. Boyd, Ph.D., Revlon Health Care Group, Tuckahoe, New York

Ian Brick, Ph.D., Institute of Clinical Pharmacology Ltd., Dublin, Ireland

Allen E. Cato, M.D., Ph.D., Burroughs Wellcome Company, Research Triangle Park, North Carolina

Lester Chafetz, Ph.D., Warner-Lambert Company, Morris Plains, New Jersey

Robert S. Cohen, M.S., Berlex Laboratories, Inc., Wayne, New Jersey

Linda Cook, Burroughs Wellcome Company, Research Triangle Park, North Carolina

Austin Darragh, M.D., F.R.S.H., Institute of Clinical Pharmacology Ltd., Dublin, Ireland

John J. Donahue, Ph.D., Hoffmann-La Roche Inc., Nutley, New Jersey

---

**Present affiliation:* E. R. Squibb and Sons, Inc., Princeton, New Jersey

Theodore I. Fand, Ph.D., Warner-Lambert Company, Morris Plains, New Jersey

Richard E. Faust, Ph.D., Hoffmann-La Roche Inc., Nutley, New Jersey

William Feinstein, Ph.D., Whitehall International, Inc., New York, New York

Marion J. Finkel, M.D., U.S. Food and Drug Administration, Rockville, Maryland

Dan M. Hayden, Quincy Research Center, Kansas City, Missouri

Angela R. Holder, LL.M., Yale University School of Medicine, New Haven, Connecticut

Keith H. Jones, M.D., Merck Sharp & Dohme Research Laboratories, Division of Merck & Co., Inc., Rahway, New Jersey

Robert J. Levine, M.D., Yale University School of Medicine, New Haven, Connecticut

Gary M. Matoren, M.B.A., M.P.A., Editor-in-Chief, Clinical Research Practices and Drug Regulatory Affairs, Middletown, New York

William F. McGhan, Pharm.D., Ph.D., College of Pharmacy, University of Arizona, Tucson, Arizona

Michael Montagne, Ph.D.,* College of Pharmacy, University of Kentucky, Lexington, Kentucky

E. S. Neiss, M.D., Ph.D., Revlon Health Care Group, Tuckahoe, New York

F. J. Novello, Eli Lilly and Company, Indianapolis, Indiana

Jack Robbins, Ph.D., Arnold and Marie Schwartz College of Pharmacy and Health Sciences, Brooklyn, New York, and Schering Corporation, Kenilworth, New Jersey

Paul H. Roberts, M.D., Revlon Health Care Group, Tuckahoe, New York

---

**Present affiliation:* School of Hygiene and Public Health, Johns Hopkins University, Baltimore, Maryland

Martin C. Sampson, M.D., F.A.C.P., Sampson, Neill & Wilkins Inc., Upper Montclair, New Jersey

Nelson H. Schimmel, M.D., Schering Corporation, Bloomfield, New Jersey

Eric C. Schreiber, Ph.D., The University of Tennessee Center for the Health Sciences Medical College, Memphis, Tennessee

Don C. Stark, M.S., Mead Johnson Pharmaceutical Division, Evansville, Indiana

Richard L. Steelman, Ph.D., McNeil Pharmaceutical, Spring House, Pennsylvania

Glen L. Stimmel, Pharm.D., University of Southern California, Los Angeles, California

Barry Strumwasser, Berlex Laboratories, Inc., Wayne, New Jersey

Thomas W. Teal, McNeil Pharmaceutical, Spring House, Pennsylvania

Donald D. Vogt, Ph.D., University of Kentucky, Lexington, Kentucky

Albert I. Wertheimer, Ph.D., University of Minnesota, Minneapolis, Minnesota

CONTENTS

INTRODUCTION

Over the years, the process of new drug discovery has evolved and matured from the days of sporadic but important or astute observations made by individuals on the therapeutic benefits of certain naturally occurring medications or substances with desirable pharmacologic activity to the present state of a highly developed art and science in which a concerted effort is made, particularly in the pharmaceutical industry, to discover, develop and bring to the marketplace new drugs for the treatment of a wide variety of disease states in human beings. While some medications from the "good old days" certainly were effective or contained truly active ingredients (e.g., cinchona bark for the treatment of malaria; foxglove in the treatment of cardiac arrhythmia), undoubtedly many purported therapeutic agents and nostrums were, in fact, ineffective. Indeed, a certain number of such preparations must have contributed toxicity rather than efficacy and likely exacerbated the disease state, increased morbidity, or shortened life. Biological science, whether we like it to be so or not, is a rather inexact science because of the inherent variability among individual animals. Such differences are really not surprising when one considers the incredible array of biological processes that are ongoing at any given moment in the human body, including, but not limited to, variability in blood flow, digestive processes, intestinal function, enzymatic levels in tissues, presence of stimuli or inhibitors, transfer of oxygen through the lung, etc. The biochemical sciences developed along a quantitative path some time ago and the sciences of animal pharmacology and toxicology have also been brought to a rather high state of reproducibility in the recent past, particularly through the use of inbred animals and careful control of environmental conditions. Because of the inherent complexity involved in conducting truly controlled studies in human beings, however, the process of clinical assessment of drug effects has lagged behind experimental laboratory processes and has been brought to a reasonably high state of control only in the recent past.

Although certain well-meaning, albeit uninformed groups argue vociferously that new drugs and therapeutic modalities can be developed and their true value proven in small mammals and lower animals, all informed and experienced research investigators and physicians know only too well that this is a laudable objective that, at the present state of knowledge, represents only wishful thinking. The only way that one can truly establish the efficacy of a new drug or a new treatment regimen in man is to study that drug or treatment regimen in human beings and to study it adequately and by properly controlled methods. To this end, a truly impressive science has developed in the recent past in which careful attention to the design of protocols, selection of patients, randomization techniques, collection of data and large-scale computerized analyses of results has led to an improved quality of new drug development, albeit at a price in both out-of-pocket costs and time. As the cost of the development of a new therapeutic agent escalates into the multimillions of dollars (some estimates putting this cost as high as $70 million per drug brought to the marketplace), the pharmaceutical industry, the government, and society in general must strive to be certain that effective and safe drugs reach the patient at the most rapid, safe pace possible. The medical profession, the government, the pharmaceutical industry and society must strive more diligently than ever before to set the standards against which "benefit-to-risk" will be measured. It is, in my opinion, reprehensible to delay, withhold, or abandon an effective therapeutic agent that can provide significant amelioration of morbidity or, better yet, cure of disease because a decision cannot be made whether the associated toxicity (of which there always will be some) justified the therapeutic benefit. Society has accepted the concept of risk in everyday life (e.g., auto accidents, smoking, swimming) and it must be educated to the fact that all agents with high pharmacological activity must be expected to show *some* types of side effects in some patients.

While the ultimate measure of efficacy is the double-blind, controlled clinical trial, academicians, government regulators, and pharmaceutical scientists must all wrestle with the question of when the use of a double-blind, controlled study may lead to ethical problems on the one hand or to an unnecessary drain on limited and precious resources on the other.

Because of the magnitude of the importance of these problems to society, it is imperative that we attempt to answer them with an evolving standard of performance that has, as one of its objectives, assurance of the forward motion of new drug development while, at the same time, giving equal assurance of protection to the experimental subject or patient ultimately receiving the drug. Throughout this process, clinical and preclinical research scientists in the pharmaceutical industry will play an intimate role. Hopefully, the contents of this book will assist those professionals who must deal with such weighty problems in formulating both the questions and the decisions that must be reached.

Charles G. Smith
Vice President, Research and Development
Research and Development Division
Revlon Health Care Group
Tuckahoe, New York

1
PHARMOGENOLOGY: THE INDUSTRIAL NEW DRUG DEVELOPMENT PROCESS

E. S. Neiss and T. A. Boyd
Revlon Health Care Group
Tuckahoe, New York

I. PHARMOGENY

The era of drug discovery, development, and application to patient care of these past 50 years has had profound, positive effects on humanity. This era is shared by rapid advancement in communications, transportation, agriculture, and other technological changes, of which all cannot be held in the same esteem as the salutary contributions of the pharmaceutical industry.

Economic forces together with growing regulatory requirements have so modulated the talented efforts of industrial pharmaceutical scientists that a special system or process of drug research and development has evolved. This chapter and much of this book deals with *pharmogeny*, the genesis or origin of drug products. As such, this text is one of *pharmogenology*, the study of the process involved in the genesis of new drug products. It is not surprising that these neologisms have not yet attained complete acceptance, as they violate a basic principle of English, namely, that all roots of a word should derive from the same language. However, perhaps violation of tradition in this case makes these terms even more appropriate—for, in fact, what is referred to as pharmogenology is a great mélange of sciences and technologies and represents a unique interdisciplinary amalgam of scientific efforts that may well deserve to be designated by a word which is itself a hybrid.

Patients who benefit from a drug encounter a dosage form and recognize it only as a medicine prescribed by their physician. The medical consumer has become accustomed to having a wide variety of medicines available to treat the signs and symptoms of their diseases. These drug products are the basis for a business called the pharmaceutical industry.

As in all other aspects of commerce, drug products must compete for the customers' attention and selection. For the pharmaceutical industry, the physician is the customer. Those drugs that meet the needs and wants of the physician who acts on behalf of patients will be prescribed and purchased and so achieve an increasing market share and become economic successes. Those drug products that do not satisfy the needs of the physician and patient will soon lose their market share. Since the aim of business is to continue to attract and progressively satisfy more customers, it behooves the pharmaceutical company to clearly identify its targets both qualitatively and quantitatively.

II. MARKET PLANNING

It may come as a surprise to some readers to find that this overview of the industrial new drug development process begins by first addressing the concept of market planning. This may be related to the misperception generally shared by the public that a better product, or one with new attributes, will automatically compete successfully with available products with lesser attributes.

Without proper promotion and marketing this does not usually happen. In medicine, where the busy and preoccupied physician is the customer, it is often quite difficult to convince him to change prescribing habits from that with which he is comfortable and confident. Indeed, success of a new drug product is much more likely if it is the first such product available for a given indication. When this is not the case, market research can indicate where a need exists for new drug products which better meet the patient's needs and better satisfy the physician's wants with respect to the benefits provided by the product and the risks attendant to it (1).

Market research groups have a variety of techniques of market analysis which help to define those therapeutic areas in which more effective or safer drugs are needed. This identification is really the first step in the industrial new drug development process. While several pharmaceutical companies may recognize similar opportunities, the history of the firm, its present drug product profile and mix, the nature of its sales force and technical representatives, and corporate strategic planning all contribute holistically to the selection of the areas to which research and development (R&D) resources should be applied.

If, however, R&D is to produce totally novel or unique modes of therapy, it is less appropriate to defer to market analysis for direction. This is because market research tends to be historic in nature and can measure only the existing markets and the trends seen in them. Medically oriented industrial research personnel are in a better position to recognize therapeutic discontinuities, heterogeneity of disease states requiring more specific therapeutics, and opportunities for drug product applications that are innovative. Once such an innovative approach is identified, an in-depth market analysis should be done to attempt to assess the concept's commercial feasibility.

A spirit of joint enterprise and a mutual appreciation of the roles and capabilities between the R&D community and the marketing community is essential if potential new drugs are to be translated into articles of therapeutic commerce. In the well-managed pharmaceutical company, such collaboration will allow both for the generation of new drug products in response to technical market analyses and for a major commitment to new and unique approaches to disease. For instance, market analysis of the agents used to treat glaucoma might not have predicted the major success of timolol maleate solution, and the sales of therapeutics for peptic ulcer might not have identified a billion dollar market following the introduction of cimetidine hydrochloride.

Just as market planning groups must be staunch advocates of their proposals based on sound analytical techniques, so also the industrial scientists and managers must champion their novel approaches or goals if major advances are to be realized. In the end, the enthusiasm of both groups must be aroused and all within the pharmaceutical company must be persuaded of the soundness of the direction chosen. As the research process moves forward and as Chemical Lead Compounds advance to Biological Lead Compounds, thence to Clinical Candidates, Investigational New Drugs (INDs), and finally to an approved drug product for commerce, the challenge of convincing the physician-customer that a better drug product is now available is prodigious and equal in effort to all of that which preceded it in its development.

The final step in the industrial new drug development process is to convince the medical community to prescribe the new medication for the patients for whom it is intended. This difficult task is eased if the marketing organization has participated in the drug development process from the beginning.

III. DRUGS ARE CHEMICALS

A well-organized and operated R&D community can usually attain the desired goals of a carefully constructed marketing plan. Providing they are reasonable goals, it is not really a question of whether these goals will be attained but rather when. The journey for a new drug from the chemist's bench to the physician's prescription pad is one that requires some 10 years and presently costs about $70 million. Of each 10,000 compounds synthesized by the scientists in the search for totally new drug products, only one has the biological, pharmaceutical, and clinical attributes to survive all the way to the marketplace (2).

The drug development process is a terribly expensive one; its costs inexorably increase as governments require progressively more preclinical and clinical studies to assure the safety of new drugs. There is no such thing as a totally safe drug. Drugs are chemicals that manifest desirable biological and therapeutic attributes, but they also have concomitant and attendant undesirable attributes, designated "side effects" or "toxicity."

A good drug has a favorable ratio of benefit (or therapeutic effect) to risk (or adverse effect) for the desired clinical situation. A drug is a highly sophisticated instrument which can be used correctly and effectively, or poorly and with adversity. Greater increases in testing will not make these potent chemicals safer. Only their proper use by well-informed clinicians who monitor their patients' responses to them can provide reasonable safety. Unfortunately, the rare and unanticipated anomalous adverse effects usually could not have been discovered even by excessively large batteries of premarketing studies and testing. Their discovery remains in the domain of postmarketing surveillance by a vigilant medical community and epidemiologists (3). Despite these facts, governments continue to require growing numbers of extensive and expensive preclinical and clinical studies in an attempt to diminish risks attendant to new drug development. This has resulted in the extraordinary costs of basic research and subsequent development which are comparatively greater in the pharmaceutical industry than in other technologically oriented businesses. An obvious tenet of a research-based pharmaceutical company is that it must make sufficient profits to pay for R&D and other costs of business. This is accomplished by the introduction into therapeutic commerce of a continuum of profitable new drug products.

IV. PRIORITIZATION

Applied or directed science requires targets that are clearly identified and of sufficient temporal longevity. Unclear or ephemeral or moving targets are counterproductive and demoralizing to the R&D organization. The planning function of a company must have its positions and recommendations carefully analyzed and critiqued by the company executive body. The endorsed or revised plans and positions are subsequently communicated fully and clearly to the R&D management. Failure to do this has led to inappropriate allocation of R&D resources in some companies with much loss of time and productivity.

When the process of industrial drug development is optimal, a committee to interface the marketing management and R&D management regularly addresses the corporate drug product needs and opportunities in a clear and precise manner. That body should be convened several times yearly to address expectations or required changes in priorities as a function of changes in business or findings of the R&D programs. The existence of such a mechanism is an important component of the industrial drug development process, as the committee provides the needed direction for optimal resource utilization by R&D management. The prioritization and program review process does not preclude innovation or creativity by the R&D community. It does, however, specify where that community should be heading. How R&D gets there—and explicitly how to define "there"—is a task given to the research community. For instance, if a market planning group and the company executive management stated their desire for a new drug approach to the prophylaxis and treatment of bronchospastic diseases, the scientists of the R&D community might suggest a biological and chemical program around:

1. Cyclic nucleotide phosphodiesterase inhibitors
2. β_2-Adrenergic agonists
3. Modulators of airway hyperreactivity
4. Inhibitors of the release of mast cell mediators
5. Leukotriene synthesis inhibitors or leukotriene antagonists

or other approaches. In any R&D community resources are limited and the choosing of one or possibly two approaches to the therapeutic target is required to assure a timely success. Such a strategic orientation is integral to the drug development process and its importance cannot be overstressed.

V. SCREENING

Ever since its origin in antimicrobial research, there has been a great emphasis on the screening approach to drug discovery in industry. This route appears to have first been taken in a systematic way by Paul Ehrlich, who recognized the principle of selective toxicity. He demonstrated that certain vital stains, such as trypan red, were antiprotozoal, and he subsequently screened long series of chemicals and stains for their activity against spirochetes and protozoa. This led to his synthesis, in 1907, of the organoarsenical arsphenamine (Salvarsan) (4). This revolutionized the medical treatment of syphilis and introduced the concept of chemotherapy. It also established a paradigm for drug discovery wherein a biological model is used to facilitate the study of the biological effects of large numbers of chemicals. Indeed, practically all antibacterial, antiprotozoal, anthelminthic, and antifungal drugs have been discovered by this time-honored screening procedure. In more recent times, the screening approach to drug discovery has been held in lesser repute by those who believe that more sophisticated methods exist for drug discovery through fundamental approaches to the basic sciences and better understanding of biological mechanisms. While this may eventually be the case, at present and for some years to come, new drug discovery is still very much a screening process.

It is axiomatic that pharmacological actions are observed as alterations in the function of cells, tissues, organs, or entire organisms by mechanisms involving chemical interactions. The mechanism of action of most drugs appears to be at the enzymatic, coenzymatic, or prosthetic group level, and they function at receptor sites through their chemical affinity and intrinsic activity. While screening in chemotherapy originated in a search for molecules with a greater proclivity for intoxicating invading microbial cells rather than host cells (selective toxicity), screening in other aspects of drug discovery led to a search for compounds which affect enzymes, membranes, or receptor sites in vitro, or tissue and organ models relevant to various states of pathophysiology or disease processes.

Many medicinal agents have survived from antiquity; their genesis, no doubt, lies in the random search by the scientifically naive for natural substances with the ability to heal or ameliorate pain. A meaningful pharmacopoeia of primitive but nonetheless useful medicines characterizes every historical culture (5). Each pharmacognosy reflects the environmental opportunities and the prevalance of certain diseases. Some medicaments appear to have been discovered and used in antiquity, then lost to subsequent civilizations, only to be rediscovered at later times. The point is that the random screening process has led to drug discovery through the ages and still contributes importantly today.

This is not to say that all is chance in drug discovery. It clearly is not. But chance and serendipity do indeed each play an important role in new discovery. Discovery of the unanticipated by the prepared mind contributes more than some of us impressed with the elegant and theoretical methods of science would like to admit. For this reason, the biological scientist must always be alert to the unexpected and be prepared to grasp an opportune finding. Scientific pragmatists will, as did Paul Ehrlich, follow up chemical leads like the special affinity of certain dyes for microorganisms. But they will also study the effects of other unique chemical structures on the system in question in an attempt to discover yet another approach to the same goal.

The first procedure is designated "biologically directed chemical synthesis and screening," or lead following; the second is nondirected "random screening," or lead seeking. Because of the enormous growth of our understandings of mechansim involved in biological systems, the contemporary approach to industrial drug discovery involves, primarily, the biologically directed chemical approach. The astute medicinal chemist will still request broad screening of molecules that are intermediates in the sequence that leads to the biologically directed target compound. A very real part of the drug discovery process is an appreciation for and acceptance of scientific discoveries made fortuitously. The retrospective integration of such events into a sophisticated rational plan for discovery is one of the pleasures of the drug development aficionado.

VI. LEAD FOLLOWING

The use of the more common biologically directed chemical synthesis approach is most likely to result in yet another drug with many properties in common with others already known or being developed in competitive laboratories. This is because of the use of relatively similar batteries of biochemical and pharmacological test systems among contemporary laboratories. New drugs are sometimes discovered by introducing into these same screens compounds (such as the chemical intermediates) whose structures suggest no reason (based on the medicinal chemistry literature) for them to demonstrate the desired biological properties. Indeed, when this happens a novel chemical lead is found and further structural modification may result in a therapeutic agent with some different and possibly more useful properties.

It was in a manner akin to this that Dr. L. Sternbach of Hoffmann-La Roche discovered the activity of the precursor benzodiazepine-4-oxide structures that led to his synthesis of chlordiazepoxide (Librium) and subsequently to diazepam Valium) (6). These drugs, whose great medical and economic successes are renowned, were discovered because a molecular rearrangement product of a substituted quinazoline-3-oxide was submitted for pharmacological screening, despite the fact that the entire series of the target compounds had thus far lacked interesting biological activity. Discovery of activity in the transformed molecule led to modifications that enhanced biological activity and greatly impacted the world of therapeutics.

Since most drugs discovered in the past 50 years are a result of the old (but very successful) modification approach, it is particularly regrettable that it is in vogue to refer to it by the pejorative term "molecular manipulation." This approach is sine qua non to the industrial new drug development process and is justified by its successes in the discovery of drugs that ameliorate the pain and suffering of millions of patients.

Now that proper honor has been paid to some earlier and still viable approaches to drug discovery, it is important to put the subject in contemporary perspective.

The design of selective, enzyme inhibitors is currently one of the most exciting approaches to the development of new drugs (7). As in all seemingly new things, a history exists for enzyme inhibition as a basis for drug action. Examples of enzyme-inhibiting drugs include allopurinal, physostigmine, penicillin, methotrexate, theophylline, digitalis, methyldopa, indomethacin, and captopril. In most situations, the fact that the drug was acting through enzyme inhibition became clear only after the fact.

Captopril is a good example of a successful attempt to moderate disease by prospective evaluation of enzyme inhibition using various natural and synthetic peptides. Angiotensin-converting enzyme (ACE) is a dipeptidyl carboxypeptidase which cleaves the C-terminal dipeptide from the nonvasoactive decapeptide angiotensin I to form the octapeptide angiotensin II, an extremely potent vasoconstrictor. An industrial drug development program directed toward finding inhibitors of ACE at the Squibb Institute for Medical Research resulted in the discovery of the important new antihypertensive agent captopril (8). The approach was based on analogy with the active site construction of a well-studied zinc-containing metalloenzyme, carboxypeptidase A. A previously reported potent inhibitor of carboxypeptidase A was used as a model for a directed chemical discovery program that eventuated first in Teprotide (a nonapeptide ACE inhibitor) and, subsequently, in the orally active antihypertensive drug captopril. This is a good example of the high degrees of selectivity and specificity that can be achieved by a program of systematic chemical structure modification using enzyme inhibition as the titrator of activity.

Industrial pharmaceutical laboratories are now approaching the search for potential therapeutic agents via chemical synthetic programs directed at the creation of selective inhibitors of enzymes involved in metabolic processes as well as tissue regulatory substances of profound effect. For example, many laboratories have expanded their interests in prostaglandin metabolism to include all of the newly elucidated arachidonic acid metabolism. Inhibitors and modulators of lipoxygenase, cyclooxygenase, phospholipase A_2, thromboxane synthetase, prostacyclin synthetase, and leukotriene synthetase will likely have important roles in the therapeutics of diseases as apparently unrelated as asthma, myocardial infarction, and peptic ulcer.

Depending upon the corporate strategy for its participation in the drug market, an R&D community can establish from those enzyme models a drug discovery program aimed at lipoprotein lipase or proteolytic enzyme inhibitors affecting hypertension (ACE inhibitors or renin inhibitors), blood coagulation, or fibrinolysis. Inhibition of a protease in spermatocytes prevents their access to the ovum and represents an enzyme-based drug approach to contraception. The opportunities seem endless. The number of enzymes is vast; since enzymes are ubiquitous in the cells and tissues of an organism, inhibition of their activity can have sequelae other than might be immediately obvious from the nature of the reaction catalyzed. Thus, the study of these potent agents in vivo, to establish the extent of their effects in the presence of all other enzymes, substrates, tissues, and homeostatic mechanisms is an essential part of the drug development process. Such studies usually follow the progression: (1) in vitro enzymatic testing, (2) in vitro tissue biochemistry tests, and then (3) in vivo enzymatic and pharmacological evaluations.

VII. FEEDBACK

Just as the biochemistry laboratories must feed information back to the medicinal chemists to help in directing their plans for further synthesis, so also must the pharmacologists contribute to the feedback process in some very important ways. For instance, we still lack meaningful enzyme models for many diseases or for the modulation of organ function. In such situations, there is no substitute for measuring effects of the new compounds on isolated tissues. In addition, pharmacology can define the selectivity of the effect, measure the magnitude of the effect in a dose-response relationshp, evaluate the effect in animal models of the disease (e.g., the genetically hypertensive rat as a model for human hypertension), and study general behavioral effects. The significance of the general behavior evaluation is obvious when one considers that feedback from the biochemists might lead the chemists to believe that they have a structure with good potency and activity as, for example, an ACE inhibitor; perhaps pharmacological data showed selectivity and potency in the spontaneously hypertensive rat test. However, if the drug also caused animals to become lethargic, it might then not have been as attractive as it might have seemed before this fact was known. In the industrial drug development process, the pharmacologist complements the biochemist in providing to the medicinal chemist a more complete picture of the biological properties of a new drug candidate in animals as whole organisms, with reflexes and compensating homeostatic mechanisms in place. For instance, the pharmacologist will advise the chemist as to whether a compound is absorbed by the indicated route of administration, and such information puts the activity and potency in proper perspective.

The specific tests performed in isolated tissues, organ systems, or in whole animals are, of course, a direct reflection of the pharmaceutical company's therapeutic goals as earlier delineated. This is a most important concept which often eludes the uninitiated. The discovery of interesting and possibly therapeutically significant biological activity is a summation and evaluation of selected studies or tests done in the departments of microbiology, biochemistry, endocrinology, and pharmacology. These studies, tests, and screens are established to determine whether the molecules synthesized by the medicinal chemist have activity that will make them of interest for further study or for chemical synthetic analoging. However, this information is, of necessity, restricted by the nature of the biological tests, and these tests are restricted to the expressed areas of therapeutic interests and capabilities of the individual pharmaceutical company.

VIII. RESOURCES

Few (if any) pharmaceutical R&D communities have *all* of the resources required to do *all* that the scientists wish to do in the vast panorama of biomedical opportunity. Under responsible R&D managements, resources are allocated and used to support a program of drug discovery and development specifically in the areas of interest agreed upon with the company marketing people and executive management. Therefore, it is not uncommon in the pharmaceutical industry to have biological screening and study restricted to a single or just a few areas of therapeutics. If, for example, a given company's defined interests are in the field of cardiovascular and gastrointestinal therapeutics, it could be that molecules passing through their screens might be the best possible drug for mental depression, or glaucoma, or systemic fungal infection, or psoriasis, or viral infection, but they will remain undiscovered for these uses. Because of this exposure, it is not uncommon in the industrial drug development process for companies to set up contract agreements to put into their screens compounds made by another company's chemists and vice versa. This is done, of course, when the two companies have different immediate therapeutic market goals and, therefore, a qualitatively different spectrum of biological screens. Such agreements usually include a licensing or perhaps a codevelopment option. Arrangements of this nature have resulted in drug discovery and are a testimonial to the continued value of broad random screening. They provide as well a further value to the involved firms and eventually to the broader medical community.

IX. THE CHEMIST

Success or failure in the industrial new drug discovery and development process very much depends upon the competence of the firm's medicinal chemists. Such chemists vary greatly in their imagination, creativity, industriousness, perseverance, and understanding of biological data. The pharmaceutical company with innovative, goal-oriented medicinal chemists who can function with a team in a collaborative approach toward a difficult or long chemical synthesis, for example, is in a favorable position to achieve its goals.

Medicinal chemists are, of necessity, interdisciplinary people who must be conversant with the biological implications of their structures as well as needing to be expert in organic chemistry and intimate with the literature of medicinal chemistry. The capacity of such chemists to propose new structures and areas of proposed synthesis always transcends, by far, their abilities to prepare these compounds at the bench. To have notebooks full of ideas of compounds for synthesis provoked by conversations, consultants, the literature, and a fecund imagination is the characteristic indicator of excellence in medicinal chemistry. A good medicinal chemistry management allows for a reasonable exercise of this individuality in a synthetic program that might be addressed communally. The most productive chemistry management maintains a complete sense of the specific goals at hand, and, with a restraining maturity derived from earlier goal attainment and collaborative experience, keeps its team focused on the accepted targets.

Communication of screening results back to the medicinal chemists by the biological scientists may be the weakest point in many pharmaceutical new drug development programs. This is particularly unfortunate in that the very success of drug development is directly proportional to the openness, intensity, and timely nature of this communication. Expressed another way, the failure of the industrial new drug development process is assured by inadequate, tardy, and desultory communication of test results by the biological scientists to the chemists.

The medicinal chemist initially prepared 1-4 g of a new compound for biological screening and then goes on to another synthetic project. If there is any activity of interest, within about 4 weeks he may be asked to prepare an additional 5-25 g of the material. The significance of the biological data derived from the first few grams must be completely conveyed to him by the biologists to allow for informed communication with his management as well as for medicinal chemistry communal evaluation of the possible meanings of that activity from the perspective of chemical structure. These intellectual exercises in coupling variations of chemical structure with biological activity (referred to as structure-activity relationships, SAR) are of immense value to systematizing drug development (9). They allow the chemist to connect the present experience with all former experience of which he and his colleagues are aware in such a way as to suggest the most reasonable changes that should be made on subsequent synthetic compounds in order to enhance or otherwise appropriately modify the biological activity. Often, but not invariably, this exercise results in the preparation of more active or more potent compounds and sometimes in less toxic drugs. To make it happen, however, the ongoing communication of biological findings and their interpretation must be received and considered by the chemist.

X. THE PRECLINICAL TRAIL

A compound with activity of sufficient interest to warrant a request for resynthesis is usually considered a Chemical Lead Compound and forms the basis for an SAR chemical program and drafting a formal document entitled "Record of Invention." In the desirable case where biological evaluation of a member of the SAR family suggests it is an important compound which, when studied in full pharmacological depth, continues to be of high interest, it is considered to be a Biological Lead Compound. Biological Lead Compounds stimulate the patent application process, which is based on the original Record of Invention referred to above. Such compounds are prepared by both radiolabeled synthesis for drug disposition studies and cold synthesis, in at least a 600-g quantity, by a separate scale-up chemistry group in order to do drug safety studies and for preliminary pharmaceutical evaluations. This very desirable situation, the formal identification of a Biological Lead Compound, is not as common an event as a pharmaceutical company management might like. It can happen only as a consequence of and in proportion to the availability of resources to support the broad scope of required effort.

A Biological Lead Compound is advanced to a Clinical Lead Compound status when satisfactory results are obtained in radioisotope drug metabolism studies, preliminary animal safety studies, and when an adequate analytical chemistry and pharmaceutical profile is available. Procedures vary among companies. Most will propose to advance a Clinical Lead Compound to the subject of an Investigational New Drug Application (IND) when the complete biological profile is consistent with the desired therapeutic activity and potency and when at least 2 weeks of toxicological studies with full histopathology is available from two animal species. More conservative firms await 30-day (interim sacrifice) histopathology results from rat and dog studies together with 90-day clinical observation of those two species in an adequate study done in compliance with the Good Laboratory Practices (GLP) regulations. The advantage of the latter drug safety evaluation program lies in the greater confidence that the firm, the initial clinical investigators, and the volunteers or patients can have in the safety of the new drug. It can also permit the drug to be given for several weeks to volunteers in the Phase I and II clinical studies and for up to 3 months administration when the 90-day animal studies are completed. For many clinical indications up to 3 months of exposure in patients is adequate to indicate the utility of the new drug and the degree of its tolerance and safety. Such a clinical exposure is usually a sufficient basis upon which to make management judgments with respect to the extremely costly, long-term animal safety studies that are required. These may include carcinogenicity studies in animals and segments I, II, and III reproduction studies. This judgment is made by the R&D management community and is implemented with executive management approval.

XI. THE BIOLOGIST

The Biological Lead Compound becomes the subject of intense investigation by the scientists in the drug disposition and metabolism disciplines. The use of a radiolabeled compound in animal studies at an early stage in development of the Biological Lead Compound facilitates an early and economical appreciation of the degree and rate of absorption of the drug and an indication of possible metabolism (10). An elimination or balance study is done in the rat and the dog and is designed to determine the comparative extent and routes of drug elimination after oral and intravenous administration of labeled compound. The amount of radioactivity eliminated in the urine and feces, as well as respiratory $^{14}CO_2$, is determined. An estimate can be made of the degree of oral absorption of the compound by comparing the radioactivity detected in the urine of animals dosed intravenously or dosed orally. Using radiolabeled drug, the distribution of radioactivity in the various tissues of animals after oral administration of compound is also commonly studied. This distribution is measured at various periods of time after dosing to determine the tissue elimination kinetics of radiolabel and to determine whether radioactivity (unchanged drug and its metabolites) is concentrated in any tissue.

An ABLE (absorption, blood level, elimination) study is also commonly done in the dog as part of the early drug development process.

This study determines blood concentrations of total radioactivity and the extent and routes of elimination of radiolabeled compound after oral and intravenous administration. The amount of radioactivity eliminated in urine and feces is determined. From data in studies with animals dosed orally compared to similar data from animals dosed intravenously, an estimate of the degree of gastrointestinal absorption can be determined, both by comparison of urinary excretion of radiolabeled drug and by comparison of the blood radioactivity concentration-time curves (serum or plasma concentration plotted with respect to time after dosing) for total radioactivity.

By application of analytical tools, such as thin-layer chromatography, on the urine obtained from rats and dogs dosed with radiolabeled drug, a metabolite profile in these species is available for subsequent comparison with that seen in urine from humans dosed with the drug. This may help identify the animal species which has a metabolite profile most similar to the human. Such data are sometimes useful in determining the proper species for long-term animal drug safety studies. It is rational to choose to do long-term studies in a species which has a drug metabolism similar to that of man. The situation is complicated by judgments as to whether the metabolite patterns should be similar in urine or serum and by concern with the respective quantities of the common metabolites and possible species-associated differences in kinetics. Other elimination and ABLE studies in yet different species are sometimes necessary in order to find animals with humanlike metabolite profiles and so attempt to support the choice of species selected for drug safety studies if they are other than the rat and the dog. For the most part, 2-year chronic studies are practically always done in the mouse and rat despite these sophisticated approaches.

Plasma protein binding may play an important role in the disposition of a compound especially where renal and/or hepatic dysfunction exists. The drug development process requires evaluations to determine the extent of protein binding, to determine whether it plays an important role in therapeutics, and to evaluate whether interactions with drugs might occur through the protein-binding mechanism. Other important drug disposition studies include investigations of the effect of chronic dosing on drug half-life (enzyme induction), the isolation and structural identification of metabolites, and the correlation between the serum concentration of drug and pharmacological effect (pharmacodynamics). More detailed aspects of the contributions of drug disposition studies in the drug development process are covered elsewhere herein by E. C. Schreiber (see Chapter 14).

The development of an analytical assay for the new drug in biological fluids is an extremely important and significant part of the drug development process. The assay method should have sufficient selectivity and sensitivity to define with confidence compartment (serum, urine, etc.) drug concentrations as a function of dose and time after dosing. Automating the analytical method facilitates industrial drug development by allowing for the efficient and cost-effective analysis of literally thousands of blood and urine samples from patients under many different conditions of investigation. This activity constitutes one of the most important ongoing interfaces between preclinical and clinical scientists in the pharmaceutical industry drug development process. This interface, like that between the biologists and chemists, can be rife with problems that challenge and perplex the most competent of research managers. It is absolutely essential to the success of the drug development process, however, that functional systems be established that reasonably satisfy both biological and clinical scientists. While clinical pharmacology studies can be and often are done in the absence of drug analyses in biological fluids, such studies are clearly of greater value to all concerned when drug concentration values are available. Similarly, drug disposition scientists do extensive and very important development work in which they compare metabolite profiles and kinetics of a drug in animals with that of humans.

Increasing numbers of basic scientists are joining the contemporary clinical pharmacology department, and because of their training they are quite comfortable and conversant with pharmacokinetic calculation, biopharmaceutical mathematics, and statistical techniques. They prefer to have the raw analytical data from biological fluids of volunteers dosed in their studies returned to them for processing and reduction to findings. The drug disposition scientists, however, do not lightly accept a role of reference analytical testing laboratory without proprietary interest in the data. Thus, research managers commonly experience a polarization between the disciplines of drug disposition and clinical pharmacology which have been quite compatible historically. This not uncommon situation is now frequently being resolved by delegating full responsibility for the processing of data acquired by clinical pharmacologists from volunteers and patients to that department. It is not uncommon for a separate section created within the drug disposition discipline to now service clinical pharmacology, drug safety, and even drug disposition scientists when the great volume of analytical samples requires highly sophisticated "crank-turning" by someone. In this case, drug disposition scientists "maintain ownership" of the processing of bioanalytical data from all species save *Homo sapiens*. The latter data are reserved for the clinical pharmacologists, who have the prerogative of collaborating with other well-trained and competent drug disposition scientists or moving the drug forward separately. The industrial pharmaceutical development process requires a sensitive and vigilant research management, and awareness of this drug disposition/ clinical pharmacology interface is an excellent opportunity to facilitate the success of the process.

XII. THE PHARMACEUTICAL SCIENTIST

While drug disposition scientists develop the isolation and identification processes that allow for analyses in biological fluids, a different group of chemists have earlier established the chemical and analytical procedures used to define the bulk drug substance and its dosage forms. These analytical chemists work very closely with the medicinal chemists on one aspect of the drug development process and with the pharmaceutical development scientists on another quite different part of the process (11).

Scientists in analytical chemistry have the responsibility of establishing, without equivocation, the elemental composition of a new compound, its purity, and its chemical configuration or absolute spatial structure. By use of techniques such as high-pressure liquid chromatography, the analytical chemist can isolate the new drug in an ultimate state of purity. With an armamentarium of tools from the basic infrared spectrometer through more advanced devices such as the mass spectrometer, nuclear magnetic resonance spectrometer, and optical rotary dispersion spectrometer, the analytical chemist proves the chemical nature of the drug. Through x-ray diffraction crystallography, the structure of drug substances can be certified to the most skeptical of colleagues. These chemists develop analytical profiles as well as analytical methods for new drug substances. Since this methodology must be stability-indicating, it must be established that the assay method is specific for the drug entity intended to be measured and does not include in the measurement any of the degradation or decomposition products of the new drug substance. This requires the isolation and identification of decomposition products from degradation studies done on the new drug.

As all synthetic substances made by a sequence of chemical reactions contain traces of reactants from the earlier steps, analytical chemists must also develop, define, and establish an impurity profile of the new drug. In order to better understand the new drug, they do a series of physicochemical studies which include solubility and stability (as a function of oxygen, water, heat, light, and pH), pK_a, partition coefficients, decomposition profile, thermal effects, crystal properties (e.g., polymorphism), and kinetic studies of degradation processes. The analytical chemists prepare and report raw material specifications and establish the first certified reference standard for the new drug. As the development process moves forward, they test and release bulk drug for the preparation of clinical supplies as well as for animal safety studies. They also later test and release the first batch of product intended for clinical evaluation.

In collaboration with pharmaceutical scientists, analytical chemists conduct comprehensive stability studies of bulk drug as well as samples from developmental batches of clinical drug product supplies. Analytical chemistry maintains an important presence throughout the entire drug development process, for it must test and release all subsequent batches of active ingredient and formulations. It seems obvious—but it is far too important to go without saying—that all of these detailed tests and studies in analytical chemistry, as in all of the collaborating sciences in this industrial drug development process, must prepare regular, rigorous, and timely reports on every aspect of their work. These reports form a "paper trail" that allow any scientist, manager, or regulatory investigator to trace back to its origin any material or data and to be satisfied that the work was done professionally, that the material is exactly as represented, and that the data are unequivocally accurate, reliable, and true.

In the past, some R&D managements have experienced great difficulty and embarrassment by failing to establish the lineage of the data in regulatory filings. Open and clear communication of data, with statements of their significance, replication of results, and validation of methodology is not only part of the drug development process but it integral to all science. Any disregard shown to so basic a tenet of science can only imperil the goal of the entire drug development process.

XIII. BIOPHARMACEUTICS

At one time the pharmacy in a drug firm was a place where powders were converted into tablets or capsules and that was all. Now, with a better understanding of biopharmaceutical concepts, an area with an enormous recent literature, the contributions of the pharmaceutical scientist are looked on with much appreciation. It is now accepted that the actions of a new drug substance depend not only upon its intrinsic pharmacological and biochemical properties but also very much upon its ability to reach target sites in the body. "Biopharmaceutics is the study of the factors influencing the bioavailability of a drug in man and animals and the use of this information to optimize pharmacologic or therapeutic activity of drug products in clinical applications"(12). Drug delivery systems are now designed and formulated not only to enhance the amount of administered drug that reaches the systemic circulation but to target the drug more specifically to its site of action. In approaching the optimal formulation and drug product presentation, the pharmacist now considers:

Solubility at various pH values and in a variety of solvents
Stability in the dosage form
Dissociation constant (pK_a)
Crystalline properties (polymorphs)
Dissolution rate
Partition coefficients

Biological transmembrane permeability
Enzymatic stability
Gastrointestinal drug interactions
Hepatic first-pass effects

among other variables.

The research pharmacist or pharmaceutical scientist works in close collaboration with many disciplines. Not only does pharmacy interface with analytical chemists on the stability evaluation, but it also interacts with marketing, process chemistry, medicinal chemistry, drug disposition, drug safety, scale-up chemistry, clinical pharmacology, clinical research scientists, biometricians, and electronic data processing people.

An example of optimizing the absorption of an insoluble compound is the work done with the antifungal antibiotic griseofulvin (13). In an early application of biopharmaceutical intervention, it was shown that serum concentrations of the antibiotic were doubled by preparing capsules of this insoluble drug from bulk materials ground to very small particle size. Since only the drug which reaches the general circulation has a chance to exert its intrinsic desirable biological effects, it was then possible to use capsules with finely ground drug at half the dose of original antibiotic of larger particle size. The literature is now rich with examples wherein reduction in particle size has augmented drug absorption and hence bioavailability.

The concept is not one that can be considered routine, as there are compounds which also accrue undesirable properties by micronization, such as loss of wettability or increased electrostatic charge resulting in consequent handling difficulty. Besides altering particle size to affect the degree and rate of absorption of the drug substance, modulation may be achieved by proper selection of crystal form or salt derivative.

Preformulation pharmaceutics begins with studies of the physicochemical characteristics of the new drug substance and identification of interactions with typical dosage form excipients. Close collaboration with the analytical chemists is essential here to avoid extensive duplication of effort; although generally there is significant overlap in this early work, the overlap represents a reasonable scientific control through replication of results.

In addition to studies of the stability of the drug substance at specified temperatures and humidities, its stability upon exposure to light at various conditions of pH is examined. As drug dosage form data are accumulated and evaluated, the drug substance and requisite excipients are examined for optimization in laboratory-scale processing and from the perspective of adequacy for scale-up to pharmaceutical pilot plant size batches and ability to run on large-scale, high-speed manufacturing equipment. Continuing evaluations of tablets include tests of hardness, tablet disintegration time, friability, moisture content, and dosage form uniformity. Other dosage forms require other tests to ensure their quality (e.g., clarity of solution for injection, or particle size distribution for an inhalation aerosol).

The speed in which a tablet breaks up is called its disintegration time, which is measured in a standardized test. Along with particle size, disintegration time is an important factor in the subsequent solubilization (dissolution) of the active component. Dissolution is prerequisite to absorption and thus bioavailability. A goal of the pharmaceutical scientist is to formulate the new drug substance so that absorption is optimized, with concomitant controlled increase in therapeutic action. In this pursuit, a number of apparently equivalent formulations generally become available for testing. The dissolution rate is used to screen these solid dosage formulations to decide which one is most likely to have maximal bioavailability. The ultimate test, of course, is to challenge an animal species (and optimally man) with the designed oral dosage forms and to determine, by serum concentration measurements, which formulation has the best total bioavailability and time-course profile. Very importantly, the dissolution time (from an appropriately designed procedure) can often be shown to correlate to bioavailability in humans (14). If the correlation can be well established, it is not necessary to run resource-intensive bioavailability studies on each manufactured batch of drug product. Rather, the dissolution rate test can be relied upon as an important measure upon which the bioavailability of production batches is indirectly assured.

Biopharmaceutic data and pharmaceutical technology are used to design quality into the dosage form; the modern dosage form must be bioavailable, stable, and readily and reproducibly prepared. Scale-up studies are performed to ensure that the quality which was evaluated during clinical trials will be present in the commercial dosage form.

Biopharmaceutical studies not only ensure an optimal formulation for the eventual patient but also attest to and ensure the availability of the new drug substance to the tissues of animals in toxicity evaluations. Drugs are rarely distributed selectively to tissues. The same physicochemical characteristics responsible for directing a drug substance to tissues where desirable biological interactions occur also results in the distribution of the drug to other tissues where unneeded, undesirable, or adverse effects may occur. The qualitative and quantitative characterization of these effects on tissues and organs and their role in mortality in animal studies are the responsibility of drug safety evaluation scientists, a group which includes toxicologists and pathologists.

XIV. DRUG SAFETY EVALUATIONS

Of each 10,000 new compounds prepared by chemists and screened by biologists, only about 10 have sufficiently interesting activity in the biological models of disease and are also considered sufficiently innocuous in the preclinical drug safety evaluations to advance to clinical investigation. The toxicity testing program is an animal-intensive one of very great expense. Governments and the public have worried about the possibility of another disaster to humans due to undetected toxicity ever since the thalidomide misfortune. A result of this concern has been a very substantial increase in the number, variety, and length of animal tests required to support the safety of a new drug substance.

As stated earlier in this chapter, there is no such thing as a totally safe drug. Drugs are biologically potent chemicals capable of doing much good when used properly in disease states for which they have been demonstrated to be effective. Increasingly more extensive and more stringent animal toxicity testing effectively precludes the availability of potentially useful therapeutic agents for which elucidation of minimal and idiosyncratic adverse effects requires close patient monitoring by scrupulous and meticulous clinicians. It is likely that the pattern will not change and that governments will continue to require progressively more evidence of safety before permitting full-scale clinical investigations of new drugs.

A. Acute Toxicology

The industrial drug development process encompasses a toxicity evaluation program which is founded on and coordinated with biological and pharmaceutical investigations (15). The most commonly mentioned toxicity test is the LD_{50} (the estimation of a dose that is lethal to 50% of the animals in a test situation); it is also probably the least useful because it is a measure of acute single-dose toxicity and its real need is now being seriously questioned. In such an acute toxicity study a potent poison can be readily identified. Corrosive chemicals, organic solvents, and irreversible enzyme inhibitors having very low LD_{50} values are rarely considered for further drug development. The acute toxicity study is usually done in two or three species and by several routes of administration; in addition to identifying frank poisons, it provides limited information about the dose ranges that might be appropriate for the toxicity evaluation of longer duration or for initial human use. An interesting ancillary use for the LD_{50} is that one might obtain the first indication of the extent of gastrointestinal absorption by comparing LD_{50} values after oral and parenteral administration.

After a single dose or escalating single oral doses of a drug to different rodent species or to dogs, the animals are observed for signs and symptoms of toxicity such as vomiting, sedation, tremors, ataxia, convulsions, and respiratory changes. Should any such signs be observed, they may serve as clues to the potential adversity attendant to the use of the new drug, and additional, more specific studies might be suggested. The death of animals shortly after dosing is testimony of its potency as a biological poison. Death a day or more after dosing is also of concern because this indicates more occult organ damage induced by the drug.

The identification of target organs or tissues that are selectively sensitive to the drug under study is particularly important in understanding and assessing the toxic potential of the drug. The degree of toxicity in animal studies that is considered acceptable as well as the number and kinds of animal studies done are related to the eventual clinical indication and the treatment already available. Drugs with a low therapeutic ratio but with life-saving potential (such as certain agents used in oncology or refractory collagen diseases) may be evaluated in only animal studies of short duration (1-2 weeks or a once-weekly dosing for 6-8 weeks) as compared to the more usual drugs for which studies of 1, 3, and 6 months duration are done. The U.S. Food and Drug Administration (FDA) has prepared guidelines for animal toxicity studies that are deemed appropriate to support various kinds of clinical study (see Table 1) (16). These guidelines have provided an ordered basis for current procedures in the industrial drug development process.

B. Chronic Toxicology

The research management of pharmaceutical firms is greatly challenged during each new drug development project to determine the correct temporal sequence in which to place various aspects of the toxicological and clinical evaluations. If chronic (1- or 2-year) toxicity studies are begun too early, major commitments of personnel and funds may be lost should it be found that the new drug substance is insufficiently well tolerated in humans or that it does not have the desired therapeutic activity at clinically or economically acceptable dose levels.

Shorter (e.g., 1-month) toxicity studies may not be sufficient to detect a drug's potential for producing a serious adverse effect. If such effects are not discovered until the chronic toxicity study is ongoing or complete, a considerable investment of resources applied to the parallel clinical development is also lost and greater risks are involved in the early study patients. Thus, programming the proper sequences of drug safety studies for each drug is an exercise in risk and resource management which is an important aspect of industrial pharmaceutical new drug development.

Chronic toxicity studies in animals are classically 1 year in length and are usually conducted in rats and one nonrodent species, using male and female animals of each species. The animals are dosed via the major route to be used in patients. During this prolonged study, animals mature and grow under the stress of chronic influence of the drug, a fixed diet, and an unnatural and restraining environment. These latter effects, together with the degenerative organ changes

Table 1 Synopsis of U.S. Food and Drug Administration Guidelines for Animal Toxicity Studies

Drug category	Duration of human administration	Clinical study phase	Subacute or chronic toxicity
Oral or parenteral	Several days	I,II,III,NDA	Two species; 2 weeks
	Up to 2 weeks	I	Two species; 2 weeks
		II	Two species; up to 4 weeks
		III,NDA	Two species; up to 3 months
	Up to 3 months	I,II	Two species; 4 weeks
		III	Two species; 3 months
		NDA	Two species; up to 6 months
	Six months to unlimited	I,II	Two species; 3 months
		III	Two species; 6 months or longer
		NDA	Two species; 12 months (nonrodent), 18 months (rodent)
Inhalation (general anesthetics)		I,II,III,NDA	Four species; 5 days (3 hr/ day)
Dermal	Single application	I	One species; single 24-hr exposure, followed by 2-week observation
	Single or short-term application	II	One species; 20-day repeated exposure (intact and abraded skin)
	Short-term application	III	As above
	Unlimited application	NDA	As above, but intact skin study extended up to 6 months
Ophthalmic	Single application	I	Eye irritation test
	Multiple application	I,II,III	One species; 3 weeks daily application as in clinical use
		NDA	One species: duration commensurate with period of drug administration
Vaginal or rectal	Single application	II	Local and systemic toxicity
	Multiple application	I,II,III,NDA	Two species; duration and number of applications determined by proposed use
Drug combinations		I	LD_{50} evaluations
		II,III,NDA	Two species; up to 3 months

associated with aging and endemic chronic infections, tend to confound the toxicological picture and the evaluation of tissue pathology. Considerable expertise is needed to distinguish nondrug from drug effects. The separation of nondrug effects from drug effects is aided by maintaining a control group of each species that comprises animals of both sexes. The control group is exposed to the same conditions as those of the drug-dosed animals, except for the active drug treatment. At the end of chronic toxicity studies, as is the case for the acute and subchronic studies, all animals are sacrificed and the tissues of all major organ systems are examined for gross and microscopic pathology.

All toxicity studies in animals are done to elicit toxicity. Doses are increased sufficiently or extended over sufficient time to permit the identification of "target organs" for toxicity. These might be considered sites of selective toxic actions that occur at tissue concentrations not generally destructive but that occur only in animals which are generally debilitated because of the drug. Such selective organ toxicity is considered an amplification of the real-world animal-drug interaction and is considered only a coarse guide to the type of toxicity for which the clinical investigators of the new drug must be ever-vigilent and suspicious.

Organ changes that occur in most dosed animals in a defined cause-and-effect relationship at a dose that does not impair the general welfare of the animal are specific organotoxic effects. These uncommon organotoxic effects are worrisome and require considerable reflection by R&D managements deliberating whether to advance such a drug to clinical investigation. In such cases, it is customary to have a group of experts and consultants evaluate and deliberate the relevance and significance to the human of all animal toxicity data. Organotoxic effects are often species-specific and, not uncommonly, occur at doses that are many orders of magnitude greater than the proposed human therapeutic dose level. Nonetheless, organotoxic effects are a clinical reality whose existence must be detected early to protect and preserve the patient's welfare.

C. Reproduction Studies

The thalidomide misfortune underscored the necessity to determine a drug's potential to affect either parents or offspring during reproduction. Since no single experimental study can examine the entire reproductive process, animal reproduction studies are done in three segments, each involving a different phase of reproduction (17).

Segment I requires the drug dosing of male rats for 60-80 days prior to mating to assess spermatogenesis and general gonadal function.

Female rats are dosed for 14 days prior to pairing, and half of these are sacrificed in midpregnancy to evaluate the number and state of embryos. The other females are allowed to litter and are studied for litter size, survival of pups, and gross anomalies. Thus, Segment I studies provide information concerning drug action on the entire reproductive process including teratogenesis, late stages of gestation, parturition, lactation, and weaning.

Segment II is the formal teratology study in which it is determined whether a drug has a potential for embryotoxicity and/or teratogenicity. For this purpose drug dosing is restricted to the period of organogenesis in pregnant mice, rats, or rabbits. The fetuses are surgically removed approximately 2 days prior to anticipated parturition and are subjected to gross examination and to special soft tissue section and skeletal examination techniques for abnormalities.

The purpose of the Segment III study is to determine if adverse effects of the drug occur when it is administered during the last third of pregnancy and through the period of lactation. This study is particularly relevant to the safety of a drug intended for chronic use in pregnancy because it delineates effects of drug on late fetal development, labor and delivery, lactation, and neonatal viability and growth.

The Segment II study and the female portion of Segment I are completed before a new drug substance may be administered to women of childbearing potential. Before Phase III clinical trials may commence, acceptable data must be made available to the FDA from all three segments of the reproduction studies.

The last drug safety evaluation done on behalf of a new drug that successfully overcomes the hurdles of clinical trials is the carcinogenicity study. This is last because currently the animals are dosed for about 2 years. As selective processes have resulted in healthier strains of mice and rats from laboratory animal vendors, it is now sometimes necessary to dose these animals for as much as 30 months to achieve a target effect on morbidity and natural "background" or control tumor incidence.

It is necessary to do a full dose-ranging study prior to selection of the dose levels used in the carcinogenicity studies. This is done to establish an upper dose that is sufficiently high as to be intoxicating yet not so great as to induce excessive mortality and thus decrease the high dose population necessary to elicit and observe the incidence of tumorgenicity. It usually requires a full year for comprehensive tissue preparation, examination, and reporting after the study is completed and the animals have been sacrificed. The histopathology studies are best performed by experts in animal pathology.

The modern industrial drug development process requires electronic data processing support and substantial statistical capability in order to evaluate and report the enormous data bases that drug safety evaluations produce. While electronic data processing support is commonly provided to the clinical operations of contemporary pharmaceutical firms, it is not generally recognized that the data base in the drug safety disciplines is equal to that from clinical studies or even greater in magnitude and requires the same sophisticated remedy.

XV. THE CLINICAL TRIAL

It has often been said in jest by those involved in the new drug clinical development process that "their product is paper." It must indeed seem so to the industrial clinical scientists involved in drafting many revisions of protocols, correspondence with principle investigators, investigators' new drug brochures, clinical study reports, summaries of the relevant literature, summaries of the full Phase I program (and Phase II and III), reports to management, comprehensive reports on clinical adverse events or reactions, and the clinical optional expanded summary for a New Drug Application (NDA). The writing is, indeed, prodigious but constitutes a most important and essential part of pharmogeny.

As most of this book deals specifically with various aspects of the clinical research process as it is employed in the pharmaceutical industry, this overview chapter will treat that aspect only briefly.

The clinical development of a new drug really begins with a formal research management decision to do so.

A. The Area Team

A number of important steps routinely precede this management decision. The recommendation to advance a Biological Lead Compound to a Clinical Lead Compound is made by a multidisciplinary scientific group. In many pharmaceutical firms such eclectic groups of scientists are assembled monthly as area teams to review and assess the laboratory findings on compounds of interest. Area teams are topical and are customarily dedicated to a specific area of therapeutics or diseases. For example, the scientists of a cardiovascular diseases area team might address themselves to novel approaches to drug discovery for hypertension during one period of years and to innovative drug treatment for the atherosclerotic diseases in a subsequent period of years. The choice, of course, is directed as described earlier by the corporate marketing strategic plan. It is the scientists of the area team that design and evaluate relevant biological screens and evaluate the data they generate. This same group identifies the Chemical Lead Compound and suggests to the chemistry department ideas for improving on that lead. The chemists who participate in the area team communicate such discussions to their own management and develop full synthetic programs responsive to their management.

The scientific area team designates a Biological Lead Compound when a critical mass of biochemical, pharmacological, and early drug safety evaluation results generates high interest in the chemical. At this point the involved scientists formalize plans for further in-depth biological study of the lead and stimulate an agenda for pharmaceutical, drug disposition, analytical chemistry, and toxicological evaluations that more completely defines this new agent. Area teams regularly develop such Biological Lead Compounds, but it is much less common when the consensus of their deliberations on the expanded preclinical data base causes them to recommend to management that the subject of their intense interest merits advancement to a Clinical Lead Compound status. This designation is tantamount to the proposal that the chemical be the subject of an Investigational New Drug filing (IND) and that the research management facilitate the development of all data on the potential new drug required for such a regulatory filing.

The details of this development process have been presented earlier. The area team presents its drug candidate in a formal document; and subsequent to this promulgation a major seminar on the drug is addressed to the company general R&D community, R&D management, invited consultants, market planners, and executive management. The intramural drug monograph is an anthology of all that is known about the new agent and presents the findings of every scientific discipline. Also described in the text is the patent status, a profile of competitive drugs, a market plan, and an outline of the clinical operating plans for evaluations essential to creating an approvable NDA or international regulatory filings. If the drug monograph and the drug conference satisfies the questions and expectations of R&D management, the decision is generally made to proceed to IND preparation.

B. The "First in Man" Committee

At a defined point in the IND preparation process, a specific group meets to consider the question: "Are there now adequate, sufficient and appropriate data to recommend the investigation of this new drug in man?" In some pharmaceutical firms, this is the responsibility of a group of research managers referred to as the First in Man Committee. These physician-scientists and basic scientists who are also professional R&D managers review the drug safety evaluation data, as well as drug disposition, pharmacological, biochemical, chemical, analytical, biopharmaceutical, and all other relevant data and decide if the available data base is sufficient to advance the new drug substance to a well-defined clinical pharmacology study in volunteers.

Should the data base on the drug be found to be insufficient or to raise some question, the First in Man Committee attempts to define the additional preclinical studies required to satisfy the questions in order to advance the drug to study in humans. If the committee concludes that company data demonstrate that it is inappropriate to advance the new drug substance for evaluation in humans under any circumstances, the substance is deleted from further consideration and all preclinical work with it is terminated. A positive conclusion results in the filing of the IND.

C. Phase I Clinical Studies

When a sponsor, usually a pharmaceutical company, wishes to begin the clinical trials process in the United States, a Notice of Claimed Investigational Exemption for a New Drug, (referred to as the IND and identified as Form FD 1571) is submitted to the FDA (18). The IND requirements are summarized in Table 2. A detailed description of these requirements is contained in the Code of Federal Regulations, Title 21, Part 312.1. A plan or outline of proposed clinical investigations is required under Item 10. Phases I and II are defined as clinical pharmacology, and Phase III as broad clinical trial.

Phase I studies are the initial evaluations of the new drug substance in humans. In industrial drug development, it is usual to enlist normal (healthy) volunteers to participate in these early studies. The primary reason for selecting normal volunteers for Phase I studies is that the individuals selected must be healthy in all respects and clinical laboratory test results, vital signs, electrocardiograms, and whatever else is monitored must be in a normal range prior to drug treatment. Any abnormalities in these tests found subsequent to treatment with the drug can be reasonably attributed to drug effect. If the initial studies are done in patients with disease, the clinical laboratory tests and other evaluations are likely to be outside the normal range in some respect before the investigational drug is administered. Further, the effect of the drug on existing but occult organ system impairment cannot be determined in advance. It would not be obvious which of the observed effects are due to preexisting disease with concomitant organ dysfunction and which are assignable to the investigational drug.

The Phase I studies establish human tolerance to the drug, absorption and elimination kinetics, blood concentrations as a function of time after dosing, the metabolic profile, and pharmacological effects at various doses. If, however, a drug has dramatic biochemical or pharmacological effects specifically tailored to address a specific disease state, Phase I studies are done only in persons with such a disease state.

D. Phase II Clinical Studies

Phase II studies, are, of necessity, done in persons with signs or symptoms of the disease for which the drug is intended. This is because the purpose of this evaluation is to gain evidence of efficacy and to establish the proper dose and dosing interval. In this Phase II clinical program, more safety and tolerance data are acquired in patients with disease which the drug is intended to treat. Physicians expert in doing Phase I and II trials are referred to as clinical pharmacologists. Such persons must be well trained in human pharmacology, toxicology, biopharmaceutics, and pharmacokinetics and be cognizant of government regulations relating to human research.

Table 2 Contents of an IND

Item 1:	Name, chemical structures, dosage form, and route(s) of administration of the new drug
Item 2:	List of all components of the drug entity, including reasonable alternates for inactive components
Item 3:	Quantitative composition of the drug entity, including reasonable variations that may be expected during the investigational stage.
Item 4:	Source and preparation of new drug substances used as components; this includes manufacturing processes for new drug substances(s) and dosage form
Item 5:	Methods, facilities, and controls used for the manufacturing, processing, and packing of the new drug; the establishment and maintenance of appropriate standards of identity, strength, quality, and purity
Item 6:	Preclinical pharmacology, toxicology, and drug metabolism data; available clinical data if drug used previously (e.g., in another country) or is a combination of previously investigated or marketed drugs
Item 7:	Informational material to be provided to investigators; this includes a copy of the labels to be on the drug containers identifying the drug as investigational and a clinical monograph describing the drug, possible utility, prior investigations, and known hazards, contraindications, side effects, and precautions
Item 8:	A statement of the training and experience required of investigators
Item 9:	The names and credentials of the monitors and investigators' responsibilities regarding record keeping, informed consent, and supervision of subjects
Item 10:	Outline of the clinical investigation, including specification of phase involved: Phases I and II (clinical pharmacology) or Phase III (broad clinical trial)
Item 11:	Agreement to notify FDA if investigation is discontinued and why
Item 12:	Agreement to notify investigators if investigation is discontinued or an NDA for the investigational drug is approved
Item 13:	Completed only if sponsor wishes to sell rather than distribute test drug free to investigators; the reason for the need to sell must be explained

Table 2 (cont.)

Item 14:	Agreement not to ship drug or use in humans until 30 days after receipt of IND by FDA
Item 15:	An environmental impact statement when requested
Item 16:	Statement that all nonclinical laboratory studies comply with Good Laboratory Practices (GLPs)

E. Phase III Clinical Studies

Phase III is the program of broad clinical trials which is needed to determine whether a drug is appropriately safe and effective and has the therapeutic attributes required to satisfy the needs expressed by the market analysis. The immense cost of pharmaceutical development is particularly evident when these large-scale clinical trials are undertaken. It is possible, albeit undesirable, of course, that the Phase III evaluations of the new drug will not demonstrate any advantage over currently available marketed drugs. Faced with this reality, a pharmaceutical company may choose to accept the loss of its significant expenditures to that point rather than to increase them by pursuing a poor competitor into therapeutic commerce. For most drug candidates there is, unfortunately, no certain path other than the Phase III experience that delineates the actual therapeutic scope or profile of a new drug. Compounding the costs even further are the completion of the full reproduction studies and the very costly carcinogenicity studies, as well as the initiatives that must be taken in chemical and pharmaceutical manufacturing, in parallel with the Phase III studies. It is no wonder that figures of the order of $70 million are cited as the average cost for each successful new chemical entity drug product approved for marketing by the FDA. By way of contrast, it is of interest to note that the average R&D cost for a new drug brought to market during the 1950s was only about $1.5 million (19).

The objective of all of the clinical trials is to produce clear and well-documented evidence that the new drug candidate is effective ("substantial proof of efficacy based on adequate and well-controlled clinical investigations") and is safe when used in the manner intended. A close collaboration between pharmacists and the company clinical monitors is essential to design rigorous methodology of "blinding" (all dosages made to look alike), packaging (bottles, blister packs, etc.) and labeling for these "well controlled" often "double-blind" clinical studies. It is not unusual for thousands of patients to be involved in a massive effort to provide the required proof. The completed case report forms for each of these patients must reflect, in full fidelity, the clinical experience of the patient throughout a clinical trial. These case report forms must be complete and their contents checked for accuracy against source documents. This last step is essential because it is only the data on the case report form that the sponsor has to digest, analyze, process, and tabulate. The assurance of complete fidelity of the clinical trial to the methodology of the protocol and monitoring of the quality of the data on the case report form is the responsibility of medical scientists in the company clinical research group. Mathematical processing of compilations of data from all involved patients as reported on case reports provides the basis of the claim for efficacy and safety of the new drug substance.

These data, together with detailed reports of chemistry, pharmacology, toxicology, biopharmaceutics, manufacturing, and quality controls, are assembled in a New Drug Application (NDA) and submitted to the FDA. It is not unusual for an NDA to be significantly greater than 100,000 pages in length and for the three copies required by the FDA to weigh half a ton or more.

XVI. PROJECT MANAGEMENT

Pharmogeny as developed and practiced in the pharmaceutical industry is characterized by manifold disciplines which contribute to the process in a rather rigorous and systematic manner. The activities and reports of many scientists and disciplines must be projected, scheduled, and coordinated in such a manner as to engender a harmonious program and efficient use of resources. When this is not done, various parts of the program are out of phase and incoherent. This results in duplication of effort, significant loss of time, and unnecessary costs. To avoid this, it is now quite common to formalize this pivotal coordinative function into the separate and distinct discipline of project management (20). The persons involved in project management are often scientists with sufficient scope of experience so as to have a good functional understanding of the contribution of all component R&D groups. The coordination of the program of studies of all disciplines results in cost and time savings and contributes greatly to the assurance of the success of multiple, concurrent drug development projects. The positive impact of good project management and dynamic project coordination is difficult to overestimate. When this function is done optimally it requires electronic data processing support and an interactive system with all participating research managers. Such use of computer technology for project management is now common in the pharmaceutical industry.

XVII. THE PRODUCT

The industrial new drug development process begins with sophisticated market analyses and comes full circle when an NDA is approved. It is then that the pharmaceutical company competes for the attention of the prescribing physician and attempts to make known the attributes and significance of the new drug. If the new therapeutic entity offers advantages to the patient, and the marketing arm performs well in communicating its virtues, then the business of the pharmaceutical firm will grow and prosper.

This is well, for disease is still prevalent and all R&D communities require continuing support and resources. Past successes have left only the most difficult and most challenging therapeutic needs as the remaining targets for the industrial new drug development process. These present therapeutic needs are even now stimulating the pharmogeny of the future.

REFERENCES

1. J. E. Schnee and E. Caglarcan, Economic structure and performance of the ethical pharmaceutical industry. In *The Pharmaceutical Industry* (C. M. Lindsay, Ed.). Wiley, New York, pp. 23-40, 1978. [See also *Drug Development and Marketing* (R. B. Helms, Ed.). The American Enterprise Institute for Public Policy Research, Washington, D.C., 1975.]
2. W. M. Wardell, The history of drug discovery, development, and regulation. In *Issues in Pharmaceutical Economics* (R. I. Chien, Ed.). Heath, Lexington, Massachusetts, pp. 3-11, 1979. [See also a more quantitative analysis in W. M. Wardell, M. Hassar, S. N. Anavekar, and L. Lasagna, The rate of development of new drugs in the United States, 1963-1975. *Clin. Pharmacol. Ther. 24*:133-145 (1978).]
3. W. M. Wardell, M. C. Tsianco, S. N. Anavekar, and H. T. David, Postmarketing surveillance of new drugs. *J. Clin. Pharmacol. 19*:85-94, 169-184 (1979).
4. P. Ehrlich, Pro and contra Salvarsan. *Wien Med. Wochschr. 61*:14-19 (1910).
5. K. H. Beyer, *Discovery, development and delivery of new drugs*. Spectrum, New York, pp. 14-15, 1978.
6. L. H. Sternbach, The benzodiazepine story. *Prog. Drug Res. 22*:229-266 (1978).
7. T. I. Kalman, Enzyme inhibition as a source of new drugs. *Drug Dev. Res. 1*:311-328 (1981).

8. D. W. Cushman, H. S. Cheung, E. F. Sabo, and M. A. Ondetti, Development and design of specific inhibitors of angiotensin-converting enzyme. *Am. J. Cardiol. 49*:1390-1394 (1982).
9. F. Hartley, Recent examples of chemicals as new drugs. *Chem. Britain July*:300-304 (1966).
10. L. E. Martin, The role of pharmacokinetic and metabolic studies in new drug discovery. *Chem. Industry May 15*:430-436 (1976).
11. G. A. Brewer, Biopharmaceutics: The role of the analyst. *Anal. Chem. 45*:702A-706A (1973).
12. Guidelines for biopharmaceutical studies in man, American Pharmaceutical Association Academy of Pharmaceutical Sciences, February 1972, p. 17 (quoted in Ref. 11).
13. R. M. Atkinson, C. Bedford, K. J. Child, and E. G. Tomich, The effect of particle size on blood griseofulvin-levels in man. *Nature 193*:588-589 (1962).
14. G. Levy, J. R. Leonards, and J. A. Procknal, Development of in vitro dissolution tests which correlate quantitatively with dissolution rate-limited drug absorption in man. *J. Pharm. Sci. 54*: 1719-1722 (1965).
15. W. B. Abrams, Introducing a new drug into clinical practice. *Anesthesiology 35*:176-92 (1971).
16. E. I. Goldenthal, Current views on safety evaluation of drugs. *FDA Papers 2*(4):13-18 (1968).
17. J. K. Lamar, Considerations of the effect of drugs on reproduction. *J. Clin. Pharmacol. 8*:9-14 (1968).
18. W. J. Gyarfas and A. Welch, The IND procedure: Assuring safe and effective drugs. *FDA Papers 3*(7):27-31 (1969). [See also proposed revisions to IND regulations and an informative preamble, prepared by the FDA, offering an overview of IND procedures as they are practiced today by pharmaceutical companies and independent clinical investigators, in *Federal Register 48* (June 9): 26720-26749 1983.]
19. V. A. Mund, The return on investment of the innovative pharmaceutical firm. In *The Economics of Drug Innovation* (J. D. Cooper, Ed.). The American University, Washington, D.C., pp. 125-138, 1970.
20. W. C. Wall, Jr., Ten proverbs for project control. *Res. Management 25*:26-29 (1982)

2
RESEARCH PLANNING AND DEVELOPMENT PERSPECTIVES

Richard E. Faust
Hoffmann-La Roche Inc.
Nutley, New Jersey

I. R&D PLANNING: APPROACHES AND CHALLENGES

The broad goal of research planning is to improve the drug discovery and development process and increase R&D productivity. Therefore, planners are concerned with many aspects of the organization and the environments affecting research operations. Pharmaceutical firms vary in their approach to planning. In some companies a fairly liberal and flexible philosophy prevails which is characterized by a somewhat less structured system to select and monitor projects and extablish priorities. At the other end of the spectrum, some organizations employ quantitative approaches and more controlled or systematized schemes to aid decision making. In most cases the approach used reflects the thinking and orientation of the senior research executive. Some may feel that unfettered and open scientific endeavors patterned after academic research will be most productive in the long run. However, in most firms today there is a movement to generate more structured, quantitative, and computerized planning techniques. One must always keep in mind when installing such control systems that in the very nature of things research is a field of endeavor in which the unknown and unexpected is always present. Therefore, it will always be impossible to assemble all existing information at the onset of any plan of action and establish an immutable schedule of events. One must be careful in guiding research activities to avoid rigidity and the destruction of the creative and innovative spirit through bureaucratic "administrivia" and "planning overkill."

Most pharmaceutical firms develop functional or operational research plans which convey the status of research activities and commitments, present a profile of research resources and their deployment, and display a projection of outputs both in the near and long term. Many plans cover three time frames: (1) near term, or 1-5 years, with emphasis usually on the first year; (2) midterm, or 5-10 years, which is a reasonable time frame to include most compounds in early development stages; and (3) long term, or 10-20 years or longer, reflecting the more exploratory or basic research programs sometimes described as "new horizon research."

What makes research planning difficult? What are some of the trends or patterns that present challenges in managing the research process and ensuring success? What are some emerging concerns in the drug discovery and development cycle today?

1. In most R&D organizations there are more project possibilities than can be carried out under existing personnel and funding resources. Therefore, project/program selection systems and priorities become important.
2. Increasingly, the modern pharmaceutical research organization is composed of numerous specialists and subspecialists from various disciplines, each bringing their expertise to the discovery and development process. It is essential that these experts come together in an integrated and productive manner.
3. An effective research group needs to generate a portfolio of projects and activities that will ensure a balance between near-term and long-term output. All too often pressures tend to force effort toward the shorter range and more immediate output, which may jeopardize the future.

4. Environmental forces both within and outside the corporation affect research operations and the management of the R&D process. Research administrators must recognize this trend and develop appropriate responsive strategies.
5. The essence of research is characterized by uncertainty, risk, and the unknown. Although through the use of various planning, monitoring, and analytical techniques we can "plan" and "control" research, we must also provide within the system for flexibility, serendipity, and the unexpected.
6. Research is carried out by people. The R&D management process must be concerned with essential motivational factors that are critical when managers seek to encourage creativity and a productive scientific milieu.
7. A successful R&D operation must be concerned with the entire process from innovation and drug discovery through clinical evaluation and New Drug Application (NDA) approval. A balanced admixture of resources must be constructed so that some vital link in the NDA generation chain is not weak or inoperative.

II. THE RESEARCH ENVIRONMENT

The research operation in the modern pharmaceutical firm is affected by two environments—the external (that outside the corporation) and the internal (that within the firm) [1]. The positive forces or constraints generated by these interlocking and interdependent systems influence research success and ultimately the flow of new drugs to the patient. Important external factors include: (1) economic conditions, such as the rate of inflation and the cost of energy and capital; (2) numerous social forces, such as the growing influence of the consumerist movement and price pressures on pharmaceutical products; and (3) the impact of a whole host of regulations that have increased the costs and time frames associated with drug discovery and development. Within the corporation the research operation must compete for funds with other operations and be responsive to the probing concerns of corporate planners, financial experts, and marketing personnel.

The pharmaceutical and biomedical researcher faces many new challenges as economists, lawyers, budget planners, diplomats, technology assessors, politicians, and consumer representatives study the research process and seek answers to productivity, cost/benefit, and risk/benefit considerations. If these concerns are not addressed adequately and realistically, they may have an inhibiting impact on scientific work and the milieu for discovery and technological innovation. Therefore, research personnel need to communicate with those groups outside and within the corporation who are questioning R&D activities and resource allocations and who influence research funding and support. Since the ultimate consumer of the output of research operations is the patient, the clinician can play an important role in this communication process.

III. THE DRUG DISCOVERY AND DEVELOPMENT PROCESS

A. Two Critical Stages

The costly drug development cycle in the United States now extends from 10 to 13 years. In this complex process two critical stages may be identified (Fig. 1). The first is the selection of a lead compound for development, a stage which involves primarily interaction among biologists, chemists, and pharmacologists. Close coordination and rapid recycling of information is essential, so that candidate compounds can be screened quickly and guidance provided to the chemical staff. Although the medical department is not usually involved at this stage, the counsel of clinicians can be another valuable input in the drug screening and selection process and should be encouraged. The second critical step occurs after extended pharmacological and toxicological testing and consists of the evaluation of the compound for the first time in man (Phase I). There is a growing trend in certain therapeutic categories to combine Phase I and Phase II studies whenever possible and this is veiwed by many as a favorable development. These early clinical trials confirm animal and other preclinical observations and indicate whether or not the compound is active in man and has a reasonable therapeutic ratio.

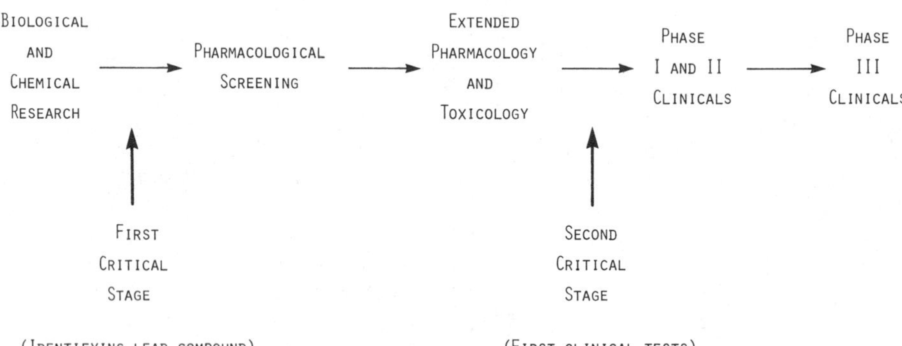

Figure 1 Two critical stages in the drug discovery and development cycle.

For the successful discovery and rapid development of new pharmaceuticals we need to accelerate the movement of new, valuable therapeutic agents to and through these two critical stages. We need to foster more creative investigations enabling us to understand better the significance of various observations in animal studies. Research on new experimental methods and models that are more predictive of success in the clinic should be encouraged. More research resources should also be directed to the study of various basic biochemical processes initiating the degenerative "wear and tear" diseases, so that compounds can be created to prevent or ameliorate undesirable changes.

At the second critical stage of the R&D cycle, the movement of compounds into man, we need skilled and observant clinical pharmacologists. At this stage and also as compounds move into broader Phase III trials the phenomenon of "clinical serendipity," which is the accidental or unexpected discovery of various clinical effects, can be an important factor in the generation of new therapies. Many historical examples of such drug usage may be noted. Recently, however, we have seen propranolol broaden its approved use from an antiarrhythmic in 1967 to an antianginal in 1973, to an antihypertensive agent in 1976, and to an antimigraine agent in 1979. It is now also being used to prevent second heart attacks. Since such utility, sometimes discovered through unexpected clinical observations, has been the method

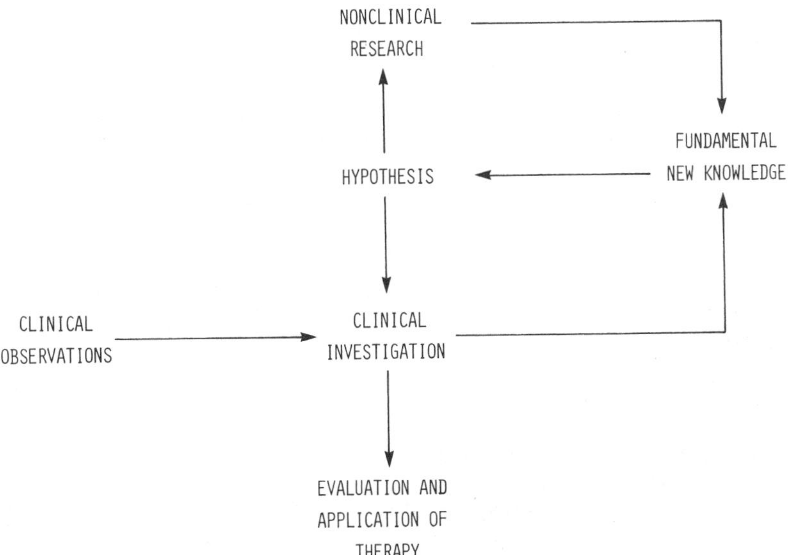

Figure 2 The role of clinical investigation in drug research and discovery.

by which an impressive array of new drugs and drug uses have been generated in the past, one can conclude that drug "discovery" is not restricted to the early phases of the R&D process but extends to all stages including postmarketing. Figure 2 illustrates the role of clinical observation and investigation in the evolution of new drugs [2]. Observations in the clinic may generate impetus for specific clinical programs aimed at providing new therpeutic applications. Clinical observations and studies may also stimulate fundamental knowledge about the behavior of drug substances and related pharmacological phenomena which results in new or revised hypotheses that trigger preclinical research or entirely new clinical investigations. In these ways the astute and observing clinical pharmacologist can be a prime factor in the discovery of utility in novel drug moieties and in designing new drug substances.

B. Changing Patterns in Drug Discovery

The nature of the drug discovery process has gradually changed over the past two decades. Many new developments and trends have had an impact on how research is being conducted and how exploratory programs are being carried out in the search for new pharmaceuticals. For example, developments in bioanalytical methods have had a profound impact on the assessment of drug action and therapeutic ratios [3]. Through radiometric, radioimmunoassay, and the newer enzyme-immunoassay techniques, the sensitivity, specificity, accuracy, reproducibility, and speed of assays have enabled workers to detect and study quantities of drugs in the body at nanogram and picogram levels. Such precision has merit, but it simultaneously opens up a Pandora's box concerning the significance of such quantities to drug utilization and safety. Along these lines, the race among research groups to be the first to discover new biologically active peptides may well be decided not so much by creative genius as by who can make the best use of high-performance liquid chromatography. Instruments such as the automated amino acid analyzer, the sequenator, and solid state synthesizer are being used widely for research on proteins and peptides that may have therapeutic value. The recent breakthroughs in the opioid peptides, characterized by the discovery of the endorphins and lipotropins, opened up new horizons in the biochemistry of analgesia and could not have occurred without such instrumentation.

The changing nature of the drug discovery process has reenforced the importance of biomedical scientists and clinicians. The three principal ways scientists search for new leads are screening, molecular modification, and the rational approach [4]. Random screening consists of synthesizing large numbers of new chemical compounds and evaluating them in various animal or other test systems with a view to finding relationships between chemical structure and biological activity.

This approach is costly and, although it has been used extensively in the past to find valuable new drugs, its relative importance is diminishing. Molecular modification involves developing compounds with optimal properties from a given type of biologically active structure. Often, the aim of this effort is to increase potency, alter the spectrum of activity, reduce side effects, extend the duration of action, or improve absorption. The strategy behind this approach seeks correlations between biological properties and structural or physiocochemical parameters. Increasingly, today, as a result of our improved knowledge of biochemical mechanisms and disease processes, a more rational approach is being used to design compounds that can be expected to interact specifically with biochemical targets or drug receptors. As this rational approach becomes even more important, the role of the clinician and his knowledge of the pathology of disease states will expand and contribute in a more significant way to drug innovation.

C. The Vital and Expanding Role of Clinicians

The cost of developing a typical new drug today can be as much as $70 million if one includes various exploratory efforts, the time value of money, and expenditures associated with compounds that do not survive the development process [5]. One assessment of the direct cost of the development of a typical "oral systemic drug for long term administration" places the figure at $24 million [6]. Table 1 describes

Table 1 Direct Clinical Costs at Various Stages of Drug Development

Stage	Activity	Cost $(thousands)
Pre-IND	Data for IND	200
Phase I	Safety-tolerance bioavailability	1,300
Phase II	Dose-range studies controlled efficacy	5,700
Phase III	Clinical efficacy long-term safety	3,700
NDA Wait	Long-term follow-up	1,100
	Claims/indications, extensions	
	Start Phase IV studies	
		TOTAL 12,000

half of these expenditures, or $12 million incurred by the clinical area, which is an indication of the large investments needed to demonstrate safety and efficacy in human clinical trials.

In many instances, however, the cost of Phase III studies can exceed those of any other stage of clinical effort. After NDA approval, substantial clinical programs may be carried out to strengthen existing claims, generate new indications, or introduce new dosage forms. Costs associated with these post-NDA activities can also be sizable. All of these cost perspectives illustrate the key role of the medical effort in the overall development of new pharmaceutical products.

In many research-oriented industries (e.g., communications, aircraft, or electronics) technicians and scientists within the firm are involved at all phases from the creation of a new product concept to its introduction. In the pharmaceutical industry, however, the critical stage of testing in human subjects is conducted by clinical researchers outside the company. Thus, in some respects the control of the program shifts and depends upon the expertise and cooperation of external clinical investigators. It depends also on the communication skills and talent of those members of the medical department within the firm. Members of the medical department must be selected carefully because of trends today in drug development that require exceptional communication and interactive skills. For example, most large multinational firms in the United States conduct many clinical trials overseas. This calls for careful planning and coordination between the various clinicians involved. Also, developing appropriate clinical protocols and processing the enormous volume of documentation involved in clinical studies requires that clinicians establish good rapport with other key personnel in the research operation, such as statisticians, data processors, clinical research associates, and project managers.

Various regulatory and related trends that impact on the latter stages of the drug development cycle or early marketing efforts increasingly involve the medical staff in some fashion. A number of examples may be noted: (1) The move to extend Good Laboratory and Good Manufacturing Practices into the medical area with the creation of Good Clinical Practice (GCP) means working closely with quality assurance personnel within the firm. (2) Some consumerist and regulatory forces have called for the approval of drugs based on comparisons with existing products. This relative efficacy position would preclude the introduction of a drug simply because there are existing products which already treat a disease "adequately." This would deny therapy to a patient who might not be able to tolerate one product but would respond to a similar but alternate product designed for the same illness. Physicians are becoming increasingly aware that since individuals respond differently to similar drug products an array of pharmaceuticals are needed so that a medication can be tailored to a particular patient. (3) The adoption of any formal postmarketing surveillance system geared to accelerate the drug approval process and establish limited postapproval utilization of a drug will probably involve a firm's medical staff in some way in the monitoring process. (4) Product liability claims against pharmaceutical firms are increasing in number and size of awards. Clinicians within the firm are often called upon to challenge such claims and provide evidence countering unjustified or exaggerated consumer complaints. (5) The proposed ban on the use of prisoners in penal institutions as research subjects represents another area in which members of a firm's clinical staff can influence public policy favorably and encourage support for research flexibility and initiatives.

In summary, the pattern that is emerging today in the discovery, development, and marketing of new pharmaceuticals favors a more significant role for clinicians and related members of the medical department.

D. The Impact of Licensing and Technology Transfer

In an effort to augment internal research capabilities and assure growth and/or diversification, many firms are seeking new products and research leads from external groups [7]. Various modes of interaction may take place, including drug screening agreements, licensing of specific agents, right of first refusal, cooperative product development, formal joint ventures, or acquisition of an entire company or one of its operating divisions. In most of these actions the role of the clinician is important in the assessment of the scientific merit of the compounds to be jointly developed or acquired. Often the physicians in the medical department of a company will be asked to evaluate data based on clinical studies conducted by others, including data generated overseas. Although personnel in quality control, manufacturing, finance, and other functional areas become involved in any technology transfer process, the three groups most directly concerned are law, marketing, and research. The scientific evaluation of acquisition candidates, however, is one of the most time-consuming and critical aspects of any such program and frequently clinical advice and judgments are crucial.

IV. PROJECT/PROGRAM SELECTION AND MANAGEMENT

The pharmaceutical R&D process consists of a number of project management steps encompassing idea generation and processing, project evaluation and selection, project monitoring and control, and project completion or termination. Each of these steps is interrelated, for after ideas are developed and projects initiated, alert managements need to monitor progress to be assured that the company is investing in those projects which will have the most beneficial effect on business success. Whenever a project encounters major technical problems or when it becomes obvious that commercial results will be significantly less than estimates used to justify the program originally, decision makers may elect to drop the project and invest funds in more promising activities.

Project management in the pharmaceutical industry is becoming more important due to a myriad of factors and trends, among them (1) the longer and costlier development cycles; (2) the complexity of modern science, which requires a large critical mass of experts and skills; (3) an emphasis on near-term productivity rather than longer-term exploratory research; (4) a multitude of external forces, including mounting regulatory requirements and consumerist pressures; (5) the shortened patent life following new product introduction; (6) the growing impact of licensing and technology transfer groups; (7) the concern for "orphan" drugs; (8) the influence of marketing and financial inputs as R&D productivity is scrutinized; (9) the shift of certain research overseas and a more global approach to project development; and (10) R&D management concern for maintaining a motivated and creative staff in an era when research funding is being reexamined and science itself is being questioned for relevancy in today's culture.

Of the various activities involved in the management of R&D, perhaps the most challenging is the selection of projects and programs. The judicious deployment of research resources is vital to the success of any firm. In the final analyses, the payoff from good project/program selection is a reduction in the time between initial discovery and eventual commercial introduction of attractive innovations. The chance of success and profitability is enhanced and at the same time expenditures of money and other resources on unprofitable project activities are minimized.

Project/program selection decisions are often influenced by a number of important considerations such as: (1) corporate policy; (2) the size of the research budget; (3) the type of research, i.e., exploratory, development, or defensive; (4) technical and scientific skills available; and (5) competitive activity. In a study of the research planning methodologies and approaches of six leading pharmaceutical firms, the selection process was found to be influenced by a number of scientific, marketing, and organization factors [8]:

Scientific Factors

1. Interrelationship with other research activities . . . synergistic advantages or competitive with other programs.
2. Probability of achieving project objectives.
3. Time required to achieve project objectives.
4. Impact on balance of short- and long-term programs within research.

5. Estimated cost of the project in the coming year and to completion.
6. Utilization of existing research talent and resources.
7. Value as a means of generating experience and gaining a technical expertise in a field . . . a foundation for future research activities.
8. Need for critical mass of expertise and activity to ensure progress.
9. Elasticity of resource input and probable output relationships.
10. Patentability or exclusivity of discoveries from project.
11. Competitive research effort in the area . . . in academic and government research centers.

Marketing Considerations

1. Projected sales and profits from effort.
2. Relationship to need as reflected by current state of consumer satisfaction.
3. Status and efficacy of current competitive products or means of meeting consumer need.
4. Compatibility with current marketing capabilities and strengths.
5. Influence of new competitive products under development.

Organization and Other Elements

1. Relationship to activities at other research centers or units within the company.
2. Timing of project with respect to other activities in marketing, research, etc.
3. Manufacturing capabilities and needs.
4. Prestige and image value to the company.
5. Effect on organization esprit de corps and attitudes.
6. Impact of governmental, public opinion, and other environmental pressures.
7. Alternative uses of scientific personnel and facilities if project dropped after a few years.
8. Moral compulsion to develop drugs meeting medical need but having low or no profit potential.

All of these various elements are not of concern in every decision influencing project selection and priorities, but the compilation does represent a check list of some of the factors which are often reviewed, sometimes intuitively, before decisions are finalized.

V. PERSONNEL MANAGEMENT AND MOTIVATION

At a time when many resources (e.g., energy, capital, or equipment) are escalating in cost, it is important that personnel be as fully motivated and productive as possible. Yet, all too often the environment for research is not ideally conducive to the motivation of the scientist nor sensitive to those special forces that foster innovation, enthusiasm, and creativity. A host of external groups have increased their surveillance of the pharmaceutical industry as they seek information and influence over various research operations and strategies. Some consumerist groups, for example, have advocated that no human testing be undertaken until all animal safety tests are completed, a move which would have added several years to the development time for pharmaceuticals and their availability to physicians. The frequent charge that "drugs are too expensive" is a negative feedback that certainly does not encourage the scientific investigator.

Within the corporation many groups compete for available funds and often the research function is the first to be cut when sales and profits patterns are not encouraging. It is somewhat of a paradox that when sales and profits are increasing rapidly management readily expands R&D operations, whereas in less favorable periods R&D is the first functional unit that is examined for possible "belt tightening." This can be discouraging to most members of the scientific staff, who tend to feel that in difficult times the rational approach would call for increased research efforts to generate new and improved products to reverse the negative pattern.

A successful research manager is aware of these external and internal forces that are affecting the morale and motivation of his scientific staff. He frequently seeks the counsel of personnel specialists who examine these relationships and explore insights on how to improve esprit de corps. Today, many organizational specialists are studying the outstanding success of Japanese industry and the role of the Japanese worker in this outcome. Japanese management is apparently skilled at inspiring a sense of purpose in workers. There is a general recognition that no matter how outstanding a manager may be in technical and other ability areas, if he is unsuccessful in motivating the average worker, his company will not achieve the necessary productivity levels and business results. In his book, *Theory Z: How American Business Can Meet the Japanese Challenge,* Ouchi provides a comprehensive analysis of the characteristics of the Japanese managerial style [9]. Some of the components of this style are rooted in Japanese culture and tradition and would be difficult to emulate. Other characteristics, however, are similar to those good management principles we have tried to foster in the United States and are, therefore, more easily pursued. Of these, one has especial relevance to pharmaceutical R&D and that has to do with the fact that successful drug discovery and development requires the input and collaboration of many individuals with different backgrounds and expertise.

There is a tendency for our managerial style to emphasize the importance of the individual and his accomplishments. We tend to have an "I,me, my" orientation rather than a more global perspective that stresses group effort and accomplishment. We often reward individual achievement rather than the group. We coordinate technology rather than people. We focus on self-centered rather than company-centered career patterns. We have formal and sometimes rigid reporting relationships that reinforce individual prerogatives and unproductive bureaucracy. Frequently the "information is power" syndrome discourages good communication and the sharing of important data or views. What results from all of this is a strong individual desire to "get ahead" above all else, which generates a lack of trust in the organization.

In order to encourage mutual trust, more widespread participation, and decision making down in the organization, as well as to develop "big picture" concepts emphasizing group effort and achievement paralleling the Japanese approach, many pharmaceutical research operations have organized task-oriented project teams. The Project Management Institute reported that in 21 of 22 pharmaceutical firms formal R&D project planning teams were functional and involved from the identification of a lead compound to NDA filing [10]. In such a task-oriented approach the traditional hierarchy is supplemented by or integrated with subunits called task forces or project teams. The distinctiveness of the subunits lies in the fact that (1) they are designed to deal with specific problems or to achieve relatively explicit goals or objectives and (2) they cut across traditional departmental and disciplinary lines in using the expertise of personnel. Because of this they provide the means for submerging the interests of the individual to the broader goals of the group and thereby encouraging mutual trust and shared responsibility.

VI. ASSESSING RESEARCH PRODUCTIVITY

A. Factors to Consider

As drug industry R&D expenditures neared $2 billion (in the early 1980s), the pressure on research and marketing groups to recoup these enormous investments has increased. A dominant concern in most pharmaceutical firms today related to the question: "How can we measure research output and productivity?" In one sense the ultimate criterion of success is the contribution that a research effort makes to the profit profile of the firm through new and improved products. Most large research-intensive firms need a major product success periodically to cover the cost of project failures and increasing R&D budgets.

In assessing "research productivity," however, we must recognize that the scientific success of a pharmaceutical firm is the result of more than the output of its R&D laboratories. Manufacturing must be able to make and marketing must be able to sell new products in an efficient and profitable manner. Therefore, all of the negative regulations and cost/pricing pressures that impact on these functions will affect how well research output is utilized and in turn the overall techno-scientific success of the business enterprise.

There are other factors that should be considered in assessing R&D progress and proficiency. For example, an analysis of activities related to research momentum and trends can provide indicators of success. A key question to be asked is: "Compared to the past, what are we doing better and what trends exist today in our research operation that support our goals, reflect research dynamism, and are indicative of progress?" Some parameters that might be evaluated are the number of (1) compounds synthesized; (2) animal screening tests performed; (3) patents filed and issued; (4) papers published; and (5) INDs/NDAs filed. In addition one can assess the magnitude of research resources directed to the support of existing products through the development of new dosage forms and/or new indications or to special studies designed to challenge unjustified negative publicity. As much as 20-25% of an R&D budget may be allocated to such "defensive" research activities. Other indicators of progress may relate to the balance between long- and short-term output and the development of a research organization that possesses sufficient flexibility to meet new challenges and rewarding opportunities quickly and effectively.

Many economists who study the drug industry tend to use the number of new single chemical entities (NCEs) as the sole measure of productivity. They ignore the other positive elements noted, as well as the valuable role of drug development in generating novel physiological and pharmacological insights, that is, the evolution of new basic knowledge broadening the frontiers of science that occurs during many drug development programs. The "storehouse of basic science" is often replenished during the course of drug discovery and development. For example, research groups in industry now exploring the arachidonic acid metabolic pathway in search of drugs having potential value in several therapeutic categories are uncovering new insights concerning the nature of this metabolic system [11].

B. Socioeconomic Benefits

One meaningful measure of research contributions relates to the positive impact of pharmaceuticals on the health and well-being of society reflected in (1) increased economic productivity through lives saved and increases in time on the job and (2) savings in medical care costs for society resulting from more rapid cures of illness and reduction of costly hospitalization and the need for constant care by physicians. Considerable historical evidence has been generated to demonstrate the socioeconomic value of new drug therapies in a number of disease areas.

In assessing the various modes of therapy and comparing different pharmaceutical approaches, the efficacy of drug products certainly must enter the decision-making process. However, improvements in economy and efficiency increasingly are being viewed as vitally important also. The traditional delivery of health care is changing today and this is affecting the mode of procuring, dispensing, and utilizing pharmaceutical products. More than ever before individuals other than the physician (e.g., formulary members, Medicaid administrators, pharmacists, and group practice officials) are playing increasing roles in the pharmaceutical selection process. Purchasing is becoming more centralized in an effort to achieve economies in procurement practices. The emergence of third party financiers of pharmaceutical products has created new decision makers with different needs and who have influence over product availability and purchase price.

Increasingly, there is need to demonstrate not only product effectiveness and safety in clinical studies but also, whenever feasible, economic benefits and advantages. Therefore, in designing certain clinical protocols or studies, clinicians and research planners should try to gather information that will measure various economic parameters, such as the following:

1. Number of days the patient is institutionalized with an average cost per day
2. Cost of administering the medication (e.g., by nurse, physician, or pharmacist)
3. The economies resulting from the reduction in the use of adjunctive drugs as a result of using the test drug
4. The number of days the patient is on active drug before he can return to his livelihood and the economic values involved
5. The economic value associated with the number of days lost from their means of support by family members who might be required to take care of patients on test drugs

C. Emerging Perspectives

The escalating cost and expanding time frame for the development of a drug has made the orientation of research efforts even more critical.

To ensure an adequate return on research investments, drug candidates must have a strong probability of demonstrating clear advantages over existing compounds or of meeting unfulfilled medical needs. Over the past 25 years many remarkable advances have taken place in the biological sciences that have not yet resulted in improvements in therapy. The virtual explosion of knowledge in biochemistry, immunoology, molecular and cell biology, and neurobiology should result in a new era of drug discovery and development in which agents are designed to interrupt a specific disease process rather than simply treating its signs and symptoms. The traditional approach of screening thousands of compounds for biological activity is gradually changing. Increasingly, the pharmaceutical research scientist is able to design drug molecules to perform specific pharmacological functions with minimum side effects. This has evolved as a result of new knowledge of biochemical processes within the cell, the body's immune response systems, the structure and function of the cell surface and receptor molecules, and a host of related biological mechanisms.

Developments in biochemistry have encouraged a greater focus on enzymes as specific drug targets. Efforts to identify, characterize, and develop inhibitors or activators for key enzymes involved in metabolic processes are increasing. The use of radioactive ligands has enabled researchers to achieve better biological characterization of various hormone and neurotransmitter receptors. The study of such receptors is yielding increased understanding of how hormones and neurotransmitters work that will permit the development of specific, potent antagonists (blockers), and agonists (mimics). Conceptual advances are occurring which will lead to novel approaches to inhibition or stimulation of immunological pathways, that is, control of individual cellular events in the generation and expression of immune responses. The study of metabolic pathways will become even more important in the search for new drug leads. As new compounds are isolated which are involved in important physiological processes, the routes of synthesis and degradation and the role of associated enzymes will provide important clues for drug development projects. Neurobiology is another exciting area and the potential therapeutic implications of synthetic peptides will grow substantially as more active peptides are isolated and peptide analogs resistant to proteolysis are synthesized. Molecular biology and gene technology have progressed dramatically and recombinant DNA technology represents one of the most exciting and potentially productive areas of biomedical research and development.

Therefore, one of the clearest trends is that the biological sciences will play an even more important role in drug innovation. The productivity of a pharmaceutical research operation will depend on how well it organizes to pursue these various exciting biological frontiers.

VII. INDUSTRY-GOVERNMENT-ACADEMIA INTERACTIONS

Drug research is now oriented toward the major disease areas, such as cancer and cardiovascular disorders, and governmental groups are becoming more and more concerned with these diseases that affect large segments of the population. Research emphasis is shifting to the tough, hitherto intractable diseases such as arteriosclerosis and coronary artery disease, infectious diseases of viral origin, disorders of metabolism such as diabetes, and the whole formidable group of serious conditions associated with aging including essential hypertension, rheumatoid arthritis, and cancer.

Biomedical research today is a giant enterprise linking universities, industry, and a host of nonprofit organizations with federal agencies in unprecedented activities and relationships. A system of government grants and contracts providing for various kinds of scientific exploration and technological development outside the government itself is a major element of this new phenomenon. While the government has not directed specific attention to the development of drug products, it is supporting much of the basic research being conducted in the pharmaceutical and related industries through the National Institutes of Health (NIH) and other federal laboratories in addition to its support of academic research; and it is from this storehouse of knowledge that fundamental new leads often emerge.

Because of the complex nature of research required to solve the serious health problems of today, there is increasing need to seek new modes of interaction and collaboration between the public and private sector with respect to biomedical research and new drug development. Society should encourage means of enhancing communication and cooperative relationships between research personnel in industry, government, and academia so that they can bring together synergistically the total resources and skills available for the greatest common good. Innovative new programs for encouraging interaction should be explored which will allay government fears that the taxpayers' money may be used to benefit a particular firm and industry fears that government involvement may jeopardize its patent position if a new drug emerges from a cooperative program.

VIII. SUMMARY

The discovery and development of new pharmaceuticals, from the synthesis of a new agent to a marketable drug, is a long and costly enterprise requiring good planning and the collaboration of many specialists. In this chapter a number of global perspectives affecting this complex process are elaborated, including the impact of environmental forces and the nature of research planning activity. As the overall research process is described, including key factors that influence its success, special attention is directed to the role of clinical investigation and investigators, as well as emerging trends and patterns that influence R&D planning, including the clinical research phases.

REFERENCES

1. R. E. Faust, Economic restraints on research and development in the pharmaceutical industry. Presented at the Conference on Regulation and Restraint in Contemporary Medicine in the United Kingdom and United States, sponsored by the Royal Society of Medicine, London, October 12-14, 1981.
2. J. A. Oates, Regulatory trends in early clinical trials. Presented at the Conference on Regulation and Restraint in Contemporary Medicine in the United Kingdom and United States, sponsored by the Royal Society of Medicine, London, October 12-14, 1981.
3. S. Udenfriend, Translation of basic biomedical research into practical applications. Presented at dedication of Bodman Laboratories for the Life Sciences, Stanford Research Institute, Menlo Park, Cal., February 17, 1977.
4. B. Berde, Industrial research in the quest for new medicines. *Clin. Exp. Pharmacol. Physiol. 1:*183-195 (1974).
5. R. W. Hansen, The pharmaceutical development process: estimate of current development costs and times and the effects of regulatory changes. In *Issues in Pharmaceutical Economics* (R. J. Chen, Ed.). Heath, Lexington, Mass., 1979.
6. M. Katz, The birth pangs of a new drug. *Drug Cosmetic Ind. 127:*40-44 (1980).
7. R. E. Faust, Acquisition/licensing strategies for pharmaceutical products. *Drug Cosmetic Ind. 112:*48-50 (1973).
8. R. E. Faust, Project selection in the pharmaceutical industry. *Res. Management 14:*46-55 (1971).
9. W. G. Ouchi, *Theory Z: How American Business Can Meet the Japanese Challenge.* Addison-Wesley, Reading, Mass., 1981.
10. C. A. Trautwein, Survey of pharmaceutical R&D planning departments. Presented at the Project Management Institute Meeting, September 29, 1981.
11. R. E. Faust, Statement presented before the Congressional Subcommittee on Natural Resources, Agriculture Research, and Environment and the Subcommittee on Investigations and Oversight, Washington, D.C., April 27, 1981.

3
HISTORICAL REVIEW

Donald D. Vogt and Michael Montagne*
University of Kentucky
Lexington, Kentucky

Substances able to exert their final action exclusively on the parasite harbored within the organism would represent, so to speak, magic bullets which seek their target of their own accord.

—P. Ehrlich, 1906 [1]

Study of molecular shape (conformation) and reactivity can aid in formulating models of drug receptors and mechanisms of drug action. Computer-assisted molecular modeling makes such studies convenient to perform.

—P. Gund and colleagues, 1980 [2]

Present Affiliation: School of Hygiene and Public Health, Johns Hopkins University, Baltimore, Maryland

I. INTRODUCTION

Within this century, there have been several paradigmatic shifts in the approach to the discovery and development of drugs: the testing of natural substances for useful properties; the clinically empirical analysis of large numbers of substances (i.e., screening) for a desirable activity in model systems; and the systematic molecular manipulation of original or parent compounds to optimize their pharmacological properties. More recently, computer-assisted molecular modeling may permit the testing and refining of ideas before laboratory work is undertaken. The contemporary paradigm of drug development evolved from Ehrlich's notion of drug-receptor interactions. The drug-receptor theory is epitomized by the elucidation of structure-activity relationships between chemical substances and physiological receptors and has been useful in explaining the activity of known drugs and more recently in designing new drug molecules [3].

During the first decades of the present century, a paradigmatic shift in therapeutics is also discernible. The traditional centuries-old drug armamentarium had been largely superseded in practice in the nineteenth century by such drug entities as the alkaloids, endocrines, and the antisera of Emil von Behring. After 1904, Paul Ehrlich's concept of chemotherapy was seminal to a new era in therapeutics; the search for "magic bullets" had begun. Through much of the nineteenth century, the pharmaceutical industry prepared dosage forms of the traditional materia medica in response to a demand by practitioners. The development of totally new drug entities reversed the relationship between medical practice and industry in which the latter would increasingly lead innovations in practice. Particularly after 1930, determination of the rational medical uses of drugs through informal experience in practice with different patients was no longer considered adequate or safe. At the same time, the drug industry was called upon to provide something more than assurance of quality; the innovators were now also creating the knowledge of drug use and transferring this knowledge to the practitioner.

One of the principal problems of the industry has been to formalize and to ensure the integrity of the clinical research process. Idiosyncratic empirical studies by independent practitioners have largely been replaced by a more tractable process of controlled clinical experimentation to provide assurances of safety and efficacy within acceptable limits. Instead of a methodic study and observation of the effects of an unknown chemical, contemporary therapeutics hypothesizes the main effects a priori and places the drug on "trial" to prove pharmacological activity (i.e., efficacy) and to determine the significance of side effects (i.e., safety).

II. GENESIS OF THE CLINICAL RESEARCH PROCESS

The development of the research process probably had its genesis in Renaissance Europe, becoming fully evolved as a self-conscious process in the seventeenth century [4]. Early Greek philosophers constructed a system of explaining natural phenomena on the basis of speculative reflection. Hippocratic medicine was based both upon a knowledge of the natural sciences of the time and, more important, upon clear and rational reasoning of cause-and-effect relationships. Emphasis was placed upon the value of practical observation of the process of disease rather than upon disease theory.

Galen was the principal authority in anatomy and physiology until the seventeenth century. It was not until this dogma was challenged by renaissance scholars that modern medical science could begin to take shape. The authority of Galen was tested by Andreas Vesalius, who demonstrated in his *De Humani Corporis Fabrica* (1543) serious discrepancies in Galen's anatomic teachings. William Harvey's *De Motu Cordis* (1628) integrated known but ineffective facts into a new and comprehensive generalization. His theory was fully supported by experimentation. Harvey was but one participant among many in a "scientific revolution," demonstrating by experiment and accessory evidence a conclusion that was diametrically opposed to traditional assumptions [5,6].

In the eighteenth century, William Withering introduced the rational clinical use of digitalis through controlled clinical experimentation. His 10-year study of digitalis and his clinical findings, based upon 163 case histories, was published in 1785. These simple clinical trials were notable for the use of unselected cases [7]. François Magendie, the nineteenth-century French clinician, was the acknowledged founder of experimental pharmacology, modern experimental pathology, and modern experimental physiology. The chemical isolation of the alkaloids and the halogens afforded the clinicians and the physiologists a number of chemically pure substances of known composition which allowed for true quantitative experiments. This had not been possible with the old extracts and raw substances with their uncertain concentrations of active principles. Working with the various recently isolated alkaloids, Magendie proceeded in a rational order: preparation of the substance from raw materials; determination of the physical and chemical properties; effects upon animals; effects upon healthy and diseased human subjects; indications; and application in various dosage forms [8].

Magendie's pharmacological research was continued by his pupil Claude Bernard. His *Introduction to Experimental Medicine*, published in 1865, synthesized the basic philosophic ideas of the nineteenth-century physiologists. He insisted upon proof and counterproof in confirming experimental data as expressions of philosophic doubt carried as far as possible [9]. It was in the Paris School of clinicians that systematic clinical studies became routine through the efforts of Pierre Louis, who introduced his numerical (i.e., statistical) method in the 1820s. Although not the first clinician to use statistics, Louis was the founder of medical, as opposed to vital, statistics [10].

The concept of clinical study in a patient population and the use of control groups in the design of experiments may have begun with James Lind. His epochal *Treatise on the Scurvy* was published in 1753:

> [I] took twelve patients . . . with scurvy. . . . Their cases were as similar as I could have them. . . . They lay together in one place and had one diet common to all. Two of these were ordered each a quart of cyder a day. Two others took twenty-five drops of elixir of vitriol three times a day upon an empty stomach. Two others took two spoonfuls of vinegar three times a day. . . . Two of the worst patients were put upon a course of seawater. Of this they drank half a pint every day. Two others had each two oranges and one lemon given them every day. The two remaining patients took an electuary recommended by a hospital surgeon made of garlic, mustard, balsam of Peru, and myrrh. The consequence was that the most sudden and visible good effects were perceived from the use of oranges and lemons, one of those who had taken them being at the end of six days fit for duty. The other was the best recovered of any in his condition [11].

A half century later, in 1798, Edward Jenner performed controlled experiments in man to assess the value of vaccination against smallpox [7]. However, the use of a control group for comparative purposes was not realized until the twentieth century.

The scientific method evolved from René Descartes' work in geometry and was given its first description in his *Discourse on Method* (1637). Descartes suggested the application of specific mathematical methods to all types of scientific inquiry, outlining a method for analysis which still dominates medical research today. The major tenets of this method are a priori reasoning and a physicalistic approach, epitomized today by quantitative, experimental designs in studying phenomena. John Graunt undertook the first attempt to interpret mass biological phenomena and social behavior from numerical data (births and deaths in London) in his treatise *Natural and Political Observations Made Upon the Bills of Mortality* (1662).

Jacob Bernoulli's *Ars Conjectandi* (1713), the first attempt at deducing statistical measures from individual probabilities, resulted in what is referred to today as the Law of Large Numbers. These first attempts at analyzing phenomena through the use of numbers, forerunners of modern statistical methods, were refined and standardized in the nineteenth century in the surge of activity and development in mathematics, philosophy, and all the sciences in general. Both Carl Friedrich Gauss's *Disquisitiones Arithmetical* (1801) and Pierre de Laplace's *Théorie Analytique des Probabilités* (1812) were instrumental in refining current knowledge and in developing statistics as a mathematical discipline with potential application to social issues and industrial problems. These works formulated a framework for number theory, presented theories of probability, and set forth a calculus for determining the occurrence of chance events and the significance of error in scientific measurement.

The real breakthrough came at the turn of this century when a number of scientists began employing statistical measures in experiments and other research studies. They saw the process of interpreting research results as a statistical exercise, with the primary purpose of attempting to determine or explain the amount of error present in their measurement techniques. The development of the theory and practice of experimental designs, led by R. A. Fisher's *Statistical Methods for Research Workers* (1925) and *The Design of Experiments* (1935), signaled a shift to planning and performing experiments with the intent of controlling for errors or chance events. The basic work on experimental designs was undertaken in agricultural research, where a variety of alternative treatments were applied to plots of land, sometimes arranged in blocks, on which a particular crop was grown. Specific measurements were made at various points in time and analyzed comparatively to arrive at a set of results concerning the impact of the treatment given. Consequently, the contemporary experimental method is described as a comparative study of a specific intervention or treatment, i.e., a drug entity, with alternative treatments or no treatment, i.e., the control group(s), involving the randomized selection and placement of cases, i.e., patients, into each of the various treatment groups.

The early use of experimental designs and statistical techniques were in agriculture and industry, where application to problems of product development, manufacturing efficiency and refinement, and quality assurance had a significant impact. The advantages of experimental designs—replication, randomization, and economy of arrangement—were well suited to industrial research. What has resulted is a shift in emphasis concerning observational errors and other irregularities in our ability to scientifically examine and measure phenomena. Prior to this century, scientists, through theories of probability and random errors, attempted to fix statistical limits within which experimental results were acceptable despite variations. The change in emphasis to the design of experiments was an attempt to make certain that the structure of the research study is logical, that it is broad enough to serve as a foundation for inference, and that every recognizable and avoidable source of error has been eliminated. Of course, the theory of experimental designs has its own flaws and inherent problems, some of which are very complex—requiring solutions through additional strategies and techniques. As Fisher states, "Experimental observations are only experience carefully planned in advance" [12].

The nature of clinical drug studies has also changed from auto-experimentation and direct screening in patients for potentially useful drugs to the multiphase testing of specific compounds, first in healthy subjects and then in those patients with the medical condition or symptomatology for which the compound was designed. Auto-experimentation, wherein a researcher serves as a subject in his experiment, has been referred to as one of the strongest and yet unappreciated traditions in medical and drug research. While some auto-experimental studies have resulted in tragic ends, many have led to important discoveries and breakthroughs, and a few have been seminal in opening and defining important categories of substances for extensive clinical research [13]. Additionally, such auto-experimentation probably had a profound influence on the individual researcher and practitioner in terms of the ethics of giving the drug to patients, the actual nature and extent of the effects present in the drug experience, and the therapeutic potential and indications of the drug in clinical therapy. Just as the individual clinical researcher has become part of a larger group of scientific specialists, so the intensely personal spirit of auto-experimentation has become a more detached and depersonalized statistical view with an emphasis on the process of research itself.

The pharmaceutical industry quickly adopted experimental designs to develop and screen new compounds, improve production, and test the drugs for therapeutic value. The full potential of experimental designs and controlled studies was realized in the 1930s and 1940s when, as part of the immense growth in knowledge and technology and the war effort, controlled clinical trials were commonly adopted as a standard way of assessing the value of drug entities. Finally, in the 1950s, the controlled clinical trial became the norm of pharmaceutical research, when the double-blind strategy of testing (i.e., keeping both patient and observer unaware of the nature of the medication being given) was adopted for supposedly eliminating problems of the placebo phenomenon and other spurious factors. Regulations in the 1960s requiring the proof of efficacy for drug entities reinforced the importance of controlled clinical experiments and made them the standard method of pharmaceutical research.

However, one should not be deluded into believing that experimental studies are a perfected method or that clinical trials are the only source of truly reliable evidence.

The objective of a clinical trial is to ensure a high probability that the better treatment is identified. Integral to this approach is the use of controls—a collection of patients who provide responses with which the effects of a specific therapy can be compared. The term "control" does not necessarily involve randomization, and a controlled clinical trial is not necessarily a randomized clinical trial. Controls may consist of patients receiving no treatment, different treatment, or the same treatment with a different dose or administered according to a different dosing schedule. When a control group is chosen by a method other than randomization, the researcher must assume that either the control and treatment groups are identical with respect to all important variables except the treatment under study or that all relevant differences can be corrected. The randomized trial is most useful and appropriate when the value of a new therapy is uncertain [14].

Recently, many authors have pointed out the less than humanistic nature of medical experimentation and the problems of statistical and conceptual design of experiments, much of which are exacerbated by ethical issues and regulations on human experimentation. In considering the failure of the basic principles of biometric science to adequately evaluate the complex response of a human subject to a treatment, Feinstein [15] has noted that the therapeutic sciences are based on hard, precise, easily measured information (e.g., laboratory values, demographic data, financial costs) from which most of the uniquely human distinctions of people have been systematically excluded. For instance, the double-blind technique masks iatrotherapeutic and placebo effects, and the controlled clinical trial does not answer completely the questions of what happens to a patient when given a treatment and which of the many things that do happen are important in evaluating the worth of the treatment. Tukey [16] has pointed out that demonstrated effectiveness, either quantitative or qualitative, is far from being perfected in clinical trials. He notes that even the term "clinical trial" has a wide variety of meanings from clinical inquiry, where some treatment is hoped to be of help to some class of patients not specified in advance and massive amounts of data are collected and analyzed, to the other extreme of focused clinical trial, where both the class of patients and the end point of therapy are specified in the initial protocol. He states that possible consequences of this confusion about what a clinical trial is include unbalanced boundaries of efficacy, the use of historical controls, and not-very-sequential designs in experimental studies.

Legislative decisions have impacted upon the form of the clinical research process. Other factors have shaped the scientific content of this highly specialized area of inquiry, both in the development of the preclinical sciences and the formalization of the controlled clinical research process.

One indication of the dynamic development of the preclinical sciences is exemplified by the dramatic restructuring of pharmacology. In 1926, pharmacology was considered to be that part of physiology which was concerned with the actions of substances other than foods upon the living orangism. Pharmacology was modestly defined by some as "all scientific knowledge concerning drugs" [17]. Toxicology was described as the detection of the effects of poisons and the diagnosis and treatment of poisoning. Almost 30 years later, the mainstream of academic pharmacology and toxicology still retained much of this global scope [18].

On the other hand, there was evidence of a new approach. In 1931, James C. Munch published a classic study entitled *Bioassays: A Handbook of Quantitative Pharmacology*. This work demonstrated, among other things, that pharmacometrics had become detached from classical pharmacology. Further, the book contained much data relating to drug evaluation. At the same time, no effort was devoted to the problem of reliability. In 1949, Leopold Ther published a booklet entitled *Pharmacologische Methoden zur Auffindung von Arzneimitteln und Giften und Analyse ihrer Wirkungsweise*. While such works were important, at this time problems were still being considered by the statistics of small numbers, up to the probability units. Later studies showed an increasing interest in statistical analysis of the biological evaluation of drugs that contributed to the control of the reliability of results.

In the 1960s a number of important publications focused upon drug evaluation. For example, Laurance and Bacharach's *Evaluation of Drug Activities: Pharmacometrics* (1964); Turner's *Screening Methods in Pharmacology* (1965); and Nodine and Siegler's *Animal and Clinical Pharmacological Techniques in Drug Evaluation* (1964) are indicative of the scope of these studies. The last-named publication paired the experimental and human methods which are suitable for quantitative pharmacological analysis of the more important actions of drugs. These and other studies did much to formalize the process of research.

The concept of drug toxicity became much broader in scope to include, by the 1960s, drug allergies, blood dyscrasias, teratogenic effects, as well as behavorial toxicity. It was not until 1959 that the first volume of *Toxicology and Applied Pharmacology* was published. The Society of Toxicology was founded 2 years later.

In the 1970s, social pharmacology began to describe a number of social factors which are important in drug-taking behaviors and experiences, many of which had been ignored to that time [19]. Prior to the experimentalistic approach developed in this century, all that physicians and scientists knew about the action of drugs was obtained from direct observation of their patients and subjects who had taken them for some purpose. It has been noted that what medical science knew of most psychoactive drugs, such as opium, cannabis, cocaine, ether, and chloroform, resulted from addicts' and other users' accounts of their addictions and drug-taking experiences [20]. However, with the advent of the currently favored sociotechnological approach, scientists and clinicians moved away from such observational data and began to perform experiments on subjects in laboratory settings. What has resulted is a dichotomy in descriptions of effects for many drugs between drug taker and researcher. This difference in approach may be one reason why pharmacology has been unable to explain many types of drug-taking experiences and the placebo phenomenon. In many instances, especially with regard to social and nonmedical drug taking, users' descriptions vary with the normative pharmacological textbook account. The users hold science and the medical descriptions in disdain and continue to experience the effects that they want to. This difference in user versus researcher descriptions of specific drug effects may also account, in part, for the continued success of many drugs used in self-medication.

A similar dramatic shift in scope occurred in another preclinical science. As late as 1920, clinical biochemistry was largely concerned with the determination of urea, uric acid, creatinine, nonprotein nitrogen, and glucose in the blood. Even here, the blood level relationship of these constituents in disease was not well understood. Both Claude Bernard and Louis Pasteur had noted the importance of carbohydrate metabolism in the living organism by focusing upon blood sugar levels in the normal and diabetic conditions and in fermentation studies. The isolation of insulin in 1920 actually shifted attention from the control of blood sugar levels toward broader problems of cellular metabolism and the influence of individual hormones upon metabolic processes. By 1950 it was clear that there was a certain unity in the interlocked chemical reactions of life processes and in the hormonal interrelationships which guide and control these processes. The relatively recent use of isotope techniques permitted the elucidation of the links in metabolic systems.

The development of the contemporary formalized clinical research process was also dependent upon the emergence of a corps of highly skilled scientific research personnel, the organization of research, the favorable economic growth of the pharmaceutical industry, and by social forces exemplified by legislatively mandated objectives. Barber [21] has noted that the education and communication processes inherent in the training of research personnel and clinicians socializes these individuals into a specific and standard way of thinking and doing things, based upon the normative mind-set (paradigm) of the status quo research establishment. The education and training that a future research chemist, pharmacologist, physician, or pharmacist receives will dictate, in most instances, how that individual will assume his or her role in an industrial or clinical setting.

Through much of the nineteenth century, American scientific education lagged behind that of Europe. The reform of medical education, neglected during the Civil War, was resumed in the 1870s. In the 1890s, the University of Pennsylvania set up a laboratory for clinical research and an associated institute of biology. In the Midwest, the University of Wisconsin and Northwestern University began regular courses in the newer preclinical sciences [4]. The first American Ph.D. in pharmacy was awarded at Wisconsin in 1902. Thirty years later, only 11 doctorates were awarded in the pharmaceutical sciences within the 1932/33 academic year in American institutions [22].

Until about 1900, American research had been generally associated with the university, with teaching and the advancement of knowledge. By 1950, foundations constituted the major source of private funds for research in the biomedical sciences. After 1913, Rockefeller boards contributed about $4 million to research in mental health, and the Rockefeller Foundation, in 1929, concentrated its medical research support in this field. This concentration of support by this and other foundations determined recruitment of research and teaching personnel and, more importantly, defined research areas. Other major foundations such as the Commonwealth, Macy, and Carnegie concentrated research funds in specific program areas as well as channeling interest at the expense of neglecting the broader spectrum of basic research [23].

III. THE DEVELOPMENT OF THE PHARMACEUTICAL INDUSTRY

Until World War I, German hegemony in fine chemicals, including pharmaceuticals, was virtually uncontested. Expansion of this science-based industry had been an important facet of Germany's rapid industrialization and economic growth, especially after the Franco-Prussian War. German universities had developed highly competent chemists capable of translating scientific advances into technological progress. Even more important, there was a highly trained core of professional chemists capable of carrying out systematic research in close association with German industry.

In general, most of the important pharmaceuticals which were produced in this country were controlled by German-owned interests and patents. An American branch of Bayer was making aspirin. Heyden, another German interest, was producing salol and certain salicylates. Saccharine, an American discovery, and phenolphthalein were manufactured here, but German chemical intermediates were used in the processes. With the onset of World War I, the halt of German imports necessitated a complete restructuring of the American chemical industry, and more particularly pharmaceuticals. The American Council of National Defense, during the summer of 1917, authorized the Federal Trade Commission to issue nonexclusive licenses to domestic manufacturers for the production of pharmaceuticals formerly protected by German patents.

American fine chemical manufacturers, including the drug makers, entered the 1920s with the proceeds from the highly profitable war years, newer plants and equipment, improved methods of production with special regard for the basic principles of chemical engineering, an awareness of economic and technical efficiency in production, low federal taxes and interest rates, a growing domestic market, and a new tariff law. The economic stage was set for the coming research revolution.

By the mid-1930s, American pharmaceuticals were within the realm of "big business." Extensive plant and equipment had been devoted to the commercial production of biological products, particularly antisera and vaccines. Almost overnight, these investments were threatened by the rapid advent of the sulfonamides. In most industries, cost-cutting innovations, process improvements, new applications, and similar changes do not ordinarily exert undue pressure upon an entire industry. In the pharmaceutical industry, the impact of the sulfonamides was profound. Over 6,000 derivatives of sulfonilamide were prepared and tested for antibacterial and toxicological activities.

The drug industry had felt the first blasts of J. A. Schumpeter's "perennial gale of creative destruction." Henceforth, innovation would be the key to industrial growth in drugs and institutionalized research and development (R&D) would be the instrument by which a continuous process of innovation would be accomplished. There remained the ever-present possibility that a new development might occur at any time—the discovery of a new and more effective sulfonamide or an entirely new antibacterial entity. This focused attention upon the problems of investment in plant and processes which might become obsolete overnight, and upon recovering costs and a reasonable profit as quickly as possible.

In the period after World War II, the 1950s were again years of active growth in the pharmaceutical industry. The general prosperity resulted in the rapid amortization of plant and equipment. Rapid growth, characterized by a stream of new products, enabled pharmaceutical manufacturers to increase production, raise wages, maintain prices at a remarkably even level, pay attractive dividends, provide internal funds for modernization and expansion, and more significantly, finance R&D. One of the dominant economic strengths was the accumulation of liquid capital.

The Internal Revenue Code of 1954 had an important effect upon the rapid growth industries, particularly drugs. Tax reductions built faster and stayed higher in those instances where new equipment and processes were rapidly introduced. The flow of new products was essential to keep pace with the rapid increase in capital stock and vice versa. The rate of capital investment increased with innovational change and the rate of profit also increased.

The National Institutes of Health reported that the amount spent in 1952 for medical research was $173 million. Of this, the federal government spent $73 million (half in its own laboratories and the other half in the universities and medical schools). In that year the industry expended $60 million for medical research ($57 million in its own laboratories and $3 million in the medical schools). As early as 1958, the industry had more than doubled R&D expenditures. Of the $128 million allocated to R&D, only $2 million came from government [23]. Internally funded R&D had become largely institutionalized.

IV. THE REGULATORY PROCESS

Social expectations were considerably heightened by the advent of the "miracle drugs"—sulfonamides, penicillin, and succeeding antibiotics. The dramatic quantum leaps in all areas of scientific endeavor during and after World War II appeared to many to signal the beginning of an age when all things were possible for science to accomplish—given sufficient support and the right focus. At the same time, serious doubts were being raised concerning the social responsibilities of the drug industry.

In general terms, prior to the 1930s, patients looked to their physicians for assurances of safe and efficacious therapy. The major pharmaceutical companies had largely assumed the responsibility for providing assurances of the quality of the raw material in their finished products, uniformity of potency, efficacy in terms of traditional therapeutic standards, stability, and elegance in appearance. The introduction of new chemical entities into therapeutics, particularly after midcentury, forced the industry to assume much of the ultimate responsibility for therapy. At the same time, the highly visible Congressional hearings of the 1950s and early 1960s led to a rather widely held social impression that the pharmaceutical industry was not acting in the public interest. The media were quick to capitalize upon the problems which must inevitably arise in drug therapy. The profitability that had triggered and financed dramatic R&D programs was condemned as public exploitation. Many of these doubts influenced efforts toward governmental intervention, some of which were based on revisions of legislation passed in the early years of this century.

The Pure Food and Drugs Act of 1906 was primarily aimed toward the elimination of unclean and adulterated foods from the market. Secondarily, the legislation was directed toward patent medicine abuses. Prescription drugs were subject to control but received less attention. In 1923, there was an increased intensity in the Bureau's control work, with attention being focused upon the bioassay of important drugs and their preparations with an emphasis upon adherence to standards of purity.

The Food, Drug, and Cosmetic Act of 1938 was intended to protect the public against quackery and the sale of dangerous drugs. The Elixir of Sulfanilamide tragedy triggered an administrative procedure for premarketing clearance of new drugs of uncertain safety. Section 505 of the new act forbade the introduction of new drugs into interstate commerce without U.S. Food and Drug Administration (FDA) determination that the new drug application satisfied regulatory specifications.

The passage of the Harris-Kefauver amendment of 1962 added a new dimension requiring the evaluation of a new drug's safety and efficacy. This legislation made the government an active participant in the research process, since FDA approval was required for testing procedures of investigational new drugs before testing could proceed for the filing of a new drug application. The FDA was administratively empowered to withdraw approval of a new drug application for a number of causes including questions of the drug's safety or lack of substantial evidence of effectiveness. This substantial evidence was to consist of adequate and well-controlled investigations by scientific experts to evaluate the effectiveness of the drug involved so that the drug would elicit the effect it was purported to have under the conditions of use for which it would be prescribed.

In 1970, "substantial evidence" was defined in greater detail. Now, adequate and well-controlled investigations had to include a formal test with explicit objectives, defined selection procedures for subject and control groups, methods for observation and recording, and statistical analysis. The test drug could be compared with a placebo, another drug known to be active from past studies, or with no treatment. Clinical experience was relegated to the final stage (i.e., Phase IV) of drug testing well after the commitment to manufacture and promote the drug had been made. Consequently, the trend in the 1980s is toward the conceptualization and implementation of postmarketing surveillance systems to detect the occurrence of problems with a drug's safety and efficacy.

V. CONCLUSION

Insistence upon assurances of absolute safety and efficacy is a reflection of a social quest for absolutes that extends beyond drugs to automobiles, aviation, environmental protection, and an all-encompassing philosophy of consumerism. The implicit question for the drug industry is whether industry scientists possess the social and ethical wisdom to control vast power for human ends—or will they use that power for exploitation? The clinical research process, which is probably in an evolving stage, attempts to meet this social demand in a simple and defensible manner. The influence of societal demand on research is no different in our own time than in former years: The call in the early 1970s to wage a "war on cancer" resulted again in an explosion of basic science and industrial and clinical research efforts aimed at finding a cure. This recent effort has resulted in the massive screening of new compounds for chemotherapeutic activity and has spawned a whole research focus and industry with the study of interferon and other contemporary magic bullets.

Part of this response to societal need has also resulted in the medicalization of many interpersonal and social problems and a "mystification" of drug-taking behaviors and experiences [24]. Dubos [25] has also pointed out that problems of drug effects on human ecology and that the toxicity of our medicinal agents is not truly known. He states that "absolute lack of toxicity is an impossibility, because absolute selectivity is a chemical impossibility." In this context, he notes that the full range of effects and the long-term consequences of most drugs are not known.

Finally, the recent focus on human experimentation and the ethics of drug taking will have a profound impact on the future development of new drug entities and clinical research. Informed consent, patient rights, and institutional review boards have already affected clinical research in many settings. Additionally, the use of placebos in controlled clinical trials are also being reevaluated by some in an ethical context [26]. These changes in societal concerns, governmental interventions and regulations, and individual patient rights may well alter many aspects of pharmaceutical development, including the focus of research efforts, the nature of clinical experimentation, the costs and feasibility of drug development, and perhaps the purpose of drug taking and the functions of drugs in society.

REFERENCES

1. P. Ehrlich, Address delivered September 1906, In *The Collected Papers of Paul Ehrlich* (F. Himmelwert, Ed.). Pergamon Press, Elmsford, N.Y., 1960.
2. P. Gund, J. D. Andose, J. B. Rhodes, and G. M. Smith, *Science 208:* 1425-1431 (1980).
3. C. J. Cavallito (Ed.), *Structure-Activity Relationships.* Pergamon Press, Elmsford, N.Y., 1973.
4. R. H. Shryock, *American Medical Research.* Commonwealth Fund, New York, 1947.
5. E. H. Ackerknecht, *A Short History of Medicine,* rev. ed. Roland Press, New York, 1968.
6. A. C. Crombie, *Medieval and Early Modern Science,* rev. ed. Vol. 2. Doubleday (Anchor Books), Garden City, N.Y., 1959.
7. C. D. Leake, *An Historical Account of Pharmacology to the Twentieth Century.* Thomas, Springfield, Ill., 1975.
8. E. H. Ackerknecht, *Medicine at the Paris Hospital, 1794–1848.* Johns Hopkins University Press, Baltimore, 1967.
9. C. Bernard, *An Introduction to the Study of Experimental Medicine.* Macmillan, New York, 1927.
10. F. H. Garrison, *An Introduction to the History of Medicine,* 4th ed. Saunders, Philadelphia, 1929.
11. J. H. Gaddum, *Proc. R. Soc. Med. 47:* 195 (1954).
12. R. A. Fisher, *The Design of Experiments.* Oliver & Boyd, Edinburgh and London, 1935
13. L. K. Altman, *N. Engl. J. Med. 286:* 346–352 (1972).

14. D. P. Byar, R. M. Simon, W. T. Friedewald, J. J. Schlesselman, D. L. DeMets, J. H. Ellenberg, M. H. Gail, and J. H. Ware, *N. Engl. J. Med. 295:*74–80 (1976).
15. A. R. Feinstein, *Lancet ii:*421-423 (1972).
16. J. W. Tukey, *Science 198:*679-684 (1972).
17. G. Bachmann and A. R. Bliss, *The Essentials of Physiology and Pharmacodynamics*, 2nd rev. ed. Blakiston, Philadelphia, 1926.
18. J. C. Krantz and C. J. Carr, *The Pharmacologic Principles of Medical Practice*, 3rd ed. Williams & Wilkins, Baltimore, 1954.
19. M. Montagne, The nature and meaning of drug-taking experiences. Unpublished doctoral dissertation, University of Minnesota, Minneapolis, 1981 (university microfilm No. 761-070).
20. H. W. Morgan (Ed.), *Yesterday's Addicts: American Society and Drug Abuse, 1865–1920.* University of Oklahoma Press, Norman, 1974.
21. B. Barber, *Drugs and Society.* Russell Sage Foundation, New York, 1967.
22. D. D. Vogt, M. Montagne and H. A. Smith, *Am. J. Pharm. Educ. 45:*232-237 (1981).
23. The American Foundation, *Medical Research: A Midcentury Survey* (E. E. Lape, Ed.). Little, Brown, Boston, 1955.
24. H. L. Lennard, L. J. Epstein, A. Bernstein, and D. C. Ransom, *Mystification and Drug Misuse*, The Jossey-Bass Behavorial Science Series. Jossey-Bass, San Francisco (copublished with Harper & Row, New York), 1971.
25. R. Dubos, in *Drugs in Our Society* (P. Talalay, Ed.). Johns Hopkins University Press, Baltimore, 1964.
26. H. Brody, *Placebos and the Philosophy of Medicine: Clinical, Conceptual, and Ethical Issues.* University of Chicago Press, Chicago, 1977.

4
LEGAL AND ETHICAL PROBLEMS IN CLINICAL RESEARCH*

Robert J. Levine and Angela R. Holder
Yale University School of Medicine
New Haven, Connecticut

---

*Portions of this chapter are excerpted or adapted from Robert J. Levine, *Ethics and Regulation of Clinical Research*. Urban & Schwarzenberg, Baltimore, Maryland, 1981, with the permission of the publisher.

I. HISTORICAL PERSPECTIVE

Modern concepts of the ethics and regulation of clinical research began after World War II when the Nazi physicians' atrocities became known to the world. The ten principles now known as the Nuremberg Code [1, p. 285] were written by the judges who presided over the war crimes trials of the Nazi physicians. These principles emphasize that the voluntary consent of any human subject in research is absolutely essential. In 1964, the World Medical Association published the Declaration of Helsinki [1, p. 287], which emphasized, among other things, that clinical research should be conducted only by medically qualified persons and that clinical research on human beings cannot be undertaken without their free and informed consent. If they are legally incompetent, consent of the legal guardian should be procured. With this Declaration establishing for the rest of the world what was assumed to be standard practice in the United States, politicians and the public alike were shocked by reports of research done without the consent of subjects in this country only a few years after the United States had led the charge against the Nazis at the War Crimes Tribunals.

The first federal policy statement requiring committee review of research supported by the federal government was issued by the Surgeon General of the U.S. Public Health Service in 1966. Much earlier, many medical schools and hospitals, including the Clinical Center at the National Institutes of Health (NIH), had created committees for peer review of research involving human subjects done within their institutions. In 1974, the U.S. Department of Health, Education and Welfare (DHEW; presently the Department of Health and Human Services, DHHS) published regulations requiring institutions applying for research funds to establish local committees for review of those studies to be funded by DHEW and required that those committees have the expertise to review the legal and community relations aspect as well as the scientific and medical aspects of research. Also in 1974, Congress passed the National Research Act [2], which established the National Commission for the Protection of Human Subjects of Biomedical and Behavioral Research (henceforth call the Commission). Based upon the Commission's recommendations, as well as public comments thereon, the Food and Drug Administration (FDA) and the DHHS published regulations which became effective in July 1981 [1, p. 259]. These regulations require Institutional Review Board (IRB) review for most federally funded research, specifying composition and procedures for IRBs, directions about the elements of informed consent which must be provided to any research subject, as well as other things.

Although legal requirements are different from ethical principles. at least in theory, the legal and ethical issues involved in research on human subjects are coextensive. What the Nuremberg Code and Declaration of Helsinki set forth as ethical norms are articulated, in much greater detail, in federal regulations.

II. CASE LAW

There is very little case law on questions of common law liability in the context of medical research. Contrary to the popular belief that patients are treated like guinea pigs when they become research subjects in a teaching hospital, the absence of "malresearch" cases suggests that research subjects tend to be treated more carefully than ordinary patients, are much more closely observed and monitored, and, apparently, are almost never involved in situations involving medical negligence.

The legal doctrine of informed consent evolved in this country beginning with a California case in 1957, solely in the context of medical practice [3, pp. 225-259]. Although there have been a few cases in which research subjects or their guardians complained that they did not understand either that they were participating in research or the risks of the research in which they were participating, the number of these is extremely small in proportion to the thousands of cases in which patients alleged that physicians did not negotiate informed consent to medical therapy. The only genuine malresearch cases which we can find are those in which a physician undertook to develop a new procedure of some sort without review by an IRB. In the few cases in which a subject in an organized research project claimed lack of informed consent, the physician has prevailed in court.

The boundaries between "standard" and "innovative" medical practice are quite difficult to define in some cases. Courts have attempted to deal with this issue by making it a question of fact to be determined by a jury. Typical jury instructions appear to be based on the concept of the "respectable minority of the medical community" standard for determining whether due care was used in selecting among alternative therapies. If the therapy was innovative, the defendant in a malpractice case may be the only physician who has ever used it. In that situation, establishing the innovation as legitimate according to the "respectable medical minority" standard may be difficult. Thus, when a new procedure is undertaken at a hospital or medical school where innovation is a customary activity, where a written protocol is submitted to and approved by the IRB, and where consent is negotiated in a manner also approved by the IRB, in any later suit by the patient-subject the peer review system has within it a method of establishing evidence of having met the standard of the "respectable medical minority." If the physicians on an IRB conclude that a procedure has a reasonable ratio of risks to hoped-for-benefits, even if the procedure has never been tried on a human subject, there is evidence that "the reasonable physician" considers it a "reasonable alternative therapy." On the other hand, if a physician invents a new procedure or other treatment independently, uses it on a patient, and injury results, there is no such evidence before the fact that his or her peers would find the innovation reasonable. In this situation a court may say, as the court said in a 1767 English case [4]: "Then it was ignorance and unskilfulness in that very particular to do contrary to the rule of the profession.... For anything that appears to the court, this was the first experiment made with this new instrument; and if it was, it was a rash action and he who acts rashly acts ignorantly."

Although IRBs are mandated by federal regulations to protect the rights and welfare of research subjects, in future cases involving research injuries their utility value in the defense of investigators may well be demonstrated. In addition to the ethical responsibilities for informing a patient or research subject about proposed interventions, documentation of informed consent, as it is applied in the context of medical research, is actually designed to protect the investigator rather than the subject [1, p. 98]. In standard medical practice, patients almost always sign very general consent forms. Most hospitals have but one surgical consent form, for example, in which the specific procedure is written in on a blank. The consent form states only that "the risks and hazards of the above procedure have been explained to my satisfaction." By contrast, consent forms for research explicate the risks of each procedure and each drug that is to be administered. Almost all courts would hold that a signature on such a document cannot be disavowed on grounds that the subject did not read or understand what he or she had signed. As long as the specific risk that materializes is listed on the consent form, any subject suing the investigator in almost any jurisdiction would be held to have assumed the risk and thus would be unable to recover damages.

In litigation arising out of the routine practice of medicine, the definition of informed consent used by the courts is very simple. It is the duty to warn a patient of the nature of a proposed procedure, its hazards and possible complications, and its likely benefits. When a physician knows that an alternative treatment exists, he or she is obliged to advise the patient to that effect. The patient then has the right to decide which course of action, if any, he or she prefers. As one court [5] stated succinctly, "the informed consent doctrine is based on the proposition that every competent person in the final arbiter of whether or not he gets cut, by whom he gets cut, and where he gets cut. A patient has a sovereign choice of whether he will submit to surgery in the course of diagnosis and treatment, and in order to make this choice meaningful and realistic, the doctor is under a legal duty to disclose to a patient any serious risks involved in the contemplated surgery and the alternatives available to him, including the risks from nontreatment."

By contrast, DHHS and FDA regulations include a much more complex set of "elements" of informed consent. There are eight elements that must be provided to prospective subjects for all research; these eight must also be included in consent forms. In addition, there are optional elements of informed consent that must be added if, in the judgment of the IRB, they are necessary for any particular research protocol. The eight required elements are [1, p. 265]:

1. A statement that the study involves research, an explanation of the purposes of the research and the expected duration of the subject's participation, a description of the procedures to be followed, and identification of any procedures which are experimental;
2. A description of any reasonably foreseeable risks or discomforts to the subject;
3. A description of any benefits to the subject or to others which may reasonably be expected from the research;
4. A disclosure of appropriate alternative procedures or courses of treatment, if any, that might be advantageous to the subject;
5. A statement describing the extent, if any, to which confidentiality of records identifying the subject will be maintained;
6. For research involving more than minimal risk, an explanation as to whether any compensation and an explanation as to whether any medical treatments are available if injury occurs and, if so, what they consist of or where further information may be obtained;
7. An explanation of whom to contact for answers to pertinent questions about the research and research subjects' rights, and whom to contact in the event of a research-related injury to the subject;
8. A statement that participation is voluntary, refusal to participate will involve no penalty or loss of benefits to which the subject is otherwise entitled, and the subject may discontinue participation at any time without penalty or loss of benefits to which the subject is otherwise entitled.

III. ETHICAL PRINCIPLES

A *fundamental ethical principle* is a general judgment that serves as a basic justification for the many particular prescriptions for and evaluations of human actions. It is fundamental in that, within a system of ethics, it is taken as an ultimate foundation for any second-order principles, rules, and norms. A fundamental principle is not derived from any other statement of ethical values. The Commission identified three fundamental ethical principles as relevant to the ethics of research involving human subjects: *respect for persons*, *beneficence*, and *distributive justice*. The norms and procedures presented in regulations and ethical codes are derived from and are intended to uphold these fundamental ethical principles (discussed in the following subsections). These principles pertain to human behavior in general; it is through the development of norms that they are made particularly relevant to specific classes of activities such as research and the practice of medicine.

A. Respect for Persons

The principle of respect for persons incorporates two basic ethical convictions: first, that individuals should be treated as autonomous agents and, second, that persons with diminished autonomy and thus in need of protection are entitled to it. As defined by the Commission, an autonomous person is an individual capable of deliberation about personal goals and of acting under the direction of such deliberation. To show respect for autonomous persons requires that we leave them alone, even to the point of allowing them to choose activities that might be harmful unless they agree (consent) that we may do otherwise. We are not to touch them or to encroach upon their privacy unless such touching or encroachment is in accord with their wishes. Clearly, not every human being is capable of self-determination. Respect for those who are not, either temporarily or permanently, may require one to offer protection to them during their period of incapacity.

B. Beneficence

Beneficence, according to the Commission, is often understood to cover acts of kindness or charity that go beyond strict obligation. From the viewpoint of research ethics, however, beneficence is understood in a stronger sense as an obligation. Two general rules have been formulated as complementary expressions of beneficent actions in this sense: (1) do no harm; and (2) maximize possible benefits and minimize harms.

The principle of beneficence is strongly rooted in the ethical tradition of medicine. Hippocrates' epigram, "As to diseases, make a habit of two things—to help, or at least to do no harm" is, unfortunately, often oversimplified as "Do no harm (*Primum non nocere*)."

The principle of beneficence was interpreted by the Commission as creating an obligation to secure the well-being of individuals and to develop information that will form the basis of our being better able to do so in the future. However, in the interest of securing societal benefits, we should not intentionally injure any person.

C. Distributive Justice

Distributive justice is concerned with the distribution of scarce benefits where there is some competition for these benefits. If there is no scarcity, there is no need to consider just systems of distribution. Distributive justice is also concerned with the distribution of burdens, specifically when it is necessary to impose burdens on fewer than all members of a seemingly similar class of persons. Just what constitutes a "fair sharing" of burdens and benefits is a matter of considerable controversy. In order to determine who deserves to receive which benefits and which burdens, we must identify "morally relevant" criteria for distinguishing between prospective recipients. Various criteria have been proposed, e.g., persons may be treated differently on the basis of their needs, their accomplishments, their purchasing power, their social worth, their past records, or their future potentials. The criterion adopted by the Commission is in accord with the Western concept of the fundamental equality of persons (e.g., before the law) and with the very strong tradition that interprets "fairness" to require extra protection for those who are weaker, more vulnerable, or less-advantaged than others. This interpretation is reflected in such disparate sources as the injunction in the Judeo-Christian tradition to protect widows and orphans, the Marxist dictum "from each according to ability; to each according to need," and most recently, John Rawls's contractual derivation of principles of justice.

IV. ETHICAL NORMS

An ethical norm is a statement that actions of a certain type should (or should not) be done. If reasons are supplied for these behavioral prescriptions (or proscriptions), they are that these acts are right (or wrong). The purpose of ethical norms is to indicate how the requirements of the three fundamental principles may be met in the conduct of research involving human subjects. The behavior-prescribing statements contained in the ethical codes and regulations on research involving human subjects are, in general, variants of five general ethical norms. There should be (1) good research design, (2) competent investigators, (3) a favorable balance of harm and benefit, (4) informed consent, and (5) equitable selection of subjects. A sixth general ethical norm has begun to appear in some guidelines, and it appears possible that it will eventually be added to federal regulations: There should be compensation for research-related injury.

Since statements of the ethical norms in codes and regulations tend to be rather vague, they permit a variety of interpretations. Often it is difficult to know exactly how to apply them to particular cases. When faced with such uncertainties, it is generally helpful to look behind the norm, i.e., to examine the fundamental ethical principle or principles it is intended to uphold or embody. Accordingly, in the discussion of each of the ethical norms, we shall call attention to the fundamental ethical principle or principles it is designed to serve. For examples of resolution of uncertainties in the application of norms to particular cases through referring to underlying principles, see Ref. 6.

In addition to the norms, the ethical codes and federal regulations present descriptions of procedures that are to be followed to ensure that investigators comply with those norms. The most important general procedural requirement that is relevant to all research involving human subjects is review by an IRB (the IRB is discussed in detail in Chap. 23). Another general procedural requirement is the documentation of informed consent on a consent form. Its purpose is to show evidence of compliance with the norm calling for informed consent.

A. Good Research Design

The primary purpose of this norm is to uphold the principle of beneficence. If the research is not well designed, there will be no benefits; investigators who conduct badly designed research are not responsive to the obligation to do good or to develop knowledge which is sufficiently important to justify the expenditure of public funds, to impose upon human subjects risks of physical or psychological harm, and so on. The norm is also responsive to the principle of respect for persons. Persons who agree to participate in research as subjects are entitled to assume that something of value will come of their participation. Poorly designed research wastes the time of the subjects and frustrates their desire to participate in a meaningful activity.

B. Competent Investigators

This norm requires that the investigators be competent in at least two respects. (1) They should have adequate scientific training and skill to accomplish the purposes of the research. The purpose of this norm is precisely the same as that requiring good research design; it is responsive primarily to the obligation to produce benefits to society through the development of important knowledge and also to the obligation to show respect for research subjects by not wasting their time or frustrating their aspirations. (2) Investigators are expected to be sufficiently competent to "take care" of the subject. In most clinical research, this requires that at least one member of the research team be responsible for observing the subject with a view toward early detection of adverse effects of his or her participation or other evidence that the subject should be excluded from the study. The investigator should have the competence to assess the subject's symptoms, signs, and laboratory results. He or she should also be competent to intervene as necessary in the interests of minimizing any harms, e.g., prompt administration of an antidote to a toxic substance. The FDA requires sponsors of new drugs and devices to include statements of the scientific training and experience considered appropriate by the sponsor to qualify investigators as suitable experts to investigate the safety and/or efficacy of a "test article." There is, in addition, a necessity for describing the relevant credentials of each investigator who participates in the study of the test article.

C. Favorable Balance of Harms and Benefits

This requirement is derived primarily from the ethical principle of beneficence. In addition, a thorough and accurate compilation of the risks and hoped-for benefits of a research proposal also facilitates responsiveness to the requirements of the principles of respect for persons and justice. A clear and accurate presentation of risks and benefits is necessary in the negotiations with the subject for informed consent. Similarly, such a compilation of burdens and benefits facilitates discussions of how they might be distributed equitably.

Without a favorable balance between harm and benefit there is no justification for research. Much of the literature on the ethics of research involving human subjects reflects the widely held and, until recently, unexamined assumption that becoming a research subject is a highly dangerous business. To many members of the public and to many commentators on research who are not themselves investigators, the word "risk" seems to carry the implication that there is a possibility of some dreadful consequence. Yet is is much more common that when biomedical researchers discuss risks, they mean a possibility that there may be something such as a bruise after a venipuncture. Recently some empirical data have become available that indicate that, in general, it is not particularly hazardous to be a research subject. For example, the occupational hazard of the role of subject has been estimated as slightly greater than that of office secretary [1, p. 25].

The vast majority of research proposals present a burden that is more correctly described as "inconvenience" than as a risk of physical or psychological harm [1, p. 27]. In general, prospective subjects are asked to give their time to have blood drawn and the like. Studies that require removal of other normal body fluids and which are associated with risk of real harm, e.g., use of blood drawn from the heart, can usually be accomplished using individuals who require removal of heart blood for therapeutic purposes.

Ethical codes and regulations require not only that the risks be justified be being in a favorable relation to hoped-for benefits but also that they be minimized. For example, prescreening tests may be done to identify prospective subjects who ought to be excluded because they are peculiarly vulnerable to injury. The mirror image of this requirement is to maximize benefits. Justification of risk is never an isolated event. To say that the imposition of risk in the interests of research is justified presupposes that the plan to do research is also acceptable in accord with the requirements of all relevant ethical norms and procedures.

Maneuvers employed with the intent and reasonable probability of providing direct benefit for the individual subject, including all diagnostic and therapeutic maneuvers whether "investigational" or "standard," are justified differently from nonbeneficial procedures. The risk is justified by the expectation of benefit for the particular subject precisely as in the practice of medicine. One additional criterion is that the relationship of anticipated benefit to the risk presented by the modality must be at least as advantageous to the subject as that presented by any available alternative—unless, of course, the individual has considered and refused to accept a superior alternative.

It is more problematic to justify the risk presented by "nonbeneficial" procedures—maneuvers performed for purposes of research to contribute to the development of generalizable knowledge. The benefits one hopes for in this case will accrue to society rather than to the individual subject.

Both DHHS and FDA have established risk-threshold criteria. When the risks presented by research proposals exceed these thresholds, the regulations call for special procedural or substantive protections, particularly if the subjects are drawn from one of the "special populations." The major threshold is "minimal risk," meaning that "the risks of harm anticipated in the proposed research are not greater, considering probability and magnitude, than those ordinarily encountered in daily life or during the performance of routine physical or psychological examinations or tests." For further discussion of other threshold standards, see Levine [1, p. 48].

D. Equitable Selection of Subjects

This norm is derived from the principle of distributive justice, which requires equitable distribution of both the benefits and the burdens of research. There are various criteria for identifying individuals as vulnerable or less advantaged in ways that are relevant to their suitability for selection as subjects. In general, we identify as vulnerable those who are relatively or absolutely incapable of protecting their own interests. While all persons are vulnerable in some ways, some groups may be particularly so. Subsequently in this chapter (Sec. V) there is a discussion of the "special populations" that have been identified as always in need of special protections. In addition to these groups there are some others that should generally be considered at least potentially vulnerable in morally relevant fashions: (1) uncomprehending subjects such as those who are uneducated, senile, unconscious, or inebriated; (2) persons in the "sick role" who may be vulnerable in several respects; (3) those who are in a relationship of dependency to the investigator or the institution, e.g., when the investigator is a physician in a hospital and the prospective subject is a patient in the hospital, or when the prospective subject is a student in the institution in which the research is to be done; and (4) impoverished persons who are unable to secure what they consider the necessities of life without the material inducements offered to assume the role of subject.

Members of various minority groups are commonly portrayed as vulnerable. It is not, however, appropriate to treat entire minority groups as if they were homogeneous with respect to vulnerability since this adds unnecessarily to the burden of stereotypes already borne by such groups.

No ethical code or regulation proscribes the involvement of any of the aforementioned groups in any type of research activity. Rather, such involvement should be justified, e.g., by demonstrating that the condition to be studied does not exist in less vulnerable populations, or by demonstrating that the knowledge to be gained is relevant to the condition that causes the subjects' vulnerability [1, pp. 61-66]. It might also be shown that the degree of risk or inconvenience is so small as to not constitute an undue burden in the particular vulnerable population.

E. Informed Consent

This norm is designed to be responsive to the requirements of the principle of respect for persons. Informed consent is commonly portrayed as a process having two components. The first is that of informing—the transmission of information from the investigator to the prospective subject. The second is that of consenting—signified by the subject's declaration that, having assimilated the information, he or she is willing to become a subject. However, informed consent should instead be seen as a process of negotiation, that is, an ongoing dialogue between investigator and subject involving transmittal of information, explanation, response to questions, and so on. While informing and consenting are both important components of the process, there may in some cases be a need also to assess the prospective consentor's autonomy and comprehension [1, pp. 71-93].

The legal criterion for disclosure of any particular point of information in medical practice is "materiality," i.e., any fact that is material to the patient's decision must be disclosed. The determination of materiality is decided by courts according to three different standards. The traditional standard in the United States is that of the "reasonable physician." According to this, the determination of whether any particular fact should be disclosed is made on the basis of whether it is customary to do so in the community of practicing physicians. The standard that is becoming accepted by an increasing number of state courts in recent years is the "reasonable person" test. In the case of *Canterbury* v. *Spence* [7], the court held that the disclosure required was determined by the patient's right of self-determination, a right that can be exercised effectively only if the patient possesses enough information to make an intelligent choice. At least one jurisdiction has adopted yet another standard, i.e., that a risk is "material" if the particular patient making the choice or decision considers it material; this is known as the "idiosyncratic person" standard. It is not particularly realistic since, by the time of trial, the uncommunicated hazard has materialized as an injury and, thus, it would be most surprising if the patient did not claim that had he or she been informed of it treatment would have been declined. Standards of disclosure for research subjects generally conform to the reasonable person rule. The task of judging what a reasonable person might wish to know has been assigned to the IRB.

It is generally agreed that negotiations for informed consent in the investigator-subject relationship should meet higher standards than those for the physician-patient relationship. The patient should, in general, be allowed more freedom than the subject to relinquish this entitlement. Patients may be permitted opportunities to delegate some types of decision-making authority to a physician, whereas most subjects should usually not be so enfranchised. The most important distinction between the negotiations for informed consent in the two contexts is that the prospective subject must be informed that if the proposed activity includes any research component, he or she will, at least in part, be a means, and perhaps only a means, to another's end.

F. Compensation for Research-Induced Injury

Commentators on the ethics of research involving human subjects have reached a consensus that those who are injured as a consequence of participation in research should be entitled to compensation. The ethical arguments to support this view are grounded in considerations of compensatory justice; they are analogous to those used to support providing benefits for soldiers who are injured in the service of their country or provisions for workmen's compensation.

In the practice of medicine, the common law provides no compensation for injury without proof of negligence. By contrast, compensation for research-induced injury would be paid whether or not the investigator is at fault. At common law, as well, an individual's consent is ordinarily construed as an agreement to assume the burdens of those risks disclosed during the negotiations for informed consent. DHHS regulations require investigators to advise subjects of the presence or absence of compensation but do not require institutions to establish a system for compensation. Very few institutions have established plans for no-fault compensation for research-induced injury. The regulations also require a statement on the availability of medical therapy for injury; many institutions do provide free medical therapy for research-induced injury.

V. SPECIAL POPULATIONS

When DHEW published its first proposals to develop regulations providing for additional protections for especially vulnerable populations of research subjects, it designated them as persons having "limited capacities to consent." Since the Nuremberg Code identifies voluntary consent as absolutely essential, it is clearly problematic to involve subjects who lack free power of choice (e.g., prisoners), the legal capacity to consent (e.g., children), or the ability to comprehend (e.g., the mentally infirm). The Commission concluded that persons having limited capacities to consent are vulnerable or disadvantaged in ways that are morally relevant to their involvement as research subjects; consequently, the principle of distributive justice is interpreted as requiring that we facilitate activities designed to yield direct benefit to the subjects and that we encourage research designed to develop knowledge that will be of benefit to the class of persons of which the subject is a representative. However, we should generally refrain from involving the special populations in research that is irrelevant to their conditions as individuals or, at least, as a class of persons. Respect for persons is interpreted as requiring that we show respect for a potential subject's capacity for self-determination to the extent that it exists. Some who cannot consent can register "knowledgeable agreements" ("assents") or "deliberate objections." To the extent that the capacity for self-determination is limited, respect is shown by protection from harm. Thus, the Commission recommends that the authority accorded to members of special populations or to their legally authorized representatives to accept risk be strictly limited; any proposal to exceed these limits ("minimal risk") requires special justifications [1, p. 155ff].

A. Children

As a consequence of the uncertainties about the ethical propriety of and legal authority to do research on children, there has been great reluctance among investigators in the United States to carry out thoroughgoing studies to determine the safety and efficacy of drugs in children. Consequently, as Shirkey[8] observed, "Infants and children are becoming the therapeutic orphans of our expanding pharmacopoeia." Since 1962, nearly all new drugs have been required by the FDA to carry on their labels one of the familiar "orphaning" clauses, e.g., "is not recommended for use in infants and young children, since few studies have been carried out in this group...." The therapeutic orphan phenomenon is not limited to children. Very similar conditions obtain in the use of drugs in pregnant women because the fetus is seen by some as a nonconsenting subject who might be peculiarly vulnerable to the effects of a drug.

The therapeutic orphan phenomenon represents a serious class injustice. If we consider the availability of drugs proved safe and effective through the devices of modern clinical pharmacology to be a benefit, then it is unjust to deprive classes of persons—e.g., children and pregnant women—of this benefit. This injustice is compounded as follows: Were we to do Phase II and Phase III clinical trials in children as we now do in adults, the first administration of various drugs would be done under conditions more controlled and more carefully monitored than is customary in the practice of medicine. It is likely that adverse drug reactions that are peculiar to children would be detected much earlier than they are now. Consequently, either we could discontinue administration of the drugs to children or we could issue appropriate warnings to physicians who are using the drugs. Since the prevailing practice in the United States is to ignore the "orphaning clauses" on the package labels, we have a tendency to distribute unsystematically the unknown risks of drugs in children and pregnant women, thus maximizing the probability and frequency of their occurrence and minimizing the probability of their detection.

The recommendations of the Commission should go far to reduce the magnitude of the problems associated with the therapeutic orphan phenomenon. In particular, their recommendation that risks "presented by an intervention that holds out the prospect of direct benefit for the individual subject" may be considered differently from risks presented by procedures designed to serve solely the interests of research should facilitate the ethical conduct of clinical trials in children.

The Commission's recommendations have been translated by DHHS into proposed regulations [9]. Following is a review of those calling for different procedures or considerations than those that apply to all research involving human subjects.

For all research involving children, the IRB is required to determine: "Where appropriate, studies have been conducted first on animals and adult humans, then on older children, prior to involving infants...." It is also required that adequate provisions be made to protect the privacy not only of the children but also of their parents. Adequate provisions must be made for soliciting the "assent" of the child as well as the "permission" of the parents or legal guardians. The elements of the transactions involved in negotiating such assent and parental permission are essentially the same as those involved in obtaining informed consent. In general, a child with normal cognitive development becomes capable of meaningful assent at about the age of 7 years (some may be younger and some older). These children should be provided with a "fair explanation" and invited to sign a form analogous to the usual consent form. In most cases, the "deliberate objection" of a child should be construed as a veto; a child who states his or her unwillingness to participate in research should have that expression of free choice honored. A deliberate objection is to be distinguished from the expected reactions of an infant who might in various circumstances cry or withdraw in response to almost any stimulus.

The Commission assigns to the IRB the authority to determine whether it is necessary to appoint an agent—an "advocate" or a "consent auditor"—to supervise the interactions between the investigator and the children and their parents or guardians. Such agents may be assigned to monitor the assent and permission discussions, assist in the selection of subjects, and so on. Such persons might be IRB members, the child's pediatrician or psychologist, a social worker, a pediatric nurse practitioner, or any other experienced and perceptive person.

The Commission recommends that, in some cases, additional procedural and substantive protections may be required depending upon the degree and nature of risk. Research that presents to children no more than "minimal risk" (as defined in Sec. IV.C) may be conducted with no substantive or procedural protection other than those specified earlier.

If there is more than minimal risk presented by procedures which do not hold out any expectation of direct health-related benefit for the child, additional protections are required. If the IRB determines that there is only a "minor increase over minimal risk," it has the authority to approve the research protocol without appealing to a higher authority. Wisely, the Commission provides no definition of "minor increase," leaving this to the judgment of the IRB.

In order to justify research involving minor increases over minimal risk, the IRB must determine that the research is likely to yield knowledge that "is of vital importance for understanding or amelioration of the subject's disorder or condition...." Moreover, the IRB must determine that the procedure that presents the minor increments in risk "presents experiences to subjects that are reasonably commensurate with those inherent in their actual...medical, psychological, or social situations...." This requirement for commensurability reflects the Commission's judgment that children who have already had medical procedures performed upon them might be more able than are those who are not so experienced to base their assent on some familiarity with such procedures and their attendant discomforts.

If, in the judgment of the IRB, the research procedures present more than minor increments above minimal risk, the IRB does not have authority to approve it. The proposal must be referred to the federal government for review by a national ethical advisory board.

Research protocols that present more than minimal risk of physical or psychological harm or discomfort to children but in which the risk is "presented by an intervention that holds out the prospect of direct benefit for the individual subjects, or by monitoring procedures required for the well-being of the subjects" may be considered differently. In this case, the IRB must determine that "such risk is justified by the anticipated benefits to the subjects; and the relation of the anticipated benefit to such risk is at least as favorable to the subjects as that presented by available alternative approaches...." Thus, for this class of research, the standards for justification of risk are precisely what they are—or ought to be—in the practice of medicine. However, if any of the procedures done in the interests of research present more than minimal risk, they must be justified according to the standards mentioned earlier for "nonbeneficial" procedures.

In some situations, the only way a child might receive a diagnostic or therapeutic agent is by participating in the research protocol. Under such circumstances, decisions are to be made as they are in the practice of medicine. The general presumption is that a parent may make decisions to override the objections of school-age children in such cases. However, in some circumstances the objection of teenagers to decisions made on their behalf by their parents may prevail.

B. Those Institutionalized as Mentally Infirm

Proposals to do research involving as subjects those institutionalized as mentally infirm present a very complicated array of ethical issues. First, there is a problem in defining a class of persons as "those institutionalized as mentally infirm." Definitions provided by the Congress, the Commission, and DHHS-proposed regulations are inconsistent with each other [1, p. 170]. Secondly, there are often disputes over what constitutes a therapeutic benefit for this class of persons. For example, a decision to administer large doses of phenothiazines to disruptive patients in a mental institution may be challenged on several grounds. Tranquilization may, at times, be done more in the interest of maintaining order in the institution than in the interests of fostering the patients' recovery; it may be chosen as a more convenient and less expensive alternative to individualized psychotherapy—which, in some cases, may be more advantageous to the patient. The disruptive behavior itself might be interpreted by some observers as an appropriate protest against illegitimate institutionalization. Thirdly, the fact of institutionalization itself presents important problems. It limits one's autonomy and creates a dependency upon the institution and its agents. Finally, some of those institutionalized as mentally infirm are capable of consent; for those who are not, it may be problematic to find someone who is willing and able to act in the interests of patient-subjects in order to give permission to supplement their assent.

The Commission's recommendations for this class of persons are very similar to its recommendations for research involving children. Following is a description of the additional or different protections recommended by the Commission.

For all research, the IRB must determine that "The competence of the investigator(s) and the quality of the research facility are sufficient for the conduct of the research; there are good reasons to involve institutionalized persons in the conduct of research; and, adequate provisions are made to assure that no prospective subject will be approached to participate in the research unless a person who is responsible for the health care of the subject has determined that the invitation to participate in the research and such participation itself will not interfere with the health care of the subject...."

For research that presents no more than minimal risk, only one additional determination must be made by the IRB. It must determine that "Adequate provisions are made to assure that no subject will participate in the research unless: the subject consents to participation; if the subject is incapable of consenting, the research is relevant to the subject's condition and the subject assents or does not object to participation...." Persons who object to participation may not participate in any research protocols except those designed to evaluate procedures that "hold out the prospect of direct benefit for the individual subjects." In such protocols, the IRB must ensure that:

> No adult subject will participate ... unless: the subject consents to participation; if the subject is incapable of consenting, the subject assents to participation (if there has been an adjudication of incompetency, the permission of a guardian may also be required by state law); if the subject is incapable of assenting a guardian of the person gives permission (if a guardian of the person has not been appointed, such appointment should be requested at a court of competent jurisdiction) or the subject's participation is specifically authorized by a court of competent jurisdiction; or if the subject objects to participation, the intervention holding out the prospect of direct benefit for the subject is available only in the context of research and the subject's participation is specifically authorized by a court of competent jurisdiction....

The remainder of the Commission's recommendations are very similar to those it made for children. The only important differences are found in the recommendation for research in which minor increments above minimal risk are presented by nonbeneficial interventions. For such protocols, there is no requirement that the procedures be reasonably commensurate with those inherent in the subject's life situation. In addition, for this class of research, the appointment of a consent auditor by the IRB is mandatory. It is important to note that this is the only class of research activities—no matter which special population is considered—in which the necessity for a consent auditor is not left to the discretion of the IRB.

In 1978, DHEW proposed regulations providing special protections for "those institutionalized as mentally disabled." These depart substantially from the commissions's recommendations. The public response to this proposal was massive and generally disapproving [1, p. 179ff]. Since we understand that this proposal is being revised drastically, we shall not detail its several substantive deviations from the Commissions's recommendations.

C. Prisoners

At the time of this writing, the status of the law governing research involving prisoners is unclear. Following the publication of the Commission's recommendations (1976) both DHHS and FDA published regulations that sharply restricted research involving prisoners. The constitutionality of FDA regulations was challenged in a lawsuit initiated by the prisoners in the Michigan state prison system [10]. Rather than argue the case, FDA withdrew its regulations and proposed new regulations in December 1981. The new proposals permit "research on practices that have the intent and reasonable probability of improving the health and well-being of the particular prisoners chosen. Subject to the approval of the IRB, prisoners may be assigned to control groups...." Research is also permitted "on conditions particularly affecting prisoners as a class (for example, vaccine trials and other research on hepatitis, which is more prevalent in prisons than elsewhere)." However, for other types of research, e.g., Phase I drug studies, there must be a demonstration that, among other things: "The type of research fulfills an important social or scientific need, and the reasons for involving prisoners are compelling...."

There is likely to be much public debate over the "compelling reasons" standard. If this requirement appears in the final regulations, it is most unlikely that there will be any further drug development done involving prisoners as subjects. The ethical problems presented by proposals to do research involving prisoners as subjects as well as DHHS regulations governing such research have been surveyed recently by Levine [1, Chap. 11].

D. The Fetus

Is the fetus a person? If so, is it entitled to be treated as the principle of respect for persons requires us to treat persons? These were the central questions in the debates conducted before the Commission over the ethical permissibility of research involving the fetus. After hearing extreme positions on both sides the Commission adopted (implicitly) the view that a fetus is neither a person nor an object but rather a potential person. Accordingly, we are to show it respect by refraining from actions that would violate its dignity and integrity. Our actions should be directed toward fostering the well-being of each individual fetus and minimizing harm. While we must respect the authority of a pregnant woman to have an abortion, we are not to encourage abortions in the interests of doing research. Research on fetuses should be designed to develop knowledge that would be of benefit to this class of potential persons.

Unfortunately, the Commission prepared its report on research on the fetus before it had an opportunity to address the general conceptual issues in its Congressional mandate. Consequently, its recommendations embody concepts that it subsequently abandoned. For example, the Commission made separate recommendations for "therapeutic research" and "nontherpeutic research." For a discussion of the conceptual confusion created by the use of such concepts, see Levine [1, Chap. 1]. The reader should keep in mind that regulation writers and others who use these terms are, in general, thinking of specific procedures as they discuss more complex activities such as those described in research protocols.

Once the Commission had accorded the fetus the status of "potential person," the second major issue argued before it had to do with proposals to conduct nontherapeutic research either on the pregnant woman or on the fetus before, during, or after induced abortions. There was little opposition to conducting "nontherapeutic research directed toward the pregnant woman" which would "impose minimal or no risk ... to the fetus ..." if abortion was not being considered. However, when an abortion is anticipated, the issue is much more problematic. Many commentators expressed concern that a woman who has chosen to have an abortion has "abandoned" the fetus. Ordinarily we call for maternal or parental consent based on the assumption that the parents love and care for the fetus and therefore will tend to protect its interests. However, if they have signaled their intention to abandon the fetus, they might too readily do things that might harm the fetus. It was agreed that, for this class of research, the standard of "minimal or no risk to the fetus" should afford sufficient protection even to those carried by nonprotective mothers.

Proposals to conduct "nontherapeutic research directed at the fetus" produced still more complex ethical dilemmas. One particularly controversial issue was whether "fetuses-going-to-term" differed in any morally relevant manner from "fetuses-about-to-be-aborted." If so, would this difference justify proposals to do research on the fetus-about-to-be-aborted that one would not consider doing also on those going to term?

It was agreed that these two classes of fetuses did not differ in any morally relevant respect. They are similar in that they are both vulnerable subjects deserving of protection from harm; the fact that one is scheduled to die does not make it any less deserving of respect or protection. According to Commissioner Karen Lebacqz, the locus of the disagreement is over what "equal protection" or "similar treatment" means. Lebacqz argues that "similar treatment" does not mean subjecting both fetuses to the same procedure but rather putting both to equal risk [11].

Consider, for example, research designed to determine whether a drug administered to a pregnant woman crosses the placental barrier. If it does, it may conceivably damage the fetus. If the fetus is aborted within a few days, it is not likely that any harm will have been done. On the other hand, if the fetus is brought to term, there might be lifelong disability. Therefore, Lebacqz reasons that the risk to the fetus should be calculated by multiplying the risk of harm were it to be carried to term by the probability that the woman will change her mind about the abortion and carry it to term. Since less than 1% of women change their minds after they have contacted an abortion clinic, the risk to a fetus-about-to-be-aborted is about 1% that presented by the same procedure to a fetus-going-to-term. On this basis Lebacqz concludes: "justice requires that fetuses to be carried to term not be subjected to some experiments which might be done on fetuses scheduled for abortion."

Opponents to exposing the fetus-to-be-aborted to risky research procedures grounded their argument, in part, in the possibility that such exposure might deprive some women of the option of changing their minds; if they did there would be a relatively high probability of delivery of a damaged infant. To avoid this possibility, some commentators suggest that possibly harmful nontherapeutic interventions upon the fetus should be permitted only as part of or during a single operative procedure designed to terminate in abortion.

Proposals to do nontherapeutic research during the abortion procedure and on the nonviable fetus ex utero generated another heated controversy. For the prospective subjects of these types of research, death is imminent. This fact is assured through the Commission's recommendation that no such research be performed on fetuses unless they are less than 20 weeks of gestational age; survival of such fetuses ex utero was unprecedented.

Current federal regulations on research on the fetus conform in large measure to the Commission's recommendations. Some exceptions will be pointed out in the following discussion.

"Fetus" is defined as "the product of conception from the time of implantation ... until a determination is made, following expulsion or extraction of the fetus, that it is viable." Following delivery, if a fetus is found to be either viable or possibly viable, it is referred to as an infant. Accordingly, proposals to do research on the infant are to be evaluated in accord with standards developed for children.

The Commission recommended that certain classes of activities be conducted only after approval by a national ethical review body, e.g., nontherapeutic research directed toward the fetus during the abortion procedure and that directed toward the nonviable fetus ex utero. This recommendation is not reflected as such in the regulations. Rather, the regulations state that any applicant, with the approval of the IRB, may request a waiver or modification of specific requirements in the regulations. The Secretary of Health and Human Services may grant such a request with the approval of the Ethical Advisory Board (EAB). Such requests must be published in the *Federal Register* along with an announcement that public comment is invited.

The regulations further require IRBs to determine that adequate consideration has been given to the manner in which subjects are selected and monitoring the progress of the activity and intervening as necessary. The IRB is also required in its review of all research involving as subjects any women who could become pregnant to determine what steps will be taken to avoid involvement of women who are in fact pregnant when such research activities would involve risk to a fetus. Implementation of this requirement may call for a variety of responses by the IRB and investigators, depending on their judgment with respect to any risk to the fetus. In some cases it may be necessary to exclude women of childbearing age who have not been surgically sterilized; in others it may suffice to perform a pregnancy test and then review the prospective subject's plan for contraception.

The regulations [1, pp. 268-271] set forth general limitations on research involving fetuses and pregnant women. No research may be undertaken unless: (1) appropriate studies on animals and nonpregnant individuals have been completed, except where the purpose of the activity is to meet the health needs of the fetus; (2) risk to the fetus must be minimal and in all cases is the least possible risk for achieving the objectives of the activity; and (3) individuals engaged in the research will have no part either in any decision as to the timing, method, and procedures used to terminate the pregnancy or in the determination of the viability of the fetus. No procedural changes which may cause greater than minimal risk to the fetus or to the pregnant woman will be introduced in a procedure for terminating pregnancy solely for research purposes, and no inducement may be offered to terminate a pregnancy for purposes of the activity.

Pregnant women may not be involved in any research as subjects unless the purpose of the activity is to meet the health needs of the mother and unless the fetus will be placed at risk only to the minimal extent necessary. The mother *and* the father (if available) must be legally competent and must have given their informed consent. The father's informed consent need not be secured, however, if health needs of the mother are being met by the research, if his identity cannot be reasonably ascertained, if he is not reasonably available, or if the pregnancy resulted from rape. The regulations also require that no fetus in utero may be involved as a subject unless: (1) the purpose of the activity is to meet the health needs of the particular fetus and the fetus will be placed at risk to the minimal extent necessary to meet such needs, or (2) the risk to the fetus is minimal and the purpose of the activity is the development of important biomedical knowledge which cannot be obtained by other means.

Fetuses ex utero may not be involved in research unless: "There will be no added risk to the fetus ... or the purpose is to enhance the possibility of survival of the particular fetus to the point of viability. No nonviable fetus may be involved ... unless: Vital functions ... will not be artificially maintained (and) experimental activities which ... would terminate the heartbeat or respiration ... will not be employed ...."

Research activities involving the dead fetus, fetal material, or the placenta are to be conducted in accord with applicable state or local laws as they apply to research on specimens or any dead body.

REFERENCES

1. Robert J. Levine, *Ethics and Regulation of Clinical Research.* Urban & Schwarzenberg, Baltimore, 1981.
2. National Research Act, PL 93-348, 88 Stat 342.
3. Angela R. Holder, *Medical Malpractice Law,* 2nd ed. Wiley, New York, 1978.
4. *Slater* v. *Baker and Stapleton,* CB, 95 *Eng Rep* 860, 1767.
5. *Congrove* v. *Holmes,* 308 NE 2d 765, Ohio 1973.
6. R. J. Levine and K. Lebacqz, Some ethical considerations in clinical trials. *Clin. Pharmacol. Ther. 25:*728-741 (1979).
7. *Canterbury* v. *Spence,* 464 F 2d 772, CA DC 1977.
8. H. C. Shirkey, Therapeutic orphans. *J. Pediatr. 72:*119-120 (1968).
9. 43 *Federal Register* 31786-31794, July 21, 1978.
10. *Fante and The Upjohn Company* v. *U.S. Dept. of Health and Human Services* et al., Civil Action No. 80-72778, U.S.D.C., E.D., Mich.
11. K. Lebacqz, Reflections on the report and recommendations of the National Commission: Research on the Fetus. *Villanova Law Rev. 22:*357-366 (1977).

5

SOCIAL PSYCHOLOGICAL IMPLICATIONS OF CLINICAL RESEARCH

Jack Robbins
Arnold and Marie Schwartz College of Pharmacy and Health Sciences, Brooklyn, New York, and Schering Corporation, Kenilworth, New Jersey

I. INTRODUCTION

In 1979 member firms of the Pharmaceutical Manufacturers Association spent a total of over $1.6 billion for research and development (R&D) activities [1]. About 94% of this amount was spent on human-use pharmaceutical products. When one considers that the 1979 R&D expenditures were up 16% over the amount spent in 1978 and that the

1980 R&D budget was $1.9 billion, again some 16% above the 1979 figure, it becomes evident that the number of clinical research projects will undoubtedly continue to increase at a fairly steady pace.

These medical research projects, of necessity, involve experimentation on human beings, and this, of course, raises some serious ethical considerations. On the one hand we believe in the value of free scientific inquiry, while on the other hand we also believe in the dignity of the individual and his right to privacy and freedom from injury. This ethical dilemma is real and cannot be simply dismissed by pious statements about the overall good which clinical research provides to the community. Every clinical researcher must face the problem of ethics each time he devises and conducts an experiment.

At the same time, while the central issue in clinical research revolves about the personal integrity and attitudes of the investigator, we must recognize that standards of conduct must continually be redefined to meet changing conditions and needs of the society at large.

The purposes of this chapter, then, are to examine and review some of the social psychological implications of clinical research, both as they affect the subject and as they affect the scientific researcher.

II. HISTORICAL PERSPECTIVE

In 1932, the U.S. Public Health Service, in collaboration with the Tuskegee Institute and Hospital, initiated a study of the effects of untreated syphilis in man. This experiment was conducted in Macon County, Alabama, with 399 black males (experimental subjects) who had contracted syphilis and 201 black males (control group) who did not have the disease [2].

The subjects voluntarily submitted to physical examinations but were never told that they were involved in a medical experiment, nor were they informed of the risks involved. Although penicillin was already available as an effective treatment for syphilis in the late 1930s and the early 1940s, the researchers made no effort to provide this proven medical treatment to the infected subjects because they wanted to see what the outcome would be if the natural course of the disease were allowed to develop without medical treatment. The poor and uneducated subjects were never told that an effective remedy was being withheld but were led to believe that they were, in fact, receiving appropriate medication at no cost.

The Tuskegee study, then, is a clear example of exploitation by scientists of a group of powerless subjects who were neither asked for nor gave their consent for this ugly experiment.

Some 30 years later, in 1963, an experiment was conducted at the Jewish Chronic Disease Hospital in Brooklyn with 22 weak and debilitated patients who were injected with live cancer cells [3]. The purpose of this particular study was to determine whether noncancer patients would reject foreign cancer cells as quickly as patients who already had cancer. While the researchers did obtain the subject's consent in this study, they did so with some measure of deception because the patients were not told that live cancer cells were being injected into their bodies.

There are undoubtedly other such cases of unethical behavior by researchers, but the two cases cited are sufficiently clear to raise a number of ethical concerns. For example, just what should the researcher tell the prospective subject about the objectives, risks, and possible consequences of an experiment? Should there be special safeguards to protect the patient from coercion in obtaining his "voluntary" consent? What do we mean when we speak of "informed consent"?

We will explore these and other issues in the next section.

III. INFORMED CONSENT

Many experimental scientists feel that if they have obtained the consent of the experimental subjects they, the scientists, have preserved the dignity of the individual. Consent by the subject, however, assumes many variables: that the subject can deal with and comprehend the information given to him by the researcher; that the subject understands the risks and potential benefits of the experiment; and, probably most importantly, that the subject is psychologically prepared to ask the right questions of the research scientist—the authority figure in the subject-experimenter relationship.

During their developmental stages, children are dependent on older people for nurture and gratification. These older people not only provide satisfactions but also can impose punishments to control the child's behavior. Thus, parents provide the guidance for the child to develop values about himself and the world around him. The adult world in and around the home also provides the child with an increasing awareness of the various degrees of status enjoyed by different people and the varying degrees of authority and power that different people appear to exercise. As the child grows, his modes of adapting to power and authority also develop, and he learns to distinguish between the various types of authority, for example, legitimate vs. illegitimate, rational vs. arbitrary, and benign vs. malicious.

Thus, because of the experiences during their developmental stages and, indeed, in some aspects of modern business life, many people by habit submit quite readily as subjects in a clinical trial because of the prestige, authority, and perceived expertness of the research scientist. The clinical researcher, for his part, consciously and sometimes unconsciously makes use of a combination of various types of power to influence the subject to consent to the experiment. The experimenter may, by action and by word, stress his legitimate role as a doctor and insist that the subject consent to participate in the experiment because, after all, patients generally do whatever their doctor recommends.

The researcher may also attempt to communicate with the subject in the subject's own language and establish a friendly personal relationship with him, thus exerting a referent influence. Additionally, the researcher may emphasize his own extensive scientific training; and his office, furnished with numerous medical books and a variety of impressive diplomas and awards, helps to establish and legitimize his power as an expert. Finally, by facial expression and behavior, the researcher may indicate signs of approval or disapproval, as the case may be, to "help" the subject decide whether or not to consent to participate in a clinical research project.

These modes of apparent manipulation of the subject are, of course, often contrary to our own value system. Yet where the public good may be served, the research scientist may at times feel that he has no alternative but to try to obtain the subject's consent to the scientific experiment.

As we have already noted, the experimenter-subject relationship is, by nature, one of unequal status. Because of this inequality, it is relatively easy for the experimenter, either consciously or unconsciously, to make the subject feel powerless, puny, and intensely uncomfortable. It is important, therefore, that research scientists go out of their way to protect the dignity and self-esteem of their subjects.

One way of protecting the subject's self-image and making him feel that he has, indeed, been participating in a socially worthwhile experiment is to debrief the subject gently and carefully after the experiment and frankly discuss the full details of the research study, including the failures as well as the successes, in the antitipated vs. the actual findings.

Some experimenters often find it more effective to wait until all the subjects have completed their participation in the experiment before informing them of the details of the study. Then they can inform all the subjects at the same time in one large group. This method has advantages in that it saves the scientist's time, while allowing all of the subjects to meet and possibly exchange views with one another about their participation in the experiment and their contributions to the goals of the study.

With the mounting concern for the protection of subjects involved in clinical research, the relevant branch of the U.S. Department of Health and Human Services (DHHS) namely, the Food and Drug Administration (FDA), published new regulations in the *Federal Register* on January 27, 1981 [4]. These regulations, which became effective on July 27, 1981, were issued in order to clarify existing FDA requirements governing informed consent and provide protection of the rights and welfare of human subjects involved in research activities that fall within FDA's jurisdiction. Following is the gist of the new regulations ([4] Subpart B-Informed Consent of Human Subjects 50.20 p. 5):

> No investigator may involve a human being as a subject in research covered by these regulations unless the investigator has obtained the legally effective informed consent of the subject or the subject's legally authorized representative. An investigator shall seek such consent only under circumstances that provide the prospective subject or the representative sufficient opportunity to consider whether or not to participate and that minimize the possibility of coercion or undue influence. The information that is given to the subject or the representative shall be in language understandable to the subject or representative. No informed consent, whether oral or written, may include any exculpatory language through which the subject or representative is made to waive or appear to waive any of the subject's legal rights, or releases or appears to release the investigator, the sponsor, the institution, or its agents from liability or negligence.

IV. COMPENSATION OF INJURED RESEARCH SUBJECTS

There are a number of views regarding the ethical issues involved in the compensation of injured research subjects. For example, at a meeting of the President's Commission on the Study of Ethical Problems in Medicine and Biomedical and Behavioral Research, held on September 15, 1980, Holly Smith, Associate Professor of Philosophy at the University of Chicago, testified that society should compensate an injured research subject primarily on the basis of gratitude for an act of altruism [5]. Smith further argued that a research subject who had been paid to participate in a clinical trial has already been sufficiently compensated for accepting risk and should therefore not be paid if injured as a result of the experiment.

On the other hand, Albert Jonsen, a member of the Commission, noted that there are strong utilitarian arguments for compensating research subjects, because compensation encourages volunteering, which in turn helps the research process and eventually society in general [5].

Another view presented at the same meeting was that of Philippe Cardon, who had served as a member of the 1977 DHHS Secretary's Task Force on Compensation [5]. Cardon stated that a distinction should be made between therapeutic and nontherapeutic research and that only those injuries sustained during experiments not intended to benefit the patient, i.e., nontherapeutic research, should be compensable [5].

Incidentally, one definition of research-related injury was proposed by Albert Jonsen at an earlier meeting of the Commission [6]. Jonsen suggested that a patient or subject should be compensated if injured as a result of using drugs for purposes other than those for which they had been approved by the FDA. At the same time, Jonsen raised quite an interesting point: What about control subjects? Would control subjects who had received placebos be considered as "injured" if their health did not improve to the same extent as those subjects receiving the experimental drugs?

Generally speaking, then, most members of the scientific community feel that they should distinguish between altruistic research and research in which subjects may expect some therapeutic benefit. Many scientists feel that in the latter situation no compensation for possible injury should be paid. On the other hand, some experimenters feel that all people who participate as research subjects contribute to the public good and are therefore entitled to compensation in the event of injury.

The DHHS Secretary's Task Force on Compensation of Injured Research Subjects, at the meeting of the President's Commision of January 9, 1981, rejected any distinction between therapeutic and nontherapeutic research for the purposes of compensation [7]. The Task Force argued that every subject in the clinical experiment is accepting risks for the altruistic benefit of society, since the experimental subject is not in control of the proceedings and especially since the subject does not, in fact, know whether he is in the control group or whether he is receiving some new drug or undergoing an experimental procedure.

V. CONFIDENTIALITY OF RESEARCH RECORDS

During 1980, the House of Delegates of the American Medical Association formulated the first major revision of its Principles of Ethics since 1957. According to Veatch [8], the most important wording in the new Principles of Medical Ethics deals with confidentiality and the rights of patients (Principle IV). According to the 1957 principles, for example, physicians were permitted to break a confidence when required by law or to protect the individual or the community. In contrast, the 1980 wording safeguards the patient's right to a trusting relationship by assuming confidentiality "within the constraints of the law."

There are a number of ethical issues which need to be considered when reviewing the topic of confidentiality of research records. For example, what possible tangible harm might befall the subject by breach of confidentiality? Some of these harms might include possible loss of income, damage to the subject's self esteem, or perhaps damage to the subject's familial and social relationships.

One of the complicating problems in dealing with the confidentiality of patient records is that such records tend to be easily accessible to third parties, e.g., other researchers, while considerably less accessible to the subject himself. One possible reason for this situation may be the self-appointed "paternalism" of the medical profession, which sees the subject in the role of the child rather than as a coequal adult. Yet, the final decision as to sharing medical record information must ultimately rest with the research scientist, because in some instances it may very well be in the patient's best interest not to disclose the contents of his medical records.

Another problem which researchers must consider is the issue of patient confidentiality in conducting retrospective research. While deletion of patient identification from existing records would not be really practical, the patient's identity could still be masked by pooling data or, in the case of individual case reports, by use of an invented name or number to identify the specific case.

The real ethical issue is the need to balance the individual subject's right and expectation of privacy, on the one hand, with the need for information to further the cause of the health of the general public, on the other.

VI. GENETIC ENGINEERING

One of the more recent developments in clinical research is the growth of expertise in the field of genetic engineering. New techniques are being developed which will attempt to replace "defective" genes in individuals suffering from single gene diseases.

The scientific community is, however, sharply divided on the ethical and social psychological aspects of genetic engineering. For example Schmeck reports on a case of a patient with a grave form of anemia, β-zero-thalassemia major, a single gene defect [9]. The patient's physician was reprimanded sharply by the National Institutes of Health for attempting experimental gene therapy without prior approval from his university. The experimental plan was to inject a small amount of genetic material into the patient in the hope that some of the material would migrate to the bone marrow and take root. In theory, enough copies of the β-globin gene would help the patient make normal blood. Controversy concerning the research appears to have jeopardized this plan.

The scientists opposed to genetic engineering argue that such efforts at manipulating human genetics to cure disease could lead to ill-advised attempts to change heredity. On the other hand, those who favor continued gene therapy research argue that successful application of gene therapy in human beings offers the strong possibility of enormous good by reducing the suffering and death caused by genetic diseases.

VII. THE PROBLEM OF BIAS

Since clinical research involves interaction between subject and experimenter, there is always the possibility that bias will intrude into the experiment. Basically, two types of bias may occur—bias due to the characteristics of the experimental situation itself and bias due to the unintentional influence of the experimenter.

In the first form of bias, the subject knows that he is in an experimental situation, and believes that certain types of responses are expected from him. The subject is, therefore, not really responding to the clinical situation itself but rather to his interpretation of what the clinical situation actually is.

We have already noted that the experimenter-subject relationship is unequal and rather one-sided. The experimenter has both the power and the knowledge in the clinical trial, while the subject can only try to guess the true nature of the experiment. In some cases, the subject may try to determine the real objectives of the clinical trial in order to appear knowledgeable and thus make a good impression on the experimenter, or he may simply feel that if he knows the experimental hypothesis he can help the clinician by behaving in the expected manner. In some instances, however, some perverse subjects may not like the experimenter or the clinical situation itself and thus may attempt to defeat the purposes of the experiment by reporting spurious effects and symptoms.

Another ever-present source of situational bias may be the subject's belief that he must behave and respond in a socially acceptable manner during the clinical interview. Many subjects therefore behave in a docile and cooperative way and do not, in fact, report adverse symptoms or other effects of the clinical trial which they believe might upset the experimenter's perception of them as cooperative subjects.

One way of reducing bias due to the subject's subjective responses to the experimental situation is to provide all subjects with a plausible explanation of the experiment which is really unrelated to the actual purposes of the clinical trial. Thus, if the subject reports effects of the clinical trial which he perceives to support or refute the plausible purposes of the experiment, then the results of the real experimental test may not be affected.

The second form of bias may come from the unintentional and subtle influence of the experimenter on the subject's responses in order to support the experimenter's hypotheses. One way of reducing this bias, of course, is to use the double-blind technique, in which neither the subject nor the experimenter knows whether a placebo or an experimental drug is being administered.

Probably one of the most effective ways of reducing possible experimenter bias is to assemble all the subjects prior to their assignment to either the placebo or drug condition. At this meeting, the general description, the objectives, and the methodology of the clinical trial are described by the experimenter to all those assembled. In this way, the experimenter's age, sex, and personality will have an approximately equal effect on all subjects in the experiment.

In summary, then, there is always the possibility of bias in any clinical research project. With adequate care and proper precautions, however, much of the bias can usually be avoided.

VIII. SUMMARY

Past and current trends in research expenditures indicate that the number of clinical research projects involving experimentation with human subjects will continue to increase from year to year. These research projects generally call for consideration by the researcher of the delicate balance between the rights, dignity, and privacy needs of the subject, on the one hand, and value of free scientific research and potential for the benefits to society at large, on the other.

Even a cursory review of clinical research activities in the United States during the past 50 years reveals that, although most research scientists are quite ethical and considerate of the subject's welfare, a small minority of clinical researchers have on occasion conducted their studies with little regard for the subject's dignity, privacy, or long-term welfare.

We have noted that informed consent by the subject involves more than simply medical or research issues but also has to do with the social and psychological factors which affect the subject.

The researcher-subject relationship is, generally, an unequal one, with the research scientist playing the authority and power role, while the subject plays the child-dependency and powerless role. There are, however, several ways of bringing a greater degree of equality into the researcher-subject interaction. One way is to treat the patient as an adult and keep him informed of the details of the clinical trial. Another, related way is to carefully debrief the subject after the clinical trial has been completed

Recognizing the need for a guide for physicians and researchers in the conduct of clinical research, new regulations have been issued by the FDA which clarify the FDA's requirements governing informed consent and provide protection for human subjects involved in clinical trials.

There is no clear consensus among research scientists as to whether subjects who are injured as a result of their participation in a clinical experiment should be reimbursed for their injuries. The federal position however, as stated by the DHHS Task Force on Compensation of Injured Research Subjects, is that subjects should be compensated for injuries sustained as a result of a research experiment.

Another real problem faced by scientist involved in clinical research is that of confidentiality of patient records. The ethical issue here is the balance between the subject's right to privacy and the scientist's need for information in order to reach some meaningful conclusion regarding the research project.

Finally, one of the more recent developments in clinical research which has sharply divided the scientific community is the technique of genetic engineering. Some scientists are opposed to research in this area because they view it as an attempt to change heredity, while proponents argue that successful application of gene therapy may alleviate suffering and death caused by genetic diseases.

In summary, then, clinical research has several social and psychological implications. Researchers should receive better training in the ethics of research and should be made more aware of the needs, rights, and feelings of human subjects. Subjects should be kept informed about the trials in which they are participating and should be compensated for any injury that they may sustain as a result of the clinical trial. While federal regulations governing clinical research will continue to play an important role, researchers will also need more effective self-regulation rather than depending on external guidelines alone.

Clinical research has made tremendous progress in ethical conduct during the past two decades. As new discoveries continue to be made in the health sciences, with all their attendant social changes, clinical research approaches and self-regulatory procedures must also progress so as to ensure the subject fair treatment by the professional clinical research scientist.

REFERENCES

1. *Annual Survey Report, 1979-1980, U.S. Pharmaceutical Industry.* Pharmaceutical Manufacturers Association, Washington, D.C., 1981.
2. Final Report of the Tuskegee Syphilis Study Ad Hoc Advisory Panel, reprinted in Stanley Joel Reiser, Arthur J. Dyck, and William J. Curran, *Ethics in Medicine: Historical Perspectives and Contemporary Concerns.* MIT Press, Cambridge, Mass., 1977, pp. 316-321.
3. Elinor Langer, Human experimentation: New York verdict affirms patients' rights. *Science 151:*663-666 (1966).
4. Protection of human subjects: Informed consent. *Federal Register,* Part IX, Subpart B, Jan. 27, 1981.
5. Minutes, Meeting IV, President's Commission for the Study of Ethical Problems in Medicine and Biomedical and Behavioral Records, Washington, D.C., Sept. 15-16, 1980.
6. Minutes, Meeting II, President's Commission for the Study of Ethical Problems in Medicare and Biomedical and Behavioral Research, Washington, D.C., May 16-17, 1980.

7. Minutes, Meeting VI, President's Commission for the Study of Ethical Problems in Medicine and Biomedical and Behavioral Research, Washington, D.C., Jan. 9-10, 1981.
8. Robert M. Veatch, Ethics. *JAMA* *245*(21):2187-2189 (1981).
9. Harold M. Schmeck, Jr., Patients wait, but is knowledge ripe for human gene therapy. *The New York Times*, May 31, 1981.

6

CLINICAL PROJECT COORDINATION

Gary M. Matoren
Clinical Research Practices and Drug Regulatory Affairs
Middletown, New York

I. INTRODUCTION

Clinical Project Coordination is needed in the drug development process because as new research frontiers are opened up, we are faced with larger areas of the unknown in planning clinical programs.

Donald S. Frederickson, former Director of the National Institutes of Health, is the author of the oft-quoted article [1] that indicates the magnitude of the clinical research process:

> If it seems to you that the number and grandeur of proposed field trials is on the increase, I believe you are correct. And the rate will continue to grow, for several decades of intensified research activities are inevitably creating more opportunities and more pressures for exploitation of basic discoveries....Choosing among the increasing options will also pose the most difficult task for the community of experts and for society at large.

Thus, clinical project coordination must be an integral part of the project management program. Clinical project coordination, which may be called coordinative planning, is utilized to facilitate the integration of the functional units in a organization to achieve the highest quality results, e.g., be it a document or design of a protocol, for the resources available. The clinical project coordination function can provide in one place a readily comprehensible body of interrelated facts, information, forecasts, judgments, and ideas designed to contribute to the most effective decision making by research management.

Clinical project coordination is a highly effective project management concept utilized in the clinical research process. Responsibilities of clinical project coordination are interdepartmental and interdivisional in nature. In essence, clinical project coordination is responsible for long-range planning and general coordination of divisional activities related to drug registration. In this chapter the term clinical research programs is understood to mean clinical pharmacology and clinical research programs (Phase I through and including Phase IV).

Since we are looking at the clinical component of the drug development process, this chapter will focus on the role clinical project coordination plays in this process. The responsibilities of the clinical project coordinator varies from company to company. In essence they may include, but are not limited to, long-range planning, project management, which includes time, resource and cost control, coordination of timetables pertaining to clinical projects, clinical grant management, program and policy analysis, and other activities outlined in detail in Table 1.

A clinical project coordinator may be involved in such activities as planning for a corporate clinical pharmacology unit; cost analysis of clinical programs; manpower planning; budget planning; and analysis of operating and capital budgets and their impact on the clinical programs. He may also be responsible for developing a Procedures Manual for the clinical division. Such a Procedures Manual may be utilized also for training and orientation purposes for new personnel. It may consist of the following component sections:

Administrative
Document preparation
Field operations
Drug regulatory requirements
Clinical monitoring procedures
Financial
Investigational drugs

Table 1 The Responsibilities of Clinical Project Coordination

A. Maintain protocol and research report registry:
   1. Issue, when requested, to senior monitors new sequential numbers for all protocols.
   2. Record all protocol titles.
   3. Record all research report titles.
   4. Index and bind all protocols and research reports.

B. Issue, when requested, to senior monitors serial numbered specimen labels for clinical studies and maintain official registry for these labels.

C. Maintian clinical division Document Center, including project schedules, 5-year plan, protocols, reports, FDA filings, and library. Catalog and distribute these items when necessary. Provide ongoing orientation for clinical staff. Maintain Document Center index.

D. Develop annual clinical grant budget with departmental directors, clinical director, research administration, and research finance.

E. Review and recommend approval to clinical director, clinical grant requests and payments. Maintain record system for these documents.

F. Design and issue annual and monthly clinical grant forecast and narrative.

G. Undertake ongoing analysis of clinical grant budget. Undertake ongoing analysis of operating and capital budgets.

H. Maintain liaison with director of finance, R&D on matters pertaining to clinical grants.

I. Write and issue monthly divisional highlights.

J. Issue divisional quarterly report.

K. Undertake program analysis.

L. Maintain and issue clinical monitors project list.

M. Develop ongoing continuing education programs for clinical staff.

N. Develop, in conjunction with clinical information and clinical departments, investigator profile and roster registry.

O. Maintain, compile, and issue clinical operating plans, consisting of description and schedule of projected clinical studies.

P. Develop and maintain a procedures manual.

Q. Be a member of all committees that impinge on divisional activities.

R. Attend meetings of clinical departments. Help coordinate activities among these departments.

S. Keep schedules and follow-up on meeting target dates for projects.

T. Compile and issue to drug regulatory affairs annual IND and NDA progress reports. Act as divisional liaison with drug regulatory affairs.

U. Special assignments from clinical director, i.e., canvassing and planning for a hospital-based clinical pharmacology unit.

Table 1 (cont.)

V. Act as chairman and member of clinical supplies coordinating committee.
W. Issue 5-year plan for clinical programs. Revise plan accordingly.
X. Compile and issue clinical study schedules.
Y. Give lectures and tours of research facilities to visitors and new sales trainees.

Since time is money it is imperative that the clinical project coordination function be an integral part of the drug development process. Changes relating to project management can be very costly, amounting to hundreds of thousands or millions of dollars spent in additional man-hours. Since a clinical project has a well-defined time span, which may change, project coordination has to be responsible for monitering the time frame and the implications that these time changes have for the clinical program.

A review of the literature would indicate that clinical project coordination is in its infancy. Whereas project coordination or project management is utilized in defense, public works, construction, and other industries, it has only been recently applied to research and development (R&D) activities of the pharmaceutical industry. It is important for us to understand certain basic definitions. A project may be described as "a combination of human and nonhuman resources pulled together in a temporary organization to achieve a specified purpose" [2]. Clinical project coordination is but one link in an R&D project management system.

The clinical project coordination function has evolved with the growth and development of clinical research activities in the drug development process. It is logical to couple or group the clinical and support departments under one broad division. Thus data handlers and data acquirers are part of an organization which may variously be called the developmental therapeutics, medical affairs, experimental medicine, or clinical division. The acquisition and processing of clinical data are interrelated. A division responsible for the clinical programs in the drug development process may consist of the departments outlined in Fig. 1. A clinical project coordinator usually serves in a staff capacity to a division director who has responsibilities for the clinical departments that may include support functions (e.g., data processing, biometrics, medical writers, clinical research associates, and a word processing center). Thus the clinical project coordinator is the link in this cycle involving the clinical research programs of the drug development process. Appreciation of the role clinical project coordination plays in the drug development process is recognized by research management.

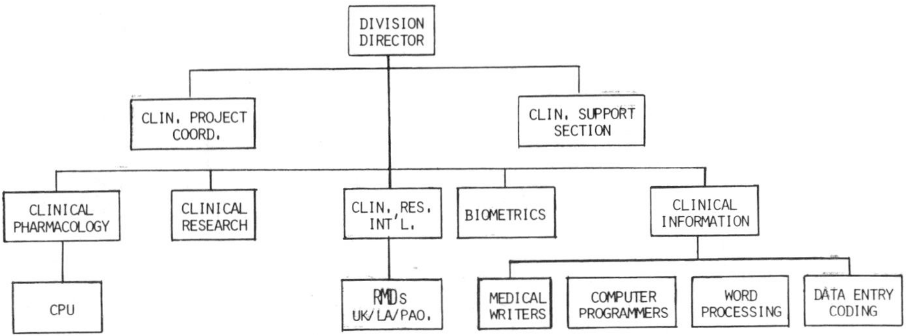

Figure 1 Organization chart: clinical division.

Throughout the component organization units (departments) involved in clinical research programs of the drug development process, clinical project coordination plays a very important role. Since drug development is a costly and lengthy program, it is important that clinical project coordination be involved at various intervals of this process.

We will attempt in this chapter to highlight the diversified activities involved in the clinical project coordination function. Because of the evolving role of the project coordinator, our focus here will be on those activities that are universal in the pharmaceutical industry. Even though the responsibilities of the coordinator and methodology may differ from company to company, the concept is the same.

The qualifications of the clinical project coordinator are important to the success of the function. The ideal clinical project coordinator should have a scientific undergraduate degree (pharmacy, chemistry, biological sciences) coupled with graduate training in planning and/or business administration and experience as a clinical monitor and/or clinical research associate prior to assuming the role of clinical project coordinator. We have not attempted to describe in detail the process sequence or the requisite organizational structure that is utilized in project management. Such organizations as the Project Management Institute (PMI) have done an outstanding job in providing a forum for individuals involved in project management.

Word processors, computer graphics, and integrated laser printing coupled with such project management programs as PERT or CPM (discussed in Sec. II, below) are utilized in the clinical project coordination function. As in many other disciplines, the impact of the computer and information science on clinical project coordination has been enormous. Utilization of the computer encourages quantification to the fullest extent possible, promoting comparable and uniform data.

II. PROGRAM AND PROJECT REVIEW

Modern project control methods are based on the premise that project network planning can be used to reasonably represent the performance sequence of a project.

—Charette and Halverson [3, p. 140]

We will only look at the role and responsibilities of clinical project coordination in program and project review as it pertains to the clinical research process. We will not go into detail on the two techniques utilized in project management. They are PERT (Program Evaluation Review Technique) and CPM (Critical Path Method). A number of excellent books have been written that address the techniques of project management. As we have indicated, project management, a concept of centralized planning and control for single or multiple projects under an individual project manager, is new to the pharmaceutical industry. The degree of sophistication of the project manager's role varies from organization to organization. The division or departments responsible for clinical programs should periodically (preferably on a monthly basis) conduct a joint program and project review. The clinical project coordinator serves as a focal point for ascertaining the availability of clinical supplies; target dates for clinical study schedules; due dates for research reports and such documents as the early registration document; and interdepartmental and divisional target dates for the various clinical documents required for Investigational New Drugs (INDs) and New Drug Applications (NDAs) of the respective clinical programs. This is accomplished by utilization of a protocol and research report registry, project management update, clinical study schedules, and attendance at monthly meetings of clinical and support departments; in addition, the clinical project coordinator is a member of the clinical supplies coordination committee and attends the formal clinical program and project review meeting on a monthly basis. He should also attend the program and project management review conducted by the vice president for research. This is the overall review process which addresses the entire drug development process. In addition to these formal vehicles, the clinical project coordinator interacts with monitors, medical writers, programmers, biometricians, and other support personnel to ascertain the variables which may impact on the target schedule.

The role of the clinical project coordinator is intertwined with the individual on the next level who is involved in project management. He usually is the director of research planning and coordination, reporting to the vice president for research, who is the overall manager of the entire research program. The success of the clinical project coordinator depends upon the acceptance of his role by line management. The coordinator is interested in achieving the project goals so the objectives of the clinical division are met on a timely basis. Benefits of an effective clinical project coordination network can be generally outlined as follows,

1. Timely and accurate project time scheduling
2. Effective resource planning
3. Planned and actual cost tracking
4. Future planning—"what if"
5. Quick recovery reaction
6. Responsibility indentification
7. Needed *and only needed* information for all levels of managemment.
8. Effective communication
9. Course of action documentation

The aforementioned list is not all-inclusive but represents those benefits which are universal to an effective clinical project coordination program.

III. THE FIVE-YEAR PLAN

Planning is the most basic function of management, that of determining a course of action.

—Michael and Stuckenbruck [4, p. 94]

The clinical project coordinator issues the 5-year plan for clinical programs. This plan is based upon the clinical pharmacology and clinical research operating plans on file in the Clinical Document Center. The 5-year plan is displayed on a magnetic planning board in the Clinical Document Center. Among the events displayed are the start of each clinical phase and the month documentation is due. The clinical project coordinator is responsible for revising and issuing the 5-year plan on an ongoing basis. The 5-year plan is a time and events schedule for each clinical program, showing important target dates. Among the events displayed are the start of each clinical phase and the month when related documentation is due. The 5-year plan should be revised once a month immediately after the major program and project review is undertaken.

The 5-year plan is an effective planning tool utilized in research management to project programs over a given time span. This enables R&D management to ascertain the manpower required for grant costs and forecast the impact of the clinical program on the entire research program. This, coupled with the Clinical Operating Plan (discussed next), is utilized to a considerable extent in the R&D planning process.

IV. CLINICAL OPERATING PLANS

> *The most important part of a project plan is an integrated work-breakdown structure, schedule and budget.*
>
> —Stuckenbruck [5, 147]

Clinical research in the United States and overseas is becoming increasingly complex, primarily because of changing regulatory requirements and attitudes in connection with demonstrating the safety and efficacy of drugs. Such matters as sound medical concept, proper experimental design, efficient data retrieval, and precise statistical design and analysis require that every clinical plan be as comprehensive and well planned as possible.

The Clinical Operating Plan is an outline of all aspects of a clinical project in a stepwise progression. The Clinical Pharmacology Operating Plan covers the early studies in man, i.e., Phase I and II or its equivalent, and the Clinical Research Operating Plan covers the broader efficacy and safety studies, i.e., Phase III or its equivalent. Each plan should include the following:

1. Background and rationale of the project
2. Purpose, including the diseases to be treated
3. Number and type of studies
4. Subject or patient selection
5. Number of subjects or patients for each study and phase and for the total project
6. Dose or dose range
7. Duration of therapy for the individual subject or patient
8. Criteria for evaluating efficacy and safety
9. Clinical grant, costs, and the time required for each phase and the total project
10. Bulk drug requirements for each study, phase, and for the total project, including active controls

When estimating the total time of a project, one must keep in mind that: the U.S. Food and Drug Administration (FDA) will require a month "hold" period after receipt of the IND before starting Phase I; time must be allowed to prepare internal documents, e.g., clinical summaries, investigator brochures, and progress reports; time must be allowed for the preparation of clinical supplies; and varying regulatory requirements for the initiation of overseas studies must be considered.

The senior monitor prepares the first draft of the clinical plan after he has considered all pertinent aspects, e.g., the drug, the diseases, and the study objectives. This may require detailed discussions with his supervisor, other members of the clinical division, clinicians or outside consultants, various basic science personnel, data processors, statisticians, corporate development, and especially potential investigators. The Clinical Operating Plan is submitted as a part of the Drug Monograph and thus must be available at the time the drug monograph is prepared. The development of the Clinical Operating Plan depends in great measure on the results of the early studies conducted in the clinical pharmacology unit.

The Clinical Operating Plan draft should undergo a two-stage review process. When the senior monitor has received all comments from the reviewers at each stage, he reviews the plan, incorporating those suggestions and criticisms which he deems appropriate. This may require clarifying discussions with one or more reviewers or a joint meeting of all reviewers to ensure that the program is understood by and acceptable to all. Final review and approval is the responsibility of the clinical pharmacology director or clinical research director, as appropriate. Each director is charged with the responsibility for ascertaining whether all pertinent suggestions and criticisms were considered in the revision process. Upon final approval, the plan is typed in finished form and forwarded to the clinical project coordinator for duplication and distribution. The Clinical Operating Plan is a combination clinical program and budget for a particular drug. This source document serves as the basis for developing the multiple studies required in moving the clinical program through the drug development process. Collectively, clinical operating plans also serve as a planning tool for research management in forecasting the cost of the clinical program.

The clinical project coordinator is responsible for distributing the final Clinical Operating Plan for each clinical program to the personnel designated for distribution by research management.

V. CLINICAL STUDY SCHEDULE

Plans aid coordination and communication, provide a basis for control, are often required to satisfy requirements, and help avoid problems.

—Rosenau [6, p. 45]

At the commencement of a new study, each monitor prepares a clinical study schedule for each clinical protocol for which he or she has project responsibility. In a multicenter study only one clinical study schedule is to be issued. The monitor maintains one copy and forwards the original to the clinical project coordinator.

It is important that the monitor consult with the clinical information, biometrics, and drug disposition staff for clinical assays (if required) in ascertaining the time framework for each open protocol. In the Document Center, magnetic boards display ongoing projects against time (weeks and months). These are listed by project name and protocol number with the phase of the study and the monitor identified. The time spans for important events, such as the clinical trial, clinical assay (where indicated), clinical information processing, statistical analysis (biometrics), and report preparation are shown. Important dates, where appropriate, may also be flagged. To provide the necessary input for displaying these events, a scheduling grid (Fig. 2) is prepared and mounted on the magnetic planning boards utilizing the bar technique (rubber or metal bar strips are utilized; a different color for each event). H. L. Gantt, an engineer, popularized the concept of bar charts some 65 years ago. It is important in preparing these study schedules that the definitions which are listed below are adhered to in developing the time sequence.

A time span for an event is represented by a line of shading extending over the period of time estimated for the event. A separate line is provided to represent each event so that an overlap may be depicted where indicated. Since all events shown may not be required for the successful execution of a protocol, only those essential for the conduct of a study are shown. Only one form is to be utilized for multicenter studies. The following are definitions for each event in the clinical study schedule:

1. *Clinical trial*—The time the drug will be studied in the clinic, calculated from the date of initial shipment of clinical supplies to the first multicenter site to when the last patient completes the trial
2. *Clinical assay*—The time necessary for analysis of biological samples
3. *Clinical information*—The time necessary for data review, encoding, and computer input and output tabulation
4. *Biometrics*—The time between the biometrics staff's acceptance of a "clean" data base to the completion of the biometric support function
5. *Report preparation*—The time necessary to write the research report

In summary, the clinical study schedule provides the following elements of time control:

1. Achievable goals vs. project deadline
2. Time control activities
3. Schedule flexibility
4. Early warning of slippage
5. Recovery from slippage

The clinical study schedule provides the clinical director and research management with a visual mechanism for monitoring the time elements in a clinical program. In addition, it serves as an important visual aid for program and project review meetings.

PAGE ____ OF ____

PROJECT NAME: ________ PROTOCOL NUMBER: ________ MULTICENTER ☐

PROJECT NUMBER: ________ TITLE: ________

MONITOR: ________

TODAY'S DATE: ________

	JANUARY	FEBRUARY	MARCH	APRIL	MAY	JUNE
CLINICAL TRIAL						
CLINICAL ASSAY						
CLINICAL INFORMATION						
BIOMETRICS						
REPORT PREPARATION						

	JULY	AUGUST	SEPTEMBER	OCTOBER	NOVEMBER	DECEMBER
CLINICAL TRIAL						
CLINICAL ASSAY						
CLINICAL INFORMATION						
BIOMETRICS						
REPORT PREPARATION						

Figure 2 Clinical study schedule.

VI. CLINICAL GRANT BUDGET AND FORECAST

> *Actual project costs must be measured to control the cost dimension and may reveal schedule or performance dimension problems.*
>
> —Rosenau [6, p. 165]

The planning process for establishing the clinical program and budget should be started some 6-9 months before the new fiscal year. There are certain distinct advantages associated with the budget planning process:

1. A more efficient organized and positive attitude toward the financial projections of the clinical grant forecast.
2. Assurance that all responsible personnel are viewing the financial objectives of the clinical grant program on the same terms.
3. The ability to coordinate and/or consolidate the diverse approaches toward earning or spending of clinical grant funds.
4. The ability to coordinate and/or consolidate diverse approaches towards financial management of the clinical grant program.
5. The possibility of analyzing how successful your financial planning for the future proved to be, and of determining what, if anything went wrong.

Also trend analysis is an effective management tool utilized by clinical project coordination in clinical grant budgeting and forecasting.

The preliminary budget is formulated by the respective clinical department directors, in conjunction with the division director, the clinical project coordinator, the vice president for research, and the director of research planning and coordination. The budget is reviewed by research administration and finance, then by the appropriate levels of divisional and corporate management. The clinical project coordinator is responsible for issuing the approved clinical grant budget to senior monitors at the beginning of each fiscal year.

The clinical project coordinator is responsible for reviewing and recommending to the division director for approval the individual grant authorization and check request forms. The clinical project coordinator ensures compliance and accuracy on grant authorization and check request forms in conformity with the divisional budget and the clinical grant forecast. The criteria for approval of subsequent payments are the following: number of patients enrolled; number of patients completed; case report forms received to date; time sequence since last payment; and other factors pertaining to productivity, quality, and performance of the investigator conducting the study. A financial management system should be maintained. A copy of each approved grant authorization and check request form should be indexed according to the clinical program, name of payee, and name of investigator. This information coupled with a forecast is utilized in monitoring the clinical grant budget.

A clinical grant forecast (including a detailed narrative) should be formulated and issued monthly or quarterly by the clinical project coordinator. The forecast is based on individual study forecasts provided by the senior monitors who have program responsibility for their respective projects. The clinical project coordinator reviews the individual forecasts. These forecasts may be modified according to past performance and future projections based on an overview of the total program.

A number of factors are associated with variance of forecast vs. the budget: the availability of clinical supplies; review and approval by institutional review boards; availability of investigators; status of preclinical studies; and priorities of the clinical departments. Also, the rate of patient enrollment is a very important factor in determining the amount of funds to be paid for a specific study. Another reason for the variance is that an investigator does not register the number of patients at the rate he indicated to the clinical monitor at the commencement of the study. Accordingly, we may have to increase the number of investigators in a given clinical program. A clinical grant forecast is always tentative in that priorities may change according to many factors. Termination or addition of clinical programs throughout the year are examples of how a forecast's projection can be altered right after release by clinical project coordination. Also the development and/or modification of a clinical operating plan will have an impact on the clinical grant forecast.

The forecast should be developed from information provided by monitors for their respective clinical programs. A clinical grant forecast worksheet (Fig. 3) should be completed by the monitor for every ongoing or projected study for which he or she has project responsibility. This is to be prepared monthly and forwarded to the clinical project coordinator at the end of the month or quarter. Each worksheet represents an individual study; the worksheet provides some of the information utilized by the clinical project coordinator in constructing a forecast and providing a documented narrative explaining the forecast.

The following guidelines should be adhered to by clinical monitors in projecting their forecasts on the worksheets:

1. Each clinical grant forecast must be completed in its entirety.
2. Grant commitment means either an approved or projected grant commitment. If a clinical monitor plans a new study, the grant commitment is either projected or approved, depending on the stage of negotiation with the investigator.
3. Payment should be projected only if funds will be utilized.

These guidelines should be utilized by clinical monitors in developing their forecasts.

PROTOCOL TITLE: ____________________

PROJECT NUMBER (RHC): ____________ PROTOCOL NUMBER: ____________

NAME OF INVESTIGATOR: ____________ INVESTIGATOR # ______

NAME OF PAYEE(S): ____________________

NAME OF SENIOR MONITOR: ____________ SENIOR MONITOR # ______

NAME OF CRA: ____________ CRA # ______

Total Number of Patients/Subjects to be Studied	# ______
Total Number of Patients/Subjects Enrolled as of This Date	# ______
Number of Patients/Subjects Completed to Date	# ______
Total Valid Case Report Forms Received In-House as of the Date	# ______
Total Dollar Amount of Grant (Including Supplemental Grant)	$ ______
Amount Paid Prior to January 1, 1982 (If Applicable)	$ ______
Total Amount to be Paid 1982	$ ______
Actual Amount Paid in 1st Quarter (January, February, March) 1982	$ ______
Actual Amount to be Paid in April, 1982	$ ______
Actual Amount to be Paid in May, 1982	$ ______
Projected Amount to be Paid in June, 1982	$ ______
Projected Amount to be Paid in 3rd Quarter (July, August, Sept.)	$ ______
Projected Amount to be Paid in 4th Quarter (Oct. Nov. Dec.)	$ ______
Projected Amount to be Paid in 1983 (If Applicable)	$ ______

Reason(s) for Change in Forecast From Previous Forecast [May 1982 (If Applicable)]

COMMENTS: ____________________

DO NOT COMPLETE

% Deviation (Variance) From Previous Forecast: ______

Figure 3 Clinical grant forecast worksheet.

The forecast is another important management barometer which enables research executives to monitor the cost of the clinical program on an ongoing basis.

VII. NUMERICAL IDENTIFICATION OF PROTOCOLS AND RESEARCH REPORTS

In order to have continuity of data flow and record keeping, a protocol and report identification system should be employed. Each study protocol (and subsequent report) should be identified by a specific number. Multicenter study protocols are to carry only one number regardless of the number of investigators participating in them. The identifying number for a protocol involving an IND consists of the IND number and a numerical suffix, e.g., the fourth protocol for a particular study may be Protocol No. XX,XXX-04. Protocol numbers should be assigned by the clinical project coordinator.

A protocol for an international study of a drug for which an IND exists in the United States is numbered as though it were a study to be done in the United States, i.e., the IND number used with a sequential numeric suffix. The identifying number of a protocol for an international study not done under an IND consists of the project number and a sequential numeric suffix. Protocol numbers for studies to be conducted solely outside the United States are assigned by the clinical project coordinator. Research reports are to have the same number as the protocol to which they correspond.

VIII. PROTOCOL AND RESEARCH REPORT REGISTRY

The clinical project coordinator should maintain a permanent protocol and research report registry. The registry is a running document

Formal Name of Protocol ____________________________

____

____

Name of Senior Monitor ____________________________

Protocol Number Issued ____________________________

Date Issued ____________________________

Issued By ____________________________

Figure 4 Protocol registry form.

Status Code	*Definition*
FIELD TRIAL COMPLETED	Clinical trial has been completed.
FINAL RESEARCH REPORT ISSUED	Indicates research report has been issued and study is closed.
HOLD	Indicates study will be implemented or protocol will be developed upon decision of director of respective clinical department.
NEVER FILED, CANCELLED	Protocol was approved on departmental basis but never implemented.
NO FINAL RESEARCH REPORT ISSUED	Indicates clinical trial has been completed but final research report has not been issued.
NO PROTOCOL ISSUED	Protocol never formalized.
PENDING	Study will soon commence.
NUMBER NOT ASSIGNED	Protocol number voided.
PROTOCOL UNDER DEVELOPMENT	Indicates protocol is being developed.
STUDY IN PROGRESS	Indicates clinical trial is in progress.
STUDY TERMINATED	Study stopped during clinical trial.
T.B.D.	Protocol will be implemented as soon as investigator is recruited.

Figure 5 Status code: protocol and research report registry.

utilized to track the protocols and research reports which are prepared by the clinical departments. Updated on a periodic basis, this registry also serves as the index for the filing of protocols and research reports in the Document Center. A protocol registry form (Fig. 4) is completed by a senior monitor when requesting a protocol number from clinical project coordination.

Document Center		Book No.	Protocol No.	Investigator			Protocol Title	Date Protocol Prepared	Date of Report	Monitor	Status
P	R			Name	No.	Perf. Eval.					

Figure 6 Format: Protocol and research report registry.

This registry serves as a reference source whereby all of the aforementioned documents are listed under their respective clinical program. Figure 5 is the status code utilized in the protocol and research registry, whereas Fig. 6 represents the outline of a sample page utilized in the registry.

IX. CLINICAL INVESTIGATORS ROSTER

The clinical project coordination unit should maintain a roster of clinical investigators who are studying, have studied, or may be available to study investigational compounds. (The roster is under continuing review for additions and deletions as appropriate.) By a simple computer procedure, potential investigators may be identified for ready reference by geographic location, specialty, qualifications, type of practice or academic standing, investigational interest and experience, and previous performance with investigational programs. Such information is derived from the clinical investigator profile form (Fig. 7) that should be completed at the time the investigator's name is added to the roster and forwarded to the data management department for processing. When a listing of clinical investigators based on any of these parameters is needed, a request should be directed to the data management section outlining the pertinent requirements. Within 3 or 4 days a computer printout providing the information requested should be available for reference.

The roster is also the source of investigator numbers that are used to identify clinical data derived from clinical trials and submitted by the principal investigator of those trials.

X. INVESTIGATOR PERFORMANCE EVALUATION

> *Reviews are the most important control tool available to the project manager.*
>
> —Rosenau [6, 135]

Clinical project coordination should be responsible for documenting the performance evaluation provided by clinical monitors. The proper completion of a clinical project depends on the performance of each investigator involved in the project. Since many of the physicians who conduct studies for the clinical division are considered again as investigators, it seems appropriate that the investigator's performance be evaluated at the termination of his/her study by the senior monitor and either by the overseas monitor or by the clinical research associates.

LAST NAME	FIRST NAME	MIDDLE INITIAL	STATUS
NAME OF INSTITUTION (if applicable)			Active (1)
			Deceased (2)
ACADEMIC OR INSTITUTIONAL RANK (if applicable)			Retired (3)
			Other (4)
STREET ADDRESS (mailing address)			
CITY	STATE	ZIP CODE	
TELEPHONE NUMBER		DEGREE	
NAME OF SENIOR MONITOR COMPLETING FORM			

SPECIALTY - (mark applicable boxes with an "X")

☐ Anesthesiology	☐ Otorhinolaryngology
☐ Clinical Pharmacology	☐ Pathology
☐ Dentistry	☐ Pediatrics
☐ Oral Surgery	☐ Physical Med. & Rehabilitation
☐ Periodontology	☐ Preventive Medicine
☐ Dermatology	☐ Psychiatry
☐ General Practice	☐ Radiology - General
☐ Internal Medicine	☐ Radiology - Diagnostic
☐ Allergy	☐ Radiology - Therapeutic
☐ Cardiovascular Disease	☐ Surgery - Colon & Rectal
☐ Endocrinology	☐ Surgery - General
☐ Gastroenterology	☐ Surgery - Neurological
☐ Pulmonary Disease	☐ Surgery - Orthopedic
☐ Neurology	☐ Surgery - Plastic
☐ Obstetrics & Gynecology	☐ Surgery - Thoracic
☐ Ophthalmology	☐ Urology
☐ Osteopathy	☐ Other

Figure 7 Clinical investigator profile.

(LEFT JUSTIFIED)
(Only if "OTHER" is Checked)

______________________________ ☐ Board Qualified
"Other" Specialty

☐ Board Certified

PRIMARY PRACTICE - (Mark Applicable Boxes with an "X")

☐ Full Time Private Practice
☐ Full Time Pvt. Pract. W/Faculty Appt.
☐ Intern in Hospital Service
☐ Resident or Fellow in Hosp. Service

☐ Research
☐ Preventive Medicine

☐ Full Time Hospital Staff
☐ Full Time Med. School Faculty
☐ Administrative Medicine
☐ Lab Medicine (Pathology)
☐
☐
☐

GENERAL INVESTIGATIONAL INTEREST OR EXPERIENCE
(Mark Applicable Boxes with an "X")

PHASE OF INVESTIGATION

☐ I ☐ II ☐ III

☐ Analgesia
☐ Anesthesiology
☐ Infectious Diseases
☐ Neuropsychiatric Disorders
☐ Cardiology
☐ Gastreoenterology
☐ Immunology
☐ Other Internal Medicine
☐ Reproductive Endocrinology
☐ Inflammatory Diseases
☐ Diabetes
☐ Nutrition

☐ Other Metabolic Disorders
☐ Neoplastic Disease
☐ Nuclear Medicine
☐ Dentistry
☐ Diagnostics
☐ Surgery
☐ Molecular Biology
☐
☐
☐
☐
☐

Prior Evaluation Form on File ☐ Yes ☐ No

☐ Previous Investigational Experience

☐ Not of Special Interest ☐ Of Special Interest

(If yes, complete information on next page)

Figure 7 (Continued)

* Name of Protocol ____________________

* Protocol Number ____________________

* Dates study conducted ____________________

* Name of Protocol ____________________

* Protocol Number ____________________

* Dates study conducted ____________________

* Name of Protocol ____________________

* Protocol Number ____________________

* Dates study conducted ____________________

* Name of Protocol ____________________

* Protocol Number ____________________

* Dates study conducted ____________________

* Name of Protocol ____________________

* Protocol Number ____________________

* Dates study conducted ____________________

* Name of Protocol ____________________

* Protocol Number ____________________

* Dates study conducted ____________________

* Name of Protocol ____________________

* Protocol Number ____________________

* Dates study conducted ____________________

Figure 7 (Continued)

Investigator's Name____________________Number________

Location____________________

Drug____________________Dosage Form________

Protocol No.____________________Indication(s)________

Reason Study Was Terminated:

☐ Completed ☐ Lack of suitable patients

☐ Loss of Interest ☐ Other (specify)

Difficulties (if any) encountered during study____________________

Overall assesment of performance ☐Excellent ☐Good ☐Fair ☐Poor

Comments and Recommendations____________________

Evaluation by____________________Date________
(CRA or Local Monitor)

(To be completed by Monitor)

Number of months scheduled for study________Number of months elapsed________

Number of case reports expected________Number received________

Percent usable ☐<50% ☐ 50%-75% ☐75%-85% ☐ >85%

Reason(s) for exclusion from efficacy analysis:

☐ Poor patient selection ☐ Loss to follow-up

☐ Failed to follow protocol ☐ Insufficient data

☐ Other (specify)

Overall assessment of performance ☐ Excellent ☐ Good ☐ Fair ☐ Poor

Comments and recommendations:____________________

Evaluation by____________________Date________
(Monitor)

C O N F I D E N T I A L

Figure 8 Evaluation of investigator's study performance.

The evaluation of the investigator's study performance form (Fig. 8) is completed at the termination of each study. In the United States, the clinical research associate (CRA) completes the top half of the form, signs it, and forwards it to the manager of the clinical research associates unit, who then delivers the partially completed form to the responsible clinical monitor. The clinical monitor completes and signs the second half of the form and forwards it to the clinical project coordinator.

Outside the United States, the local monitor completes the top half of the form, signs it, and sends it to the director of clinical research (international). He in turn completes and signs the second half of the form and forwards it to the clinical project coordinator.

The sections headed "Comments and Recommendations" should include the opinions of the senior monitor and of the CRA or local monitor regarding the suitability of the investigator for future studies.

The clinical project coordinator should maintain a file of completed performance evaluations as a reference source for the senior monitors in selecting investigators for future studies. These evaluations could be computerized in conjunction with the information on the investigator profile, providing comprehensive information on each investigator.

The clinical project coordinator should record the evaluation of each investigator and study in the protocol and research report registry.

XI. CLINICAL SUPPLIES COORDINATING COMMITTEE

In some R&D programs management has established an interdivisional clinical supplies coordinating committee. This committee has the responsibility for coordinating the scheduling of clinical orders and preparation of clinical supplies. The committee makes recommendations for the adoption of procedures to improve these activities. Among the members of the committee are the clinical project coordinator and the section head of the clinical pharmacy R&D department. The chairmanship generally rotates between these two individuals.

The clinical project coordinator acts as clinical research liaison to clinical pharmacy, and the section head of clinical pharmacy R&D acts as the clinical pharmacy liaison to the other clinical departments. Clinical supplies requests and target dates are coordinated in conjunction with R&D and clinical pharmacy personnel. The committee should:

1. Coordinate the scheduling of clinical orders and the preparation of clinical supplies to ensure the timely availability of required materials
2. Make appropriate recommendations to its respective managements for the adoption of procedures which will assist in improving activities concerning the ordering, preparation, and distribution of such materials

The committee often addresses the following issues in discharging its responsibilities:

1. Scheduling of clinical supplies
2. Standardization of packaging and labeling requests
3. Drug accountability procedure
4. Standardization of medication count
5. Labeling of medication container
6. Request for bulk clinical supplies
7. Receipt of supplies by investigators

The committee should issue a monthly report of the status of all clinical orders. This report should be distributed to all clinical departments, including directors and clinical monitors, research management, and pharmacy R&D management.

Clinical supplies are an integral part of clinical investigation. The clinical supplies coordinating committee serves as a vital link between the pharmaceutics division and the clinical division.

XII. CLINICAL PROJECT LIST

The clinical project coordinator issues on a quarterly basis a clinical project list reflecting both the domestic and the international clinical programs. This list is compiled by the clinical directors, who assign program responsibility for each clinical program to a senior monitor. The clinical directors (Department of Pharmacology, Department of Clinical Research, Department of Clinical Research/International) then send their lists to the clinical project coordinator for distribution.

The clinical project list is distributed to the preclinical, clinical, marketing and research, and executive research management directors. This information provides them with an overview of the active clinical studies being conducted by the departments responsible for the respective programs.

Inquiries of a general nature pertaining to the clinical project list and those related to nonspecific developmental therapeutic activity should be directed to the clinical project coordinator. Inquiries related to specific active projects or those which require a qualified management decision should be directed to the clinical division director.

The clinical project coordinator revises the clinical project list whenever assignments are changed or additions or deletions to the clinical program occur.

XIII. MONTHLY CLINICAL DIVISION HIGHLIGHTS

The clinical project coordinator is responsible for compiling the monthly highlights report for the clinical division. Each month the following departments and sections in the clinical division provide the clinical project coordinator with the individual highlights for their departments or sections:

Biometrics
Clinical Information
Clinical Pharmacology
Clinical Research
Clinical Research/International
Clinical Support

The individual highlights should contain the following information:

1. Status of each clinical program
2. New developments
3. Seminars and conferences attended by staff
4. Personnel appointments and promotions
5. Miscellaneous information

This outline does not preclude the inclusion of additional information which may be unique to a particular department or section to whom these highlights are distributed.

The divisional highlights which are a synopsis of the programs and activities should be distributed to all personnel in the division and a cross-section of other scientists and research managers in the R&D area.

XIV. DOCUMENT CENTER

Project success is completely dependent on adequate planning, direction, scheduling, monitoring and control.

—Charette and Halverson [3, p. 118]

The Document Center is intended to serve as an area for project planning and review. In order to facilitate this purpose, clinical plans, protocols, written reports, and clinical summaries for active clinical projects are assembled in the Document Center in looseleaf binders. The protocols and reports are grouped by project in numerical sequence (protocol number), and clinical summaries are grouped chronologically. A master title index of the contents of the individual binders is available for reference. The binders are color coded. Also filed in the Document Center are investigational drug brochures, IND filings, optional expanded summaries, workshop proceedings, and product package inserts. A copy of the protocol and research report registry is available in the center. It also contains reference texts in pharmacology and the clinical investigation process, as well as travel and airline guides and other reference information.

Magnetic planning boards displaying the 5-year plan and clinical study schedules are located in the Document Center. Because information on the division's entire clinical research program is assembled in this facility, the room must be classified as confidential and restricted to staff members of the clinical division and invited individuals. Copies of individual protocols, reports, and summaries may be made by staff personnel. The Document Center serves as a reference library for the clinical division.

XV. CONCLUSION

> *During the last three decades, the ethical pharmaceutical industry has been transformed into a research-based highly competitive industry*
>
> —Schnell [7, p. 39]

It is obvious that with this transformation clinical project coordination must be an integral part of the drug development process. We have attempted to show the important place clinical project coordination has in the R&D activities of the contemporary pharmaceutical industry.

The clinical project coordinator is a vital link in the clinical research process. The planning and implementation of clinical programs on a timely and cost-effective basis depends on the role of clinical project coordination in the clinical research process. In view of the keen competition in clinical research and the high cost of R&D, the clinical project coordination function is a necessity

In conclusion it is obvious that the following statement by Michael and Stuckenbruck [4, p. 115] is appropriate: "Effective planning is an essential ingredient for project success." The future is bright for those clinical research programs that implement the clinical project coordination function in their organizations.

REFERENCES

1. D. S. Frederickson, quoted by R. S. Gordon, Proceedings of the national conference on clinical trials methodology. *Clin. Pharmacol. Ther. 25:*629 (1979).
2. D. I. Cleland and W. R. King, *Systems Analysis and Project Management,* 2nd ed. McGraw-Hill, New York, 1975, p. 184.
3. W. Charette and W. S. Halverson, Tools of project management. In *The Implementation of Project Management* (L. E. Stuckenbruck, Ed.). Addison-Wesley, Reading, Mass., 1981.
4. S. B. Michael and L. C. Stuckenbruck, Project planning. In *The Implementation of Project Management* (L. E. Stuckenbruck, Ed.). Addison-Wesley, Reading, Mass., 1981.

5. L. C. Stuckenbruck, The job of the project manager: Systems integration. In *The Implementation of Project Management* (L. E. Stuckenbruck, Ed.). Addison-Wesley, Reading, Mass., 1981.
6. M. D. Rosenau, *Successful Project Management.* Lifetime Learning Publns., Belmont, Cal., 1981.
7. J. E. Schnell, Economic structure and performance of the ethical pharmaceutical industry. In *The Pharmaceutical Industry* (C. M. Lindsay, Ed.). Wiley, New York, 1978.

7
THE PROTOCOL AND CASE REPORT FORM

Allen E. Cato and Linda Cook
Burroughs Wellcome Co.
Research Triangle Park, North Carolina

I. INTRODUCTION

Clinical trials should be disciplined, organized research studies conducted in humans to determine the value of a therapeutic agent. Carefully defined groups of patients or subjects are selected according to proper experimental design to eliminate biases in the assessment of the outcome of treatment. Without an appropriately controlled, well-designed trial, the results of a study could be meaningless or misleading.

For example, idoxuridine (IDU) was the first clinically effective antiviral agent approved by the U.S. Food and Drug Administration (FDA). It was licensed due to its demonstrated usefulness in the topical treatment of herpetic keratitis, a leading cause of vision loss in the United States. While originally available systemically for experimental use in the chemotherapy of malignant tumors, IDU was subsequently employed empirically for the treatment of herpes simplex encephalitis, a relatively rare but usually fatal infection of the brain [1,2]. Although the drug has significant and frequent toxicity (myelosuppression in particular), at least 70 cases of encephalitis were treated with parenteral IDU before two groups independently undertook double-blind, placebo-controlled efficacy and safety trials [3].

The combined results of the two studies showed that six of the eight biopsy-proved or probable cases died despite a full course of IDU therapy, and all four of those patients in which postmortem examinations were performed had herpes simplex isolated from the brain at autopsy. In the two cases where comparisons could be made, the quantity of virus present at autopsy was equal to or greater than that present prior to therapy.

Finally, six of eight patients had clinically significant evidence of myelosuppression, and seven of eight developed pneumonia. Both studies were prematurely terminated, and further systemic use of IDU was discontinued [4].

In this chapter, we shall discuss the design of clinical trials relative to drug development. The principles of protocol design, however, remain the same no matter what the nature or purpose of the investigation. Examples of problems that have occurred during clinical evaluation of treatment are also discussed. Many of these problems might have been avoided had more careful planning gone into the initial stages of protocol design.

Whatever type of design is selected, clinical trials of some sort are clearly necessary for the evaluation of new therapeutic modalities. Two essential prerequisites for the execution of successful clinical trials are a carefully designed protocol and accompanying case report forms.

II. DESIGN OF THE CLINICAL PROTOCOL

A. Abstract and Table of Contents

An abstract is optional but can be very useful, especially when a protocol is large and/or complex. Similar reasoning applies to the Table of Contents.

B. The Introduction

The introduction of the protocol varies, depending on the phase of drug development and the relative amount of preclinical and clinical experience with the drug.

In the early stages of clinical development (Phase I, to determine initial safety and tolerance, pharmacokinetics, bioavailability and metabolism, drug interactions, and initial efficacy), the results of preclinical tests are emphasized. Included are the chemistry, pharmacology (site of action, metabolism, excretion, absorption, and degree of protein binding) and toxicology data.

As clinical experience is gained and the drug enters Phase II, in which its efficacy is tested for one or more clinical indications, the preclinical results are summarized only briefly. More emphasis is placed on summaries of safety and efficacy derived from human exposure to the test agent.

After summarizing the current status (preclinical and/or clinical) of the drug, the rationale of the study is stated briefly. This is a concise statement of the purpose of the study—of why the study should be done. The rationale, or "why" of the study, should not be confused with the objective(s), or "what" of the study. The rationale, as opposed to the objective, is a part of the overall plan for the clinical development of a drug and is therefore presented in a broader context.

C. Objective

The objective of a clinical trial is to assess the value of a drug in the treatment or diagnosis of a disease, as determined by the drug's benefits relative to its risks or undesirable effects. The objective is a crucial part of the protocol and should be presented in a clear, succinct statement detailing *what* is being investigated. It lists exactly which questions are to be answered from the study (the endpoint).

Rarely will there be a successful study with more than two or three major objectives, although there may be minor objectives as well (i.e., those objectives which may be obtained as a by-product of the study but around which the study is not necessarily designed). Objectives are usually covered in four or five statements with a minimum of modifying embellishments.

The objective should be nonbiased. A statement like "To demonstrate the efficacy of drug X as an antidepressant" predetermines the outcome, namely, that efficacy will be demonstrated. A more appropriate wording would be "To assess the effect of drug X in depressed patients relative to its safety and efficacy, if any."

In addition to determining the degree of efficacy of a compound, assessments should always be made of the nature, severity, and likelihood of occurrence of side effects and toxicity reactions. The ultimate determinant of risk vs. benefit in a new drug involves its efficacy and toxicity in comparison to currently available treatment.

The importance of stating the exact objective of a study prior to initiation cannot be overemphasized. In cases where the study objective changed as data were accumulated, serious problems have arisen with the data analysis and/or resulting conclusions.

As an example, the Anturane reinfarction trial was designed with the objective of comparing the effect of sulfinpyrazone (Anturane) with that of placebo in preventing recurrent infarction and death in patients who had had a recent, well-documented acute myocardial infarction. The study was a 2-year, double-blind, randomized, multicenter trial in about 1600 patients. The original hypothesis on which the study was designed projected that sulfinpyrazone would prevent the platelet-mediated phenomena associated with arterial disease, which in turn is thought to lead to myocardial infarction. The study results failed to show any benefit of sulfinpyrazone in reducing mortality from myocardial infarction but did suggest a benefit of the drug in preventing sudden death [5].

However, the protocol was not originally designed to look specifically at the effect of sulfinpyrazone on sudden death. The resulting shift in objective undoubtedly contributed to the FDA's refusal to approve a claim that sulfinpyrazone is effective in preventing sudden death after myocardial infarction. The FDA reviewers commented that "important aspects of the study were underdesigned, at least in view of the importance that those aspects later acquired [6].

In particular, it was felt that the definition of sudden death was defective. The FDA, therefore, recommended another, similar study. Controversy arising from the results of many clinical trials emanates from confusion over the objectives, not from the results per se [7].

D. Study Design

Consideration of the appropriate research design in the initial stages of development will protect subjects from injury or inconvenience, as well as ensure the scientific validity of the trial, since a clinical trial is only as good as its design. Generally, consultation with a biostatistician is desirable.

An example of improper study design occurred during the retrolental fibroplasia (RF) epidemic of the 1940s and 1950s. In the early 1940s, it became common practice to expose premature infants to high levels of oxygen for varying times after birth in order to increase their survival. A new disease, RF, appeared at the same time, characterized by vascular abnormalities in the eyes, progressing at the extreme to retinal detachment and blindness.

An open uncontrolled trial of therapy with ACTH (adrenocorticotropic hormone) demonstrated that only 2 of 31 treated infants became blind, whereas 6 or 7 untreated infants at another hospital apparently became blind. Therapy with ACTH became routine before a controlled trial was conducted. Ultimately, a multicenter cooperative trial demonstrated that about one-third of the ACTH-treated infants became blind, whereas blindness occurred in only one-fifth of the untreated, control infants. Furthermore, there were more deaths in the group treated with ACTH. It was later learned that approximately three-fourths of the infants showing the early changes of RF will spontaneously revert to normal without any treatment. The delay in conducting a well-designed, controlled clinical trial unnecessarily exposed many infants to risk without benefit and the discovery of the cause of the epidemic (an oxygen-rich environment) was delayed [8].

The components of a good study design are considered in the following subsections.

1. *Type of Design*

A suitable study design (open, single- or double-blind, parallel or crossover, placebo or positive control) should be chosen, considering such factors as the trial setting, institution, number of therapies, and severity and prevalence of the disease. The control group (active vs. placebo) depends on: (1) consideration of existence and value of active drugs available to treat the disease in question; (2) predictability of the natural evaluation of the disease if untreated; (3) reliability and definitiveness of diagnosis; and (4) severity and consequences of the disease if untreated.

2. *Basic Test Group*

The selection of patients is a major aspect of the design. If the test sample is not representative of the population, the results will be unreliable and of dubious worth.

3. *Number of Subjects or Patients*

There are important statistical and ethical implications to consider in determining sample size. A clinical trial should have sufficient statistical power to identify a clinically important difference between treatment groups. At the same time practical aspects such as limitation of time and availability of patients should be considered. If the sample size is too large, the trial may be judged as unethical due to unnecessary involvement of extra patients; yet a trial with a sample size that is too small may not detect clinically important differences and may lead to incorrect conclusions relative to the question(s) being asked.

Generally, the smaller the difference that is judged clinically significant and/or the larger the variability between patients, the greater the number of patients needed for a valid clinical trial. A clinical trial proposal should report the statistical methods used to determine the sample size, as well as an estimation of confidence intervals (see Sec. II.H, below).

4. *Method of Randomization*

Randomization allows the choice of treatment for each patient to be made in an independent manner, ensuring that *only* random differences can affect the treatment comparison. Results of a trial can be interpreted scientifically only if the treatment groups are comparable in all relevant aspects. And, if unknown variables are not considered, failure to achieve comparability can destroy the results of the study. Thus, the major function of randomization is to achieve approximate comparability with respect to all variables. Randomization prevents bias and contributes to equal distribution of patients and validity of statistical analyses.

5. *Duration of Study*

Under this heading are listed the following items:

a. Overall duration
b. By test agent
c. Length of washout period, if present
d. Posttreatment period, if needed
e. Duration of this study in relation to other studies and ultimate claims about efficacy and safety of the drug

(Drugs, dosages, and details of experimental procedures are given subsequently.)

E. Drugs and Dosages

This section is best presented in outline form, as follows:

1. *Name of Experimental Drug(s)*

a. Test drug
b. Active control, if appropriate
c. Concomitant therapy, if allowed, e.g., antiepileptic

Concomitant therapy deserves special attention. In the case of antiepileptic compounds, patients should enter a study with stable therapeutic plasma levels. During a study, plasma levels should be monitored frequently to determine potential drug interactions.

2. Dosage Forms

Includes strength of the unit dose(s), appearances relative to placebo or active control, and route of administration [oral, intramuscular (i.m.), intravenous (i.v.), topical, or subcutaneous (s.c.)].

3. Dosage Instructions

a. Unit dose
b. Total daily dose (single or multiple courses)
c. Number of capsules or tablets at each dose

It is preferable to keep the number of tablets or capsules constant. In a dosage titration study, this can be done by adding placebo capsules to active drug, as needed.

d. Frequency and manner of administration (fixed or flexible doses)

The rate of increasing dose is tailored to the specific drug and is usually somewhat arbitrary; may be a combination of logarithmic and geometric progression.

e. Time of administration (including relation to meals), and volume of coadministered fluid when dosing orally

In order to achieve some reliable uniformity in dosing, factors affecting the dose-response relationships have to be considered. Two of these factors—intake of fluid volume on drug absorption, and dosing in relation to timing and type of meal—are often overlooked.

Animal studies have shown that drug activity is increased in experimental animals by diluting the dose. Furthermore, absorption efficiency from solid dosage forms is often increased in humans by increasing the accompanying fluid volume [9]. For example, erythromycin stearate absorption was decreased by 43% in fasting volunteers when the accompanying fluid volume was decreased from 250 to 20 ml. When the same drug was administered with various test meals, absorption decreased by 53 to 64% [10]. Conversely, the bioavailability of propoxyphene, in solution or as a capsule, tended to decrease slightly when coadministered with a large volume of water. Additionally, overall absorption efficiency was unchanged or slightly increased when the drug was given with test meals [11]. However, results of these and other studies suggest in general that more efficient and reliable drug absorption may be obtained when an oral dosage form is taken with a relatively large fluid volume.

f. Mode of distribution to patients

The starting dose is usually tailored to anticipated pharmacological or toxic effect. The endpoint is determined by the emergence of drug effect, either pharmacodynamic or toxic.

4. *Decoding Procedures*

Instructions are given for breaking the code, if necessary, in blind studies.

5. *Disposition of Unused Drugs*

Instructions are included for the investigator as to the disposition of unused drugs.

6. *Labeling and Packaging*

Explanation of labeling and packaging of experimental drug(s) is related to drug identification and sequence of administration. If extreme caution is not used in the labeling and packaging of experimental drugs, costly errors can result and valuable time is lost. For example, in the first double-blind, placebo-controlled efficacy study of a potential antidepressant compound, some placebo capsules were mixed with the investigational drug. The error was detected through plasma and urine analyses of some of the volunteers who were randomized to receive active drug. Whereas lower doses had resulted in significant plasma levels of drug, at higher dosages drug was suddenly undetectable. Assays subsequently performed on the unpackaged test drug confirmed the mixing of active and placebo capsules. As a result of the error, the study had to be repeated and the volunteers redosed. The consequences were a time delay of approximately 3 months in the development of the drug, an additional cost of $23,000, and an exposure of volunteers to a test drug from which essentially no useful information was garnered.

F. Experimental Procedures

In this section, a series of procedures *in time sequence* is presented from initiation of the screening of subjects/patients through post-treatment.

1. *Overall Time/Events*

A time/events schedule, summarized in tabular form, is presented as a general introduction to experimental procedures. This schedule (a) provides the reader with an overall, concise summary of the flow of the study, and (b) provides the writer with a reference for the detailed instructions which follow.

2. Measurements and Evaluations

A chronological sequence of objective measurements should be used when possible. It is recommended that the details of assay procedures and associated calculations be relegated to an appendix, with reference only made to them in the body of the protocol.

Screen Phase

For the investigator to determine clearance of a subject/patient for entry into the study, screening usually includes a physical exam, history, general and specific laboratory tests, and inclusion/exclusion criteria. Excessive redundancy should be avoided in the inclusion/exclusion criteria when they are implicit in the physical examination and history.

Patient definition is of crucial importance. The inclusion criteria should be clear and specific, where appropriate, defining sex, age, and weight limitations. For Phase I studies utilizing normal volunteers it is usually wise to specify that weight should not deviate more than some 10-15% from the ideal [12].

The patient population should be suitably selected in relation to the aim of the study. A precise definition of the disease to be treated should be given, as the nature of the illness determines the way in which the treatment effects have to be conceptualized. The diagnosis and symptomatology of the disease should be objectively specified and classified so that correct assessment of the treatment outcome can be made.

In the initial idoxuridine reports described earlier, many patients reported as cured with IDU were probably incorrectly diagnosed, since they had no brain biopsy. In about 25% of patients with a clinical diagnosis of herpes encephalitis, existence of the disease cannot be proved by biopsy and culture [3].

Acceptable previous and concomitant therapy should be detailed in the selection criteria. Where practical, patients should curtail all previous drugs for at least 2—preferably 4—weeks prior to study initiation.

The exclusion criteria should specifically state the type of patient to be excluded from the study (doubtful mental status; inability to give informed consent; special groups or situations, e.g., pregnancy, drug addicts, alcoholics, children, women of child-bearing potential, frequent volunteers, geriatric patients).

A simple history may not always be sufficiently reliable for the purpose of assessing exclusion criteria. A striking example occurred in 1980, when a 23-year-old nursing student died, presumably from cardiac arrest, while participating in a National Institutes of Health (NIH) sleep experiment. Lithium was being tested in normal volunteers. However, the nursing student turned out to be far from normal. She suffered from anorexia nervosa, had experienced cardiac arrest on two previous occasions, was under both individual and group psychotherapy, and at the time of her death had a serum potassium level below normal, possibly due to self-induced vomiting [13].

History alone has proved particularly unreliable as an indicator of abstinence of alcohol. Sensitivity in one study was as low as 17% [14]. Patients with alcoholic liver disease were questioned once a week for 2 months, and urine samples were examined for alcohol. Of the patients who absolutely denied drinking for the total observation period, 24% of their urine samples contained alcohol levels above 80 mg/dl, a level associated with impaired driving [14].

The screening phase should be concluded by the statement: "The clinical interpretation of the screening results by the principal investigator and the study monitor is the ultimate determinant for clearance of a subject/patient to the baseline phase of the study." In one study using normal volunteers, 29 healthy young men were screened; only four had all laboratory data within the "normal" range [15].

Baseline Phase

A detailed description of measurements/evaluations and calculations unique to baseline should be listed here. Most of these assessments are carried over to the treatment period. Also, selected measurements taken during the screening phase may be used as baseline determinations. Thus, the time lapse between screen and baseline is preferably no longer than 2 weeks.

Treatment Phase

It is wise to review the objectives at this point, since in the first draft of many protocols the measurements and evaluations initially listed have little relation to the stated objectives.

Instructions for measurements/evaluations which are common with those of baseline phase can usually be referenced to that section, eliminating repetition of details. Specific statements on concomitant therapies, associated procedures, and general reporting of side effects are usually included in the treatment phase and can be parenthetically referenced to the appropriate sections of the data collection forms.

Posttreatment Phase

Observation after drug discontinuation is particularly important if delayed toxicity is suspected.

G. Procedures for Patient/Subject or Study Discontinuation

Instructions for steps to be taken in case of premature discontinuation of subject/patient from the study should be stated. Patient discontinuation might result from side effects; lack of effect; the patient not cooperating, e.g., lack of compliance; the disease being cured (or remitted), e.g., the test agent proving effective (medication no longer needed); or other reasons (subject moved to another city, failed to return, etc.).

Also, any special tests or evaluations to be done at the time of patient termination can usually be referenced to posttreatment assessments previously listed.

Procedures for reporting to clinical monitors of severe and/or unusual adverse reactions, along with specific follow-up procedures, should be specified.

Premature discontinuation of a clinical study by the investigator requires a detailed explanation, and such an explanatory statement should be included.

H. Statistical Analysis

The statistician can contribute to clinical research in (1) protocol design, (2) study execution, and (3) analysis of data.

Statistical considerations in protocol design include assistance in matching design to objectives, determining treatment structure (e.g., double-blind controls and dosage groups), defining response parameters and providing randomization (including stratification). Statisticians often assist in the development of efficient data collection instruments. Most importantly, after the clinical scientist prospectively determines the smallest difference of clinical significance in the parameters to be measured, the necessary sample size can be projected.

If the sample size (related to the power*) of a study is too low, there is a greater risk of a false-negative finding. In a recent survey of 71 reportedly negative randomized controlled clinical trials, over two-thirds had more than a 10% risk of missing a true 50% improvement [16]. The reason was inadequate sample size. Another survey of 612 articles from three widely read medical journals disclosed that 38% presented data on 10 or fewer patients. A single patient was reported in 13% of the articles [17].

If there are to be interim analyses during study execution, this should be considered during protocol preparation. Modifications of protocol design after study initiation may require statistical input.

A prospective plan for data analysis is crucial. The conclusions of many large, multicenter trials have often been fraught with controversy concerning the data accepted for analysis and subsequent analytical methodology [7,18,19].

Before trials begin, consideration should be given to patients that may be excluded from analysis, and any potential bias resulting thereby. Exclusions, a controversial subject, sometimes may include: noncompliance, including protocol deviations, especially the patient

---

*This is a measure of the probability of obtaining a statistically significant result for a population difference of a given magnitude.

taking restricted medications; voluntary withdrawal (e.g., the patient moving); intercurrent disease; death; treatment mix-up (wrong treatment, or treatment out of sequence); ineligibility (discovering that the patient, after receiving the test drug, does not meet entrance criteria); withdrawals due to side effects; and failure to improve or cure the disease state. The implications for study analysis will have to be considered when patients are discontinued (as some nearly always will be!).

Ultimately, the objective is to make clinical importance and statistical significance coincide, thus avoiding problems of interpretation. After 32 studies over 25 years, there is still no consensus on the efficacy of anticoagulants following myocardial infarction [20].

I. Consent/Clearance Provisions

Standard statements for institutional review, approval of clinical protocol, and informed consent agreement are presented in this section.

J. The Bibliography

A list of references should be included. Most of the references will be cited in the Introduction.

K. Appendix

Detailed assay procedures or techniques and special laboratory test procedures referenced from the body of the protocol (and the like) are presented in the Appendix.

III. DESIGN OF CASE REPORT FORMS

A. Introduction

The importance of well-designed protocols for conducting clinical trials has been emphasized. However, the protocol will not be successful unless complete and accurate records of the results are maintained. The means of recording the data obtained during clinical trials are through the use of properly designed forms called data collection forms or case report forms (CRFs). Without the maintenance of proper CRFs, the scientific and medical objectives of a well-designed protocol and carefully controlled trial may be defeated.

B. Reasons for Need

Carefully designed forms are essential: (1) as a means of checking the logistics, design, and practicality of the protocol; (2) for later processing, analysis, and interpretation of data in order that results may be accurately reported; (3) to check protocol adherence and/or investigator compliance; and (4) to fulfill FDA requirements.

1. Protocol Check

When attempting to design CRFs for a protocol, basic inconsistencies within the protocol often become apparent. For example, different areas of a protocol may call for measurements to be made at the same time, such as an electrocardiogram (ECG and urinalysis, which simply cannot be performed simultaneously. Also, the measurements requested sometimes fail to match the stated objectives.

2. Interpretation and Analysis of Data

A final medical report is the endpoint of any clinical study. In order to make calculations of a drug's safety and efficacy from a clinical study the data must be processed, usually by entry into a computer or by compilation and tabulation by statisticians, and finally interpreted and reported by the clinical monitor. The CRFs should be designed for rapid, economical, and accurate transfer into a data base for later manipulation. The importance of carefully designed CRFs can be emphasized by considering the total quantity of data which must be processed for a typical New Drug Application (NDA); see Table 1.

3. Protocol Adherence

Case report forms can assist the investigator with compliance to the protocol as well as allowing the monitor a rapid review when checking protocol adherence. For example, during the screening phase, a checklist for inclusion/exclusion criteria can be put into a form listing each requirement. The questions can be worded so that any "yes" answer, for instance, would exclude the patient from the study (Fig. 1). Demographic and other data collected during the screen phase

Table 1 Quantity of Data Characters Required for an NDA

200 data characters per CRF
25 Pages of CRFs per patient
2000 patients-volunteers per NDA
Total data characters per NDA = 10,000,000

Page ____ of ____		
DO NOT FILL IN	Cmpd. #/Name	Proj. #
	IND#	NDA#
	Protocol # — Study #	
	Dept.	Sec.

USE BLACK INK.

PATIENT'S INITIALS: ___|___|___ (F M L) PATIENT CODE No.: ________ DATE: ___|___|___ (M D Y)

	Yes	No
Is the patient known to:	☐	☐
1. have systolic pressure less than 110 mmHg and/or diastolic blood pressure less than 60 mmHg on Screen Day?	☐	☐
2. have resting heart rate greater than 140 beats per minute or less than 50 beats per minute at screening?	☐	☐
3. have presence of infection not adequately controlled by antibiotics?	☐	☐
4. have ingested within 24 hours of the first dialysis any drug known to affect blood coagulation or platelet aggregation? (information list in Appendix C of protocol).	☐	☐
5. have received nitroprusside, diazoxide, intravenous nitroglycerin, trimethaphan, anticoagulants or fibrinolytic agents with 12 hours of initiation of dialysis?	☐	☐
6. have one or more of the following cardiovascular exclusions:		
a. clinically significant stenosis of any heart valve?	☐	☐
b. acute myocardial infarction (by electrocardiographic evidence) within the past 3 months?	☐	☐
c. active angina?	☐	☐
d. stroke or other cerebrovascular diseases (e.g. transient ischemic attacks) within the past 6 months?	☐	☐
e. malignant hypertension?	☐	☐
f. suspected or proven arterial or venous aneurysm?	☐	☐

*Comments: ____________________

Based upon results of all screening procedures conducted herein, this patient is a suitable candidate for entry to the baseline phase of this study. Yes________ No________

(a)

Investigator's Signature ____________ Date ________

Figure 1 Hemodialysis study: (a) exclusion criteria checklist, screen day.

Page ______ of ______

DO NOT FILL IN	Cmpd #/Name	Proj #
	IND#	NDA#
	Protocol # — Study #	
	Dept	Sec

USE BLACK INK.

PATIENT'S INITIALS: ___ ___ ___ (F M L)	PATIENT CODE No.: ______	DATE: ___ ___ ___ (M D Y)

Yes No*

Is the patient known to: □□

1. be a consenting adult patient, 18 to 80 years of age, inclusive, who currently requires hemodialysis therapy for uremia? . . . . . □□
2. have received heparin (according to the approved, written, standard procedure of the hemodialysis unit of the hospital) during the most recent hemodialysis which was associated with:
   a. occurrence of a "major bleeding episode," defined as Bleeding Necessitating Discontinuance of Heparin? . . . . . □□
   b. verifiable heparin allergy? . . . . . □□
   c. heparin-associated severe thrombocytopenia? . . . . . □□
   d. clinically significant hematologic, cardiovascular, hemostatic, or other complications judged to be possibly exacerbated by heparinization during hemodialysis? . . . . . □□
3. be either a male, OR a female who lacks childbearing potential or who . . . . . □□
   a. is not pregnant?, and . . . . . □□
   b. will submit to blood and urine pregnancy tests within 3 days prior to receiving epoprostenol, the results of the urine test being negative with the blood test pending? . . . . . □□

*Comments: ______________________________

____

____

(b)

______________ Investigator's Signature

______________ Date

Figure 1 (b) Inclusion criteria checklist, screen day.

can be listed on CRFs. This gives the sponsoring company a permanent record of the patient's pertinent data and supplies the monitor with information from which he can spot check the patient's hospital or office records for investigator compliance.

4. *U.S. Food and Drug Administration Requirements*

Regulations of the FDA require that individual CRFs be maintained and submitted to the sponsor on a routine basis. Additionally, specific items are requested by the FDA [21]. These include: (1) the condition being treated; (2) the severity and stage of the disease; (3) the dosage, route of administration, and frequency of the medication being administered; (4) prior and concomitant medication; (5) age and sex of volunteer/patient; (6) range of normal values for the laboratory to be used by the investigator; (7) results of all clinical observations made; and (8) any adverse effects which occur.

The proposed FDA regulations [22] require that records be maintained by the investigator for the shortest of three alternate periods: (1) for 2 years following the date of NDA approval; (2) 5 years after the sponsor submits an NDA; or (3) for 2 years following the date the entire investigation is completed or terminated. The sponsors (via the monitors) are required to review the records periodically during the conduction of the study [23]. Also, the FDA maintains the right to inspect the records periodically for protocol adherence, accuracy, and compliance with FDA requirements.

In order for the investigator to assure the sponsor of the validity of the data recorded, each page of the CRF should have a place for the investigator's signature or initial and the date signed. It is usually best that a set of forms for each volunteer/patient is organized into a notebook. Instructions for filling out the forms should be carefully reviewed with the investigator and most particularly the staff who will be filling out the forms, with emphasis on the importance of accuracy, completeness, and legibility and keeping in mind the need for duplication (black ink is preferably, for instance).

C. Modular Forms

Clinical trial protocols are highly variable owing to changes in the study phase (i.e., Phase I, II, III, and IV) for which the study is directed, to the different objectives to be drawn from any study, and finally to different indications even within the same drug. Nonetheless, a great deal of similarity exists across clinical trials relative to the CRFs used. It is common now to use modular CRFs: a module is a group of related data elements, and a modular CRF is a series of modules used together in a clinical trial. Modules allow rapid and efficient design for the study-specific CRF. The basic module can be adaped to suit the individual requirements of the study.

Modular forms also allow continuous improvement in design of the modules. Once CRFs have been used in clinical trials, problems associated with the design of the forms often become apparent. The wisdom gained from actual use can be utilized to refine the design of the subsequent forms.

1. *Areas Amenable to Modular Case Report Forms*

As mentioned earlier in the discussion of protocol design (Sec. II), there are usually four distinct phases in the clinical study (Table 2). Specific information is gathered during each phase. Components can be identified which are common to phases within a single study, as well as across the majority of clinical studies. Some of the categories of CRF information for which modular forms are applicable include: demography, physical examination, history, clinical laboratory testing, test drugs and dosages, and adverse experiences.

a. Physical Exam

The physical exam provides one of the simplest examples of a category amenable to modular CRF design (Fig. 2). While the data elements may be varied, in most studies a fair degree of standardization can be achieved. If a specific entity is being extensively tested, for example, when a detailed neurological test is required, an additional CRF can be developed.

b. Drugs and Dosages

Test drugs and dosages constitute a module which can be standardized with variables in content and level of precision.

(1) Data elements. The following are data elements involved in test drugs and dosages: (a) test agent name; (b) time units (days, hours, minutes; days and hours; days or date alone); (c) time ranges of the

Table 2 Phases in a Clinical Trial

Screen	Measurements made to determine if a potential volunteer/patient qualifies for entry into the baseline phase
Baseline	Measurements made without the presence of the test drug(s) to determine qualification for entry into the treatment phase and to provide a base upon which to measure drug effects for safety and efficacy
Treatment	Measurements made during active therapy
Posttreatment	Measurements made after cessation of active therapy to determine potential delayed effects and possible reversal of therapeutic effects

Page ______ of ______

DO NOT FILL IN	Cmpd. #/Name	Proj. #
	IND#	NDA#
	Protocol # — Study #	
	Dept.	Sec.

NOTE: Please use BLACK ink

Investigator's Name	Vol. Initials	Vol. #
	___ ___ ___ F M L	

___ ___ ___
M D Y
Exam Date

SEX: ☐ Male ☐ Female BIRTHDATE: ___ ___ ___ (M D Y)

RACE: ☐ White ☐ Black ☐ Other, Specify ______________

BODY WEIGHT: __________ kg ☐ lb

BODY HEIGHT: __________ cm ☐ in

BODY FRAME: ☐ Small ☐ Medium ☐ Large

VITAL SIGNS: Systolic/Diastolic

Blood Pressure (sitting) ________ / ________ (mm Hg)

Pulse Rate (beats/min) ______________ (at rest)

Temperature (oral) __________ ☐ C° ☐ F°

Resp. Rate (breaths/min.) ______________

GENERAL EXAMINATION: (✓)	Nor.	Abn.	Are there CLINICALLY SIGNIFICANT abnormalities? ☐ Yes ☐ No If yes, please provide details below.
Head/Neck			
EENT			
Chest			
Heart			
Lungs			
Abdomen			
Liver			
Spleen			
Pelvic			
Genital			
Rectal			
Extremities			
Skin			
Neurological			
Pulses			
OTHER (specify)			

______________________ Investigator's Signature ________ Date

Figure 2 General physical examination.

above; (d) frequency of dosing or dosage number; (e) route of administration—oral, intravenous (rapid or bolus), infusion (prolonged or continuous), intramuscular, subcutaneous, rectal, or topical; (f) dosage form (tablets, capsules, ointments, salves, creams, solutions, suspensions); (g) unit of measurement (expressed as milligrams, milliliters, drops, tablespoons, grams, milligrams/kilograms or volume); (h) unit dose—quantity of drug administered to a patient at one time; (i) drug containers (bottles, vials, ampules, envelopes, tubes, jars).

Variables associated with drug containers are as follows: container(s) assigned to one patient; one container assigned to multiple patients; or container(s) unassigned.

(2) Precision of information. The above data elements can be chosen for inclusion into a CRF according to the precision of information required. Precision refers to the exactness employed in documenting the administration of drug and the time of drug administration.

High precision High precision studies (Fig. 3) are those: (i) in which every dose amount and time of administration are accounted for; (ii) which require constant direct supervision of drug administration;

Page ______ of ______

DO NOT FILL IN	Cmpd. #/Name	Proj. #
	IND#	NDA#
	Protocol # — Study #	
	Dept.	Sec.

NOTE: Please use BLACK ink.

Investigator's Name ______ Subj. Initials ______ ______ ______ Subj. No. ______

Test Drug Name ______ Period ______

Dosage Form ______ Unit of Measure ______

Route of Administration (oral, i.v., i.m.) ______

SCHEDULED DOSE ADM.				ACTUAL DOSE ADMINISTRATION					Reason for Deviation from Scheduled Dose
Day No.	Time	Dose No.	Unit Dose	Container Number	Date	Time (am/pm) Hr. Min.	Dose No.	Unit Dose	

☐ Check if no significant deviations from dosage schedule.

Investigator's Signature ______ Date ______

Figure 3 Experimental drugs/dosages (high-precision schedule).

(iii) in studies which are highly controlled; and (iv) which generally apply to bioavailability/kinetic studies and Phase I studies. In Fig. 3, the drug constants would be such items as dosage form, route of administration, and frequency of dosing unit of measurement (mg, mg/kg, or ml) and the type of dosing container; these items would be listed only once. The day number would be the number of days into the study in which dosing of drug has occurred; the dose number would be the dose within a given day; and the unit dose would be the number of drug units (such as 2 capsules or 5 ml).

Intermediate precision If an intermediate-precision dosage record is required, a form such as Fig. 4 can be used. The characteristics of intermediate-precision studies are as follows: (i) The administration of every dose of every test agent is not necessarily under the direct control of the investigator or responsible delegate. (ii) The amount

Page ______ of ______

DO NOT FILL IN	Cmpd. #/Name	Proj. #
	IND#	NDA#
	Protocol # — Study # ____ ____	
	Dept.	Sec.

NOTE: Please use BLACK ink.

Investigator's Name ______________ Subj. Initials ___ ___ ___ Subj. No. ___

Test Drug Name ______________ Period ______________

Dosage Form ______________ Unit of Measure ______________

Route of Administration (oral, i.v., i.m.) ______________

Container No.	DATES OF ADMINISTRATION		DAILY DOSE ADMINISTRATION			Reason for Deviation from Scheduled Dose
	Start	End	Unit Dose	Frequency	Total Daily Dose	

NOTE: Any change in the unit dose, frequency or total daily dose requires a new line entry. Missed days of drug administration are recorded by giving dates and entering a zero in the Total Daily Dose column.

☐ Check if no significant deviation from dosage schedule.

______________ ______

Investigator's Signature Date

Figure 4 Experimental drugs/dosages (intermediate-precision schedule).

and time of every dose of test agent administered is not recorded at the moment of administration. (iii) The time unit of measurement is at the day level. (iv) If there is no deviation from the scheduled unit dose and/or dose frequency over the scheduled range of days, only the start and end range of dates are recorded. (v) If there is one or more deviations of the unit dose and/or frequency on a daily basis, the time (dates) and daily dosage characteristics of each deviation would be recorded. (vi) On an outpatient basis with spaced visits, simple dosage records are generally required. (vii) These are usually late Phase II or early Phase III studies (such as when a range of dosages is being tested). (viii) The dosage either administered or taken is usually more variable than in high-precision studies.

Low Precision The major difference between an intermediate- and low-precision level of recording is in the dosage and time of administration information (Fig. 5). Low-precision studies have the following characteristics: (i) The investigator primarily controls the drug administration of the patient or volunteer on an outpatient basis without regularly spaced checks. (ii) Usually the dosage information is sub-

Page ______ of ______

DO NOT FILL IN	Cmpd. #/Name	Proj. #
	IND#	NDA#
	Protocol # — Study #	
	Dept.	Sec.

NOTE: Please use BLACK ink.

Investigator's Name ______ Subj. Initials ___ ___ ___ Subj. No. ___

Test Drug Name ______ Period ______

Dosage Form ______ Unit of Measure ______

Route of Administration (oral, i.v., i.m.) ______

Container Number	DATES OF ADM.		Est. Avg. Total Daily Dose	EST. DAILY DOSE RANGE		*Est. Number Missed Doses ✓ = none	Reason for Deviation from Scheduled Dose
	Start	End		Low	High		

*Missed days are included in calculations of the estimated average total daily dose and estimated daily dose range.

☐ Check if no significant deviations from dosage schedule.

Investigator's Signature ______ Date ______

Figure 5 Experimental drugs/dosages (low-precision schedule).

mitted verbally by the patient to the investigator and entered on the CRF subsequent to a patient visit, as opposed to being recorded by the patient on a daily basis. (iii) Unit measure of time is in days, with the drug administration estimated as the average total daily dose and range over a given span of days. (iv) The route of administration is generally oral or topical. (v) These studies are usually Phase III or IV.

2. *Advantages*

One of the advantages of modular CRFs is that experience gained in the field can be incorporated into future forms. It is not unusual for forms to be designed from a standpoint of data entry, without taking into account the practicality of the use under the conditions of a clinical setting. The ultimate aim of a modular approach to the design of forms is to simplify both the composition and the use of the forms. The resulting benefits include the following:

a. Time spent in CRF design is decreased.
b. Costs associated with construction and printing of forms are minimized.
c. Increase in Cross-study uniformity of content and form is increased.
d. CRF editing is simplified.
e. Errors of coding and data entry are minimized.
f. Data processing is speeded up.
g. Programming for data entry, data retrieval, and statistical analysis is simplified.
h. Generation of listings and graphic displays is more efficient.

REFERENCES

1. C. J. Breeden, T. C. Hall, and H. R. Tyler. Herpes simplex encephalitis treated with systemic 5-iodo-2′-deoxyuridine. *Ann. Intern. Med. 65:* 1050-1056 (1966).
2. D. C. Nolan, M. M. Carruther, A. M. Lerner. Herpesvirus hominis encephalitis in Michigan: Report of thirteen cases, including six treated with idoxuridine. *N. Engl. J. Med. 282:* 10-13 (1970).
3. Boston Interhospital Virus Study Group and the NIAID-Sponsored Cooperative Antiviral Clinical Study, Failure of high dose 5-iodo-2′-deoxyuridine in the therapy of herpes simplex virus encephalitis. *N. Engl. J. Med. 292*(12):599-603 (1975).
4. G. Galasso, Perspectives in antivirals. *ASM News 45*(7):353-358 (1979).
5. The Anturane Reinfarction Trial Research Group, Sulfinpyra-

zone in the prevention of sudden death after myocardial infarction. *N. Engl. J. Med. 302*(5):250-256 (1980).

6. R. Temple and G. Pledger, Special report: The FDA's critique of the Anturane reinfarction trial. *N. Engl. J. Med. 303*(25): 1488-1492 (1980).
7. G. Kolata, FDA says no to Anturane. *Science 208:*1130-1132 (1980).
8. W. A. Silverman, The lesson of Retrolental fibroplasia. *Sci. Am. 236*(6):101-107 (1977).
9. P. G. Welling, Effect of food on bioavailability of drugs. *Pharm. Int.* pp. 14-18 (Jan. 1980).
10. P. G. Welling, H. Huang, P. F. Hewitt, and L. L. Lyons, Bioavailability of erythromycin stearate: Influence of food and fluid volume. *J. Pharm. Sci. 67*(6):764-766 (1978).
11. P. G. Welling, L. L. Lyons, F. L. S. Tse, and W. A. Craig, Propoxyphene and norpropoxyphene: Influence of diet and fluid on plasma levels. *Clin. Pharmacol. Ther. 19*(5), Pt. I, 559-565 (1976).
12. U.S. Food and Drug Administration, Division of Pharmaceutics, Bureau of Drugs. *Introduction to Guidelines for Bioavailability Studies* (Drug Monograph). Revised Mar. 30, 1977.
13. G. Kolata, NIH shaken by death of research volunteer. *Science 209:*475-478 (1980).
14. *Lab Report for Physicians 2*(6):Issue 8 (1980).
15. P. Joubert, L. Rivera-Calimlim, and L. Lasagna, The normal volunteer in clinical investigation: How rigid should selection criteria be?. *Clin. Pharmacol. Ther. 17*(3):253-257 (1975).
16. J. A. Frieman, T. C. Chalmers, H. Smith, and R. Kuebler, The importance of beta, the Type II error and sample size in the design and interpretation of the randomized control trial. *N. Engl. J. Med. 299:*690 (1978).
17. R. H. Fletcher and S. W. Fletcher, Clinical research in general medical journals. *N. Engl. J. Med. 301:*180 (1979).
18. S. S. Schor, Statistical problems in clinical trials. *Am. J. Med. 55*(6):727 (1973).
19. J. Cornfield, The University Group Diabetes Program: A further statistical analysis of the mortality findings. *JAMA 217*(12): 1676-1687 (1971).
20. R. Doll and R. Peto, Randomized controlled trials and retrospective controls. *Br. J. Med. 280:*44 (1980).
21. U.S. Dept. of Health, Education and Welfare, General considerations for the clinical evaluation of Drugs. HEW (FDA) 77-3040 (Sept. 1977).
22. Obligations of clinical investigators of regulated articles. *Federal Register 43:*No. 153 (Aug. 8, 1978).
23. Proposed establishment of regulations on obligations of sponsors and monitors. *Federal Register 42:*No. 187 (Sept. 27, 1977).

8

CLINICAL DATA MANAGEMENT AND STATISTICAL DESIGN IN THE CLINICAL RESEARCH PROCESS

Joseph R. Assenzo
The Upjohn Company
Kalamazoo, Michigan

Thomas W. Teal
McNeil Pharmaceutical
Spring House, Pennsylvania

I. INTRODUCTION

This chapter emphasizes the integration of clinical data management and statistical concepts and methodology into the clinical development program. Consideration is given to program organization, planning and execution, medical objectives, and data management procedures, all of which impact directly on statistical considerations.

II. ORGANIZATION OF SUPPORT

Conceptualization, organization, and execution of clinical trials are primary responsibilities of the medical monitor. Other function which are part of this effort may include medical research coordinators and clinical research associates. The development team also will include one or more biostatisticians, computer support personnel, and clinical data processors. The size of the team will increase as development proceeds. Initial trials can be handled by the monitor and biostatistician, but Phase III multiclinic trials may require a larger formal team.

	Medical	Clinical Data Processing	Computer Operations	Statistics
Protocol Concept and Routing	✓			✓
Case Report Form (CRF) Design	✓	✓		✓
CRF Computer Acceptance Program		✓	✓	
Recruit Investigators	✓			
Distribution of Supplies to Investigator	✓			
Monitor Studies	✓			
Return of CRF	✓	✓		
Review of CRF by Science Information Division		✓		
Review of CRF by Medical	✓			
Categorization of Adverse Reactions by Medical	✓			
Encode & Encode Check CRF codes		✓		
Enter (OE) and Key Verify (KV)		✓		
Merge to Master File and Display		✓		
Check Display to Original CRF		✓		
Update Master File		✓		
Preliminary Reports - Pt. Profile, Concom. Med., ARs, Dose		✓		
Update Master File		✓		
Release File to Data Processing		✓		
Stat. Request Displays of Relevant Variables (Efficacy, Concom. Meds., ARs, Clinical Labs)			✓	✓
Stat. Reviews for Consistency & Protocol Adherence				✓
Stat. & Medical Establish List of Exclusions	✓	✓		✓
Provide Data Processing with List of Exclusions			✓	✓
Run Stat. Analysis (Variance, Mantel-Haenszel, Crosstabs, Graphs, Etc.)			✓	
Write Report and Attach Data Appendixes				✓
Type and Route for Review	✓			✓
Revise Report if Necessary				✓
Release Report for Distribution				✓

Figure 1 Key data management activities and responsibilities. Saga of a clinical trial.

Since the end product of a clinical development program is derived from data and observations, effective organization of all data management activities cannot be overemphasized. In addition to statistical design and analysis, these activities include development and/or modification of software for data collection, storage and retrieval, data encoding, and entry, data base validation and generation of listings, reports and summaries. Figure 1 lists some of the key data management tasks which must be performed and the units ordinarily responsible for them. This list, along with starting and ending dates for each task, is an effective planning and monitoring tool.

III. PLANNING OF CLINICAL TRIALS

Major planning elements include assembling appropriate background information, protocol design and development, and implementation of quality assurance procedures [1-14].

A. Protocol Design and Development

There is essentially universal agreement that protocol design is an important, perhaps even the most critical, step in the entire clinical trial process. This step is a collaborative endeavor which requires expertise in several areas. In this section some of the important statistical concepts which must be incorporated into the protocol are addressed.

1. *Objectives*

The clinical setting, experimental approach, and study plan are determined by the trial objectives. Formulation and statement of study objectives is therefore the foundation of protocol development. There are general classes of studies carried out during the development process which have very different objectives, experimental design, data acquisition, statistical analysis, and data processing requirements. These classes are methodology development, tolerance evaluation, pharmacokinetics, efficacy evaluation, therapeutic or management trials, and surveillance studies.

Clinical studies will usually have more than one objective, and since the objectives may conflict it is desirable to rank them in order of importance. This facilitates evaluation of alternative experimental methods and clinical models.

2. *Choice of Patients*

The choice of subjects or patients is influenced by the type of study and the nature of the test drug or procedure. Inclusion and exclusion criteria must be carefully specified for all types of studies. As a general rule, the inclusion criteria are more restrictive in the earlier stages of development and may be relatively relaxed in late Phase III trials. In the latter case, patient selection criteria should be based on the target population of the new therapy. Since it is inevitable that ineligible patients will be enrolled and receive the test drug, it is essential to define precise procedures for handling these violations. Excluding patients from the study entirely does not distort the results, but the methods used to handle protocol violations and dropouts can have a pronounced effect on outcome.

3. *Endpoints*

Endpoints or measurements which will be recorded during the trial also will be defined in the protocol. A small subset of these endpoints which are particularly relevant to safety and efficacy evaluation should be identified and used as design variables. This set will be used to design the trial, determine sample size, and specify data analysis methods.

4. *Judging Effects of Drug: Statistical*

Another central issue in protocol design is the approach which will be used to judge the effects of the test drug, that is, which type of control to utilize. This leads to a comparative evaluation involving either parallel or concomitant control patients, quasi-controls, historical control patients or some form of within patient control. Comparisons based on parallel controls provide the framework for the highest degree of scientific rigor and this is nearly always the preferred approach. The control group receives no treatment, an indistinguishable placebo or some form of standard therapy.

The use of historical controls may be considered in some applications such as those described by Gehan and Freireich [15]. This approach can be reserved for those situations in which parallel controls cannot be used or where a stable clinical environment can be documented. Using patients as their own controls also may have some utility, but pre- vs. posttreatment comparisons alone do not control temporal effects. However, combining this approach with a parallel control group eliminates this temporal bias and increases sensitivity of the trials.

5. *Randomization and Blinding*

Randomization is generally regarded as a fundamental requirement for the valid interpretation of treatment effects. It is a method for assigning treatments to patients in a way which guarantees that all possible assignments have an equal chance of occurring. Assignment based on chance alone assures validity of statistical significance tests, improves balance between treatment groups with regard to known and unknown prognostic factors, and removes bias from treatment assignments.

Randomization sequences are obtained from published random number tables, or they can be generated with computer programs as described later in the text.

In order to maintain a degree of balance in group sizes as the trial progresses, the randomization sequence may be forced into subsets. Each subset would be constructed to allocate an equal number of patients to each treatment group. This approach may be useful if the trial will continue over a long period of time and there is concern about temporal effects [16].

The double-blind trial, in which the clinician and patient are both unaware of actual treatment assignment, is generally recommended. This requires that the medications be indistinguishable and may therefore require some special formulation. When dosage forms of the treatments are incompatible with uniform formulation it may be necessary to use double placebos. All medication must be coded to allow identification of the actual treatment. Conditions for breaking the code should be carefully stated in the protocol.

The general rule in a double-blind trial is that as few persons as possible should be unblinded to treatment. These people should be identified and their relationship to other portions of the study should be minimal, if any, and predefined. Obviously, the patients, the treating physicians, and other medical personnel responsible for patient care and evaluation must be blinded throughout the study. In the event of a medical emergency in a patient, however, it may be necessary to the treating physician to break the code for that patient. If this happens it is important to record the reason for the breaking of the code and the outcome of the medical emergency.

In the coordinating center all persons who can affect the current and future course of the trial, that is, who have the potential of biasing the trial, should remain blinded throughout the course of the trial. For example, those physicians at the coordinating center who make decisions about study-patient care and evaluability for the study, recommend protocol changes, and classify medical events should remain blinded. All data encoders who make classification decisions should remain blinded throughout the study.

It is necessary that a few people at the coordinating center be unblinded to treatment. A physician must be unblinded to give advice to treating physicians in cases of medical emergency, for example, in the event of a potential overdose of study drug. This same physician must be unblinded to monitor safety and to report the progress and results of the study to regulatory agencies and to upper level management. A statistician at the coordinating center needs to be unblinded to determine whether the randomization is going awry and to provide the needed interim summaries and analyses to the unblinded physician and upper-level management. Interim or periodic summaries and analyses must have very limited distribution—only to those who are unblinded.

A few members of upper-level management need to be unblinded to determine whether the progress of the study is satisfactory, whether new studies should be undertaken, and whether to terminate the current study.

6. *Stratification*

The basic idea here is to classify patients according to important prognostic factors and distribute them equally across treatment groups. This is a sound procedure that can be readily applied when there are just a few important factors which could influence treatment response. Practical difficulties arise, however, when the number of factors becomes large. Excessive stratification can in fact actually lead to a reduction in balance [17-19].

Byar et al. [20] suggest that it is seldom advisable to stratify on more than three or four variables. In multiclinic trials, clinics and one or two important prognostic factors would form the strata. Randomization, with separate lists for each strata, would then be relied upon to balance other factors to the extent needed. Post hoc adjustment would still be recommended in most cases because the randomization process never leads to perfect balance.

7. *Adaptive Randomization*

The basic concept here is to increase the likelihood of comparability of the treatment groups at baseline. The procedure operates essentially as follows. An eligible patient's important baseline variables or prognostic factors are entered into the data base and compared with the baseline variables of those patients currently in the study. If the treatment groups would be more homogeneous with the patient assigned to Treatment A rather than B, then the probability of assignment would be adapted so that there was, for example a 3/4 chance of assignment to Treatment A and a 1/4 change to Treatment B. If the two treatment groups were fairly balanced, then the patient would have a probability of 1/2 of being assigned to Treatment A.

8. *Statistical Design Model*

The design should address trial objectives, make efficient use of resources, and lead to definitive conclusions. As a practical matter, the design should be as simple as possible and compatible with sound clinical practice. Statistical analysis of the data, or course, is determined by the design used. Inappropriate or faulty design will place severe limitations on data analysis and conclusions which can be drawn from the trial.

The simplest design involves two or more parallel groups and the patients are classified only by group number. This design is appropriate for a wide range of clinical investigation and is used extensively. Another useful experimental approach is the matched pair design. This is used widely in dermatology studies when there are no systemic effects resulting from topical applications. In these studies the patient is a "block" and, in the case of two treatments, will receive one treatment on each, say, arm or on bilaterally symmetrical lesions.

The crossover design has a great deal of intuitive appeal and is used extensively in clinical pharmacology studies. In the two-treatment crossover, patients are allocated randomly to two groups. The first group receives treatments in the order AB and the second group in the order BA. This approach permits comparison of both treatments in the same patients. The advantage of a crossover is related to the relative magnitude of within-patient variation to between-patient variation. Brown [21] in a comprehensive evaluation of the design, has constructed a table which gives cost efficiency of the crossover relative to the completely randomized experiment for various ratios of between to within-patient variation. Any efficiencies which can be obtained with the crossover must be balanced with the major criticism of the design: carryover effects cannot be detected in a satisfactory manner. Brown concludes that "the crossover experiment can yield great savings in cost if the assumption of no carryover effect is valid, but the design should not be used if this assumption is in doubt."

9. *Sample Size*

Although there are various methods which differ in detail, all sample size calculations are based on target values of the Type I error (α, the risk of declaring an effect which does not exist), the Type II error (β, the risk of failing to detect an effect which actually exists), estimates of experimental variation, and the effect level which the trial should detect. With all other factors constant, detection of small effects requires more patients than does detection of large effects. The usual procedure is to define with the medical monitor levels of effect which are clinically meaningful.

The power of the trial to detect meaningful differences is defined as $1 - \beta$, the probability of detecting an effect which actually exists. In practice, α is usually set at 0.05 or less and $1 - \beta$, the power, at 0.8 or more.

Sample size determination is not the definitive procedure it sometimes is perceived to be. It is an extremely important process, however, in obtaining some general ideas about patient numbers. Several values of α, β, and effect levels will usually be evaluated. It is also informative to go through this process for two or three of the most important endpoints or design variables. All this information, along with any resource constraints, is then used to determine the size of the clinical trial.

Use of measurement data rather than counts more often than not will result in a smaller total study size. For example, using time to event rather than whether or not an event occurs as an endpoint will result in a smaller study size. Use of baseline measurements as covariates or for construction of difference scores will also result in a smaller study size.

Use of measuring instruments with smaller experimental variation will result in a smaller study size. However, the cost for the increase in precision may far outweigh the reduction in cost due to the smaller sample size.

With everything else held constant it can be shown that the minimum total sample size is attained when half the patients are assigned to each of the two treatment groups. However, in late Phase III studies which seek to evaluate the drug under typical or usual clinical conditions it may be advantageous to assign patients, in a 2:1 or 3:1 ratio, to study drug and control groups.

10. Case Report Forms

The final step in protocol development is design of the case report forms. These documents are the primary means of guiding execution of the study according to the protocol. They are in fact a part of the experimental plan and must be considered an integral part of the protocol. Case report forms are the foundation of all data management activities, and a major share of the development resources are directly focused on them. This includes the recording of data in the clinic, review by the medical team, and entry of information on the forms into the data system. Important efficiencies can be achieved through appropriate formatting and sequencing which is tailored to practice in the clinic and characteristics of the data system being used.

The *case report form* (CRF) must be developed to satisfy the data collection requirements established by the protocol. Data management techniques in this respect require that the various CRF *modules* (i.e., demographic data; history of disease; adverse reaction; clinical laboratory data; concomitant medication; physician's comments; efficacy; etc.) be standardized as much as possible. Such standardization significantly improves the efficiency of CRF generation and clearly minimizes coding and data entry errors as well as simplifying computer programming associated with clinical data entry and subsequent data processing. Indeed, if the CRFs did not vary from trial to trial, essential programming could be accomplished only once.

To accomplish this goal of standardization, CRFs should be prepared on a computerized phototypesetter. Preparation of a single page of a CRF is a simple task requiring no more than several hours, since most of the modules on the CRF page can be recalled from memory and quickly modified as necessary for the particular protocol. The objective in this respect is to satisfy the medical monitor and investigator's requirements with a minimum change to standard CRF formats. However, the medical monitors' requirements are paramount.

B. Quality Assurance

Points in the data flow where observations on the CRF can be corrected, modified, or transformed should be identified and procedures developed to guarantee that these changes are consistent and well documented. Rules should also be established for classifying events which have a large judgmental component. It is preferable that those responsible for updating these rules not be involved in other aspects of the study to prevent the possible introduction of bias.

IV. DATA MANAGEMENT

Among the needs and objectives of clinical data management are the following:

1. Maximizing accuracy in data transfer from the CRFs to the computer master file [verified by quality control audit prior to submission of each New Drug Application (NDA)]. Essentially, the master file should represent the physician's intentions as well as his writing. This implies that data discrepancies in CRFs, elicited through computer algorithms, be reconciled with the investigator to ensure an accurate recording of his intended response. An example of this would be an absent decimal point in a clinical laboratory value.
2. Fast and economical transfer of the data from the CRF to the computer master file.
3. The computer master file organized in a manner readily amenable to retrieval and manipulation for various management and analytical purposes.
4. Master file and subsets of master files organized in a manner which will provide ready access for an extended period of time.
5. A clinical data system that supports management in monitoring and administering the clinical trials from which the clinical data are developed.

To achieve these primary objectives a data management organization which provides clinical data processing, computer services, and statistics is required.

Data associated with clinical trials as well as all other scientific data tend to flow naturally through these three departments on their way, for example, to an NDA or a publication if the data pertain to a marketed product. Essentially, the technical skills required in those departments consist of: (1) *clinical data reviewers*, responsible for ensuring CRF data adherence to protocol requirements, accuracy and completeness of data, identification of potential adverse experiences, and abnormal clinical laboratory test results; (2) *clinical data coders*, responsible for encoding certain of the clinical data variables such as concomitant medication in preparation for direct on-line data entry into a computer, and also responsible for the key verification or reentry of those data into the computer, where a 100% identical match is required; (3) *computer programmers*, responsible for writing the necessary programs which convert data produced on a wide variety of investigational and marketed drugs into a standard computer record for the various types of clinical data; (4) *scientific data processors*, whose primary efforts are to satisfy the data processing needs of the statisticians and whose secondary efforts deal with producing standard and specialized reports required by other areas such as management and the medical and research divisions; and (5) *statisticians*, who are primarily responsible for utilization of the bulk of the clinical data and who are ultimately responsible for statistical interpretation of those data.

The complexity of clinical data and evidence of the need for sophisticated data management systems are indicated by both the previously stated objectives and by the following brief view of the problem:

V. PERSONNEL REQUIREMENTS

In general, for example, experience indicates that in the case of a NDA consisting of data concerning 2,000 patients, each of whom has experienced an average of 3 months of therapy and for each of whom a case report form recording a physician's observations initially, at 2 weeks, 4 weeks, 8 weeks, and 12 weeks, with each visit requiring 4 pages of CRF data (a total of 20 pages of CRF data per patient), we must process about 40,000 pages of data. With each page producing roughly 300 characters, a total of 12 million data points must be processed.

Using the procedures described above, each of those case reports requires approximately 2 hours of effort in order to establish a "released for processing" master NDA data file. This is approximately *2 man-years* (50 man-weeks/year) of data review, coding, verification and display effort among the clinical data processing personnel for the 2,000 case reports. Many of those CRFs reflect data from protocols requiring but a single visit as well as CRFs from protocols requiring from one to several years of therapy. The above is considered to be a typical NDA data base in terms of clinical data *volume*.

Personnel resources required for scientific processing of those data are primarily a function of the number of protocols; secondarily, the number of individual investigators; and, finally, the number of patients. If one assumes, for simplicity, 20 protocols for the NDA, each of which contain 100 patients and each of which is in some way different from the others and requires computer programmer intervention, one can make a rough estimate of the scientific data processing time required for this NDA. This is particularly true if standardization has been maximized across protocols in such areas as CRF format, rating scales, and terminology.

Approximately one programmer week is required and approximately three scientific processor weeks are required per protocol. Thus the programmer/processor resource requirements for this NDA equals approximately 80 man-weeks. If one also adds to that resource requirement the programmer/processor effort of 20 man-weeks required to produce standard data displays such as investigator summary, patient summary, or adverse experience reports, we have identified *two additional man-years* for this NDA.

Similar resource requirements in terms of statistical support may also be identified. If we assume, 1 man-month of statistical effort per protocol, we have identified 80 man-weeks of statistical effort. With additional statistical support for the NDA in terms of project team participation, protocol CRF design and review, consultation with physicians/monitors, etc., one can easily identify an additional 64 man-weeks of effort. Thus, a total of *three statistician man-years* is required in our example.

Word processing of statistical reports, which includes preparation and circulation for review, copying, etc., requires *one additional man-year*. *Another man-year* is required for phototypesetting of CRFs and other phototypesetting support, including package insert preparation. The administrative man-hour overhead associated with these 9 man-years of effort is probably in the neighborhood of 30%. That overhead resource requirement ranges from department management through the computer terminal repair personnel which keep the functional sections in business. Therefore, to accomplish the project just described, a total personnel resource of approximately *11 man-years* is required.

In order to efficiently manage the clinical data generated by the various types of clinical trials, one must establish supporting computer systems which should be integrated with the Clinical Data System, the core system. Those supporting systems vary in complexity. Two of the most significant of those systems, necessary for efficient data management, are briefly described in the following sections.

VI. CLINICAL INFORMATION MANAGEMENT SYSTEM

This system provides management with ready access to the status of any clinical trial for which an Investigational New Drug Application

(IND) had been prepared whether or not that clinical trial is ongoing or completed. The system contains an individual record for every clinical trial established. The record contains data from various fields, each of which is the responsibility of different functional sections.

A single record exists for each investigator even though that investigator may be one of many investigators participating in a common protocol. The record is initially created by the department responsible for submitting an IND for the particular investigator to the U.S. Food and Drug Administration (FDA). Information entered into that section's segment of the record consists of, for example, the investigator's name, address, protocol identification, and title.

The second segment of the record is completed by the section responsible for supplying the investigator with clinical supplies and monitoring the course of the investigator's clinical trial. Data elements supplied by this section include the monitor assignment, investigator visit frequency, drug supplies shipment information, and other data of interest.

The third segment of the record is completed by the clinical data processing section. Data supplied by that section provides information pertaining primarily to the status of the study. These variables are, for example, number of case records received to date, number of patients per drug group, dosing regiment, protocol title, lab normals for each investigator, and status of various other events associated with the clinical trial.

From this record a variety of management reports are generated. Examples of reports typically and routinely generated are the following:

A report for each clinical research associate or monitor presents basic information about each investigator within his sphere of responsibility. The report contains a record of all previous visits or contacts made with this investigator. The report provides a method of economically scheduling visits to investigators in the future.

A second report provides answers to questions concerning the status of all investigators participating in a particular protocol or participating in a particular set of protocols associated with a particular drug.

Reports necessary to complete periodic progress or interval reports for drugs required by the FDA may also be generated routinely.

The system would also provide a historical accounting for all clinical trials.

In this system and in all associated/integrated systems, a common investigator number and common codes for other variables such as study drug and protocol identification would be used to facilitate the generation of management reports which require data from more than one system.

VII. CLINICAL SUPPLIES AND INVENTORY SYSTEM

This system contains a complete record of all batches of clinical supplies prepared for use in connection with clinical trials. These data include not only investigational and marketed drugs, but matching placebo and reference drug supplies. The data base would be queried to select the most appropriate batches of material for use in a clinical trial. It would also provide inventory control for all material shipped to individual investigators and for that investigator's supplies by individual bottle code for both double-blind and open label trials. The system, therefore, satisfies the proposed good clinical practices requirements for complete accounting in connection with these supplies. It also provides a unique benefit in that a drug consumption reconciliation report is possible. Material that is returned at the conclusion of a clinical trial is accounted for in the computer record by bottle and by tablet, prior to disposal of those supplies.

Other benefits accrue from the system, such as the ability to generate labels for the clinical supplies, properly blinded and randomized according to the statisticians' randomization requirements. That significant benefit allows clinical trials to be carried out rapidly, with a minimum of administrative effort. Labels for supplies used in the clnical trial can be randomized and printed within minutes of receiving the request.

The drug reconciliation report referred to previously provides drug consumption information on a patient-by-patient basis. These data may then be compared with both the physicians' prescription during the course of the trial and the study trial nurses' bottle-by-bottle inventory, both of which are recorded on the CRF periodically during the trial.

VIII. CLINICAL DATA SYSTEM

These systems are associated/integrated with the Clinical Data System, the core system. The Clinical Data System provides for original entry of clinical data and subsequent verification (identical reentry) of those data. Within a project, pages of data may be entered randomly, by protocol, investigator, patient visit, and study drug. Thus, pages of data from different patients participating in entirely different protocols may be entered in sequence by the same original entry coding person.

At the time of entry, every 10 pages of data are automatically stored in a minifile called a batch. It is possible to create a batch with fewer pages, even a single page, but the fixed allotment of 10 is efficient in terms of the reentry process. The data entry program should automatically assign an identification number to that batch. Subsequently the key verifier enters those same 10 pages identified by the batch number. The key verifier, after entering the batch identification number, immediately proceeds to reenter—in same page sequence—the pages contained in the batch. Any disagreement between the original entry person and the key verifier with regard to a data element should be reconciled by the key verifier, the more experienced person. When a disagreement is other than an entry error, further reconciliation of the data is required. At the end of any given day of data entry, a multitude of batch minifiles will have been created. During non-prime-time hours, the clinical data system is programmed to automically collect the minifiles on a drug-by-drug basis and to merge them into the existing master file for that protocol/drug. In both the data entry segment of the process and the merging into the master file process, various algorithms may be used to edit and range check every data element. The system, as a result, requires various data dictionaries such as drug, adverse experience, and disease/diagnosis dictionaries.

Data elements outside of a previously established range for that element are called to the attention of both the original entry and key verifier to inform them of the existence of such an "outlier." During the creation of the master drug file for a particular drug, periodic data displays should be generated by variable on a patient-by-patient basis within investigator and, in certain instances, across patients within investigator or across patient and investigator within protocol. An example of the latter would be a display on a patient-by-patient basis over all patients in a protocol of the duration of illness prior to the first day of administration of the study drug. This type data display provides the opportunity to examine unusual aspects of the data, as well as identification of patients who may not have satisfied the protocol acceptance criteria.

A second example of these data displays would be a listing in ascending order across all patients within a protocol of a particular clinical laboratory test result. Outliers are readily identified in this data display which were faithfully entered into the computer by both the original entry and key verifier but which were clearly originally in error. These outliers are reconciled with the investigator in each case.

Each file is individually released to the computer section for subsequent interim or final processing. By this time, a specific written request should have been received from the assigned statisticians for the various types of data processing. Those requests may be basic data displays used by the statistician to verify each patient's suitability for subsequent efficacy or safety analysis; sophisticated statistical analyses accomplished to compare the effects of different treatments; various hierarchical data displays which become part of the statistical analysis; data displays which are used to validate the accuracy of the computer record; and displays used to provide supporting details for various statistics appearing in an efficacy or safety report.

The Clinical Data System should be modular in nature. Those modules may consist of the following sections: patient profile, adverse reaction, dose, clinical laboratory, concomitant drug, and comments. As indicated by this set of sections, every variable on every CRF is entered into the computer master file, including comments by the investigator. These comments frequently provide insight into unusual events occurring during the course of therapy.

As indicated previously, certain computer programs should be prepared as soon as a CRF is developed which will convert those CRF data into the aforementioned standard master file format.

The various sections of the Clinical Data System may be used to produce standard reports by investigator, by drug, or by patient, dealing with, for example, clinical laboratory data or adverse experience data by patient, by drug, etc. In each patient record, in each section of the system, sufficient tag information is contained to enable association and integration of the several modules of the system to produce more elaborate reports and analyses. Certain of the reports, for example, such as a basic patient data display, would generally contain data from most of the aforementioned modules. Specifically, the patient's basic data display uses data from the patient profile, adverse reaction, dose, clinical laboratory, and concomitant medication sections. Fragmenting the master file into the various modules of the clinical data system increases the difficulty in those instances where processing requires data from various modules. However, since the data files are standardized across all drugs, programs written to produce such merged data need be written but once.

Fragmenting the master file into the various modules of the Clinical Data System also provides many benefits. The most significant of these is the ability to specifically process, within a minimum of time and computer resource, particular sections of the data base. The bulk of the data processing required for a clinical trial or, indeed, for an NDA is accomplished in connection with the patient profile section of the Clinical Data System. For this reason, the record in the patient profile section of the system should be more complete since it contains all of the demographic data and parts of the efficacy data as well as certain dose variables such as initial, average daily, high, low, and final dose and also variables associated with the patient's completion of the trial.

Additional functions associated with the above flow of data through the departments should also be mentioned in connection with good data management requirements. In no particular order, the more significant of those functions are as follows:

As indicated previously, a *protocol* is developed for each clinical trial. While that protocol must obviously satisfy the company's overall objectives in connection with the Clinical Operation Plan for a particular drug, it must also satisfy the FDA guidelines and requirements, the medical monitor and study physician, as well as the statistician's requirements. In connection with data management, however, the protocol must also be standardized as much as possible to satisfy the clinical data processing requirements. An example of this would be a similar rating scale for "overall response to therapy." Wherever possible, from one CRF to another, the terms used should be identical as well as the numeric codes assigned to the terms of that rating scale. This minimizes the difficulties associated with data screening, reviewing, and programming when establishing a master clinical data file for a particular drug.

The CRFs must be developed to satisfy the data collection requirements established by the protocol, as we discussed in Sec. III.A.10. To achieve the goal of standardization, CRFs should be prepared on a computerized phototypesetter. Phototypesetter applications range from simple management forms to master package inserts. The device may be linked to both the computer system and the multiuser word processor.

IX. INFERENTIAL ANALYSIS OF CLINICAL TRIALS

Several issues related to inclusion/exclusion must be resolved before data collection begins. Because of the complexity of clinical trials, a study may not be executed exactly as planned; thus, some of these issues may have to be reconsidered after the data collection is completed.

A basic issue is whether to analyze prescribed drug regimen or intent to treat. As more is learned about the actions of a new drug in patients, the inclusion criteria almost always broaden as one progresses from early Phase II through Phase III. For sound medical and statistical reasons, early Phase II studies are confined usually to narrow prognostic strata. The analyses of these studies almost always address the effects of prescribed drug regimen. Late Phase II and early Phase III studies which include both broadened patient and physician populations are designed to learn about the potential of a new drug. That is, they are designed to determine the safety and efficacy of the new drug when it is administered in specified fashion(s) to specified patient population(s). Therefore, for these studies prescribed drug regimen is analyzed.

Late Phase III studies are usually designed to determine what a drug might do after it is on the market. In addition to much broadened patient and physician populations, criteria for dosage regimens, concomitant therapy, and patient care are frequently broadened or specified only in a general sense. Because of these reasons as well as the randomization argument, many people hold that intent to treat must be analyzed and all randomized patients therefore must be included in all analyses. Others argue that patients who receive an inappropriate drug regimen or no study drug should not be included in any analysis. Thus, they argue that drug regimen should be analyzed. Some statisticians analyze both ways with the hope of demonstrating that both analyses result in the same conclusion. These issues should be addressed before the study begins, and the decisions should be recorded.

It must be recognized that the pre- and postmarketing experience with a new drug will be different: the patient population receiving the drug in premarketing studies may differ considerably from that receiving the drug postmarketing. Thus, many people argue that premarketing studies hardly ever predict what a drug will do after it is marketed. They further argue that a primary purpose of premarketing studies is to set a standard of therapy and to analyze the effect of a prescribed drug regimen.

Recent studies have been designed to determine the effect of drug on a disease process. For example, will it decrease the incidence of myocardial infarction in people with threatened infarction; will it decrease infarct size in those who proceed to infarction; and will it reduce short-term mortality and reduce subsequent infarction and mortality? In these kinds of trials the new drug is almost always used as an adjunct to a standard therapy. It is well-accepted that these studies should analyze intent to treat. The Anturane and Timolol studies among others are recent examples [3-8,11].

A. Inclusion/Exclusion Considerations

Entrance requirements violations, time duration between patient visits, treatment compliance, multiple courses of study drug, and events which occur outside of study treatment period are discussed here.

1. *Entrance Requirements Violations*

During the course of a clinical trial it is almost certain that at least one patient who does not meet the protocol-specified entrance requirements will be enrolled in the study. If the entrance requirements are not well defined and/or if the monitoring of the trial is shoddy, then an excess number of unqualified patients may be enrolled in the study. This will lead to many problems including disbelief in and nonacceptance of the study results.

With well-defined entrance requirements and aggressive, active, ongoing monitoring during the entire course of a trial, the number of unqualified patients who are enrolled is expected to be low. The most common solution, in this case, is to exclude the data from these patients from the analyses. The decision to eliminate data is based only on entrance requirements and not on treatment results.

If the number of patients not meeting entrance requirements is substantial, say, 50 or more in 1,000, it is worthwhile to determine whether the proportion of these patients is the same in each group and then to summarize all data for these patients separately. This helps answer questions about bias and blinding and the effect of eliminating these patients from the overall analyses.

2. *Time Duration Between Patient Visits*

All protocols specify the duration between patient visits to the physician. For example, the patient may be required to return every 90 days after the initial visit for four follow-up visits. Since patients cannot always return to the physician on the exact day specified by the protocol, the question is how much deviation is permitted before the patient is declared to have missed a visit. In determining the time interval within which a patient visit is acceptable, a decision must be made as to whether the acceptable interval should be constructed from the initial visit or from the last visit. Since the decision is based on the medical questions to be answered by the study, the medical monitor should be consulted.

Because of mistimed visits, patients are seen by their physicians more than once during the defined interval. There does not appear to be an optimum method for incorporating in the analysis the two or more sets of data for each of these patients. Some statisticians use the data from the first visit in the interval; some use the data from the last visit; others use the data from the visit closest to the protocol specified date; and still others use average (mean or median) scores. If a given side effect is being tracked, say, nausea, the most severe score is often used. The method should be specified in advance and recorded before looking at the results.

3. *Treatment Compliance and Multiple Courses of Study Drug*

Use of protocol-excluded medication, nonadherence to study medication, and multiple courses of study drug are discussed in the following subsections.

a. Protocol-Excluded Medication

To avoid confounding of effects, most protocols exclude the taking of specified medications during the course of a trial. For several reasons, however, patients do take excluded medication. If few patients take excluded medication and the use is comparable in each treatment group, the usual practice is to leave the data in the treatment groups to which the patients were assigned. Some statisticians perform more than one analysis, especially if the use of excluded medication is large (i.e., all data, data excluding noncompliers' data, and noncompliers' data). Noncompliers' data must be analyzed to determine whether noncompliance is in any way related to study medication. Patients may be using excluded medication, for example, because their study medication is either ineffective or causing unacceptable side effects.

b. Nonadherence to Study Medication
Most people argue that it is difficult if not impossible to accurately measure adherence to study medication. Thus, they will include in the analyses all patients who are not otherwise excluded, even when they know certain patients did not adhere to the prescribed regimen.
c. Multiple Courses of Study Drug
If at all possible it is best to avoid multiple treatment courses by prohibiting them in the protocol. Statistical issues here include baseline variables; potentially biased estimates of efficacy and safety; power; and interpretation.

4. *Events Which Occur Outside of the Study Treatment Period*

In this subsection we discuss the handling of data from patients who experience a medical event between randomization and the start of study treatment or shortly after the start of study treatment (early events) and from patients who have an event after study treatment has been discontinued (late events).

While it is recognized that patients should be randomized as close to the start of treatment as possible, the practical problems of implementation sometimes result in a time interval between randomization and the start of study treatment.

If the clinical trial is comparing two nonsteroidal anti-inflammatories and a patient has a myocardial infarction or dies before the assigned drug starts, then few people would argue that this event must be listed as a possible side effect. That is, it would be listed as a medical event unrelated to drug. If the clinical trial is comparing the adjunctive use of a new drug to standard therapy with standard therapy in patients who had a myocardial infarct, then a new myocardial infarct or death cannot be ignored even though it occurred before study medication had begun because the group that was receiving standard therapy only would have no minimum time from randomization before treatment began.

A patient who discontinues study on day 3 would usually be counted as a treatment failure. The data for dropouts for worsened conditions in the first X days may contain a substantial proportion of the treatment effect.

In studies where each patient's study drug dose is titrated over the first few weeks of the study, the protocol should state that efficacy variables other than dropouts will not be evaluated until the patient has reached the defined optimum dose. Side effects are evaluated from start of study.

Some events occur after a patient has stopped taking the study medication. If medical events leading to observed event started during the study, then one must count the event as medication related. For example, a patient had a bleeding ulcer which started while he was on study drug, the study drug was stopped, and the patient died 3 months later from complications of the bleeding ulcer. In this case the patient's death must be counted as a study-drug-related event.

If there is no way of determining whether medical events leading to the observed event started during the study, then there are four choices for handling events:

1. Include events in the analysis only if they occur while the patient is on study medication. There are several drugs and chemicals that have delayed reactions. Thus, this is not an optimal choice. For pre-NDA studies, however, it is a frequently used method.
2. Include events only if they occur while the patient is on study drug or during a short interval after the patient discontinues study drug (e.g., 10 blood level half-lives). This is not too much different than method 1, above, but will include all events during the time when the drug is expected to be pharmacologically active. The drug of course can affect a body system in such a fashion that the event is not noted clinically for some longer period of time.
3. Include events only if they occur during the protocol-specified study duration. In other words, if a patient is enrolled in a 2-year study, all events which occur in the patient within those 2 years will be included in the analysis regardless of how long he may have been on study drug. This method is frequently used in pre-NDA studies.
4. Include events no matter how long after study drug they occur. This method is used in large-scale, long-term intervention studies and retrospective epidemiological studies. Since there must be a cutoff for NDA filing, this method is not used to any great degree in premarketing studies.

There is no universally accepted procedure for handling early or late events. The method(s) used must be specified before looking at the results. If early events are in fact due to chance, then they should not affect the final results to any great degree. If there are several more patients in one treatment group than the second who have an early event, then one must look very critically at the study and be suspicious of bias and unblinding. For late events, especially very late events, a very critical look must be made for biased reporting of events in one group of patients.

B. Periodic and Final Analysis

The experimental designs of studies to evaluate a new instrument, assay, or experimental approach are usually one of those found in standard statistical texts [22-24], which also describe the analyses.

Design and analyses of pharmacokinetic studies are well described in Wagner [25] and Albert [26].

Some Phase I studies are conducted in an uncontrolled fashion. Analyses are within patient, based on change or percentage of change from initial value. Either a parametric, t(paired), or a nonparametric, such as Wilcoxon, statistic is calculated to test significant change from initial value [22]. Controlled Phase I studies, i.e., those where a placebo group is included, usually utilize standard designs [24]. Nonstandard designs require the writing of an appropriate statistical model, as do standard designs. The analysis is determined by the model utilized. The SAS package is helpful in the analysis of these studies. (SAS is the Statitistical Analysis System, SAS Institute.)

Standard designs are usually used for early Phase II studies. Frequently, however, the designs have to be creative and innovative to explore certain relations that cannot be tested with standard designs. In these cases, again, a statistical model must be written; and after it is written, SAS is helpful in the analysis.

Late Phase II, Phase III, and studies to determine the effect of a drug on the disease process are often described in the literature. It is essential to read this literature to gain insight and guidance on the design and analysis of these studies. (See, for example, Refs. 2-8.)

Further discussion in this section applies primarily to these aforementioned studies. Data presentation, baseline variables and other potential sources of bias, hypothesis testing and estimation, patients who discontinue study, hypotheses of no difference, problems with multiplicity and periodic analysis, and inferences and conclusions are discussed below.

1. *Data Presentation*

Because of the complexity of clinical trial data and because of the several audiences, decisions on data presentation should be made collaboratively with the medical monitor(s). Data presentation can be considered to contain five levels: case report forms (CRFs), computer data files, data listings, summaries, and statistical analyses.

2. *Potential Sources of Bias and Baseline Variables*

Assessment of possible sources of bias is based upon the properties of the sample as well as the methods used to obtain it. Randomization, of course, makes it feasible to assess possible sources of bias. Potential sources of bias which should be investigated are drifts in diagnoses and disease-staging methods, in supportive care and concomitant therapy, in follow-up procedures, in definitions of important medical events, and in protocol compliance.

After these sources are investigated, an assessment of potential bias in baseline variables is made. First, the patient population is described by treatment group and overall, and this is compared to the target population. Next, the study population that is to be included in the analyses is compared with those patients who were enrolled in the study but whose data were excluded for various specified reasons. Then, the baseline data are investigated for differences that might affect the choices of the statistical analyses. The most obvious are treatment group differences with regard to baseline prognostic variables and differences among investigators' patients.

All of these assessments help to determine whether there is a need to analyze by poststratification or to use covariance analysis, difference scores, or some other adjustment technique. Poststratification is also indicated by a significant treatment by baseline variable interaction. Interaction, in this case, means the study drug performs differently in patients with dissimilar levels of a baseline characteristic [24]. For example, a drug may be beneficial to males and be harmful or of no benefit to females. There may be significant differences in drug effects between the very young and the very old, or between mildly ill and very severely ill patients.

3. *Hypothesis Testing and Estimation*

To help construct hypotheses and an appropriate statistical model, it is best to start with a review of the summary statistics by treatment group. In doing this it is very helpful to "slice" the data set in several ways; the exact number depends on available time and resources. This is usually done on the basis of three to six prognostic variables. For a discussion of prognostic stratification, see Feinstein [27]. These summary statistics help to determine whether the responses fit a pattern. Scatter diagrams of change scores (from initial value) are very helpful in discovering interesting relationships.

A statistical model for all patient and treatment groups can be set up if the preliminary looks indicate no unusual patterns. If there are unusual patterns, a model for each patient group may have to be constructed. Because of the complexity of the data structure and because it is not known a priori which variables affect outcome, it is often not clear initially which model is the most appropriate. In this case several may be tried and then the models which were considered and why the final one was chosen must be documented.

Models should include, in addition to treatment effect, investigator effect and effect of any of the baseline variables that the first look at the data and prior knowledge indicate should be included. The model should include also at least all first-order interactions involving treatment.

The data are analyzed separately by time period when there is a significant treatment by time interaction. This is likely to occur when one of the treatments is placebo. If there is a treatment by baseline variable interaction, subgroups of patients are constructed by that variable and the data analyzed within subgroup. If there is a treatment by investigator interaction and if the possibility that investigators' patient populations differ has been ruled out, the examination of the interaction starts with the determination of whether it is of magnitude or direction. For an interaction of magnitude (each investigator's data say treatment A better than B, but for some the difference is much larger than for others), an overall statement about the presence or absence of treatment differences is made. Because of the interactions, however, no general estimate of the difference can be made. In this case many statisticians, for practical reasons, do report the overall mean difference. For an interaction of direction (for some investigators' data treatment A is better than B, and for other investigators' data treatment B is better than A), differences in patient "handling" or evaluation procedures among investigators are looked for first. If differences are found, the data can be stratified on these and analyzed within stratum. When no differences are found, each investigator can be considered as the experimental unit and an analysis of the proportion who have results favoring, say, treatment A is then performed.

In analyzing the study data it is best to avoid, if at all possible, complex modeling and adjustments. Use of these complexities raises issues of believability, appropriateness, and interpretation.

4. *Patients Who Discontinue Study*

Patients who drop out of the study are evaluated because their reason for discontinuing may be due to treatment and thus their data contain part of the treatment effect. Patients may discontinue the study due to side effects, lack of efficacy, or—as is often noted in acutely administered drug studies—because they believe they have improved sufficiently to stop the drug. Obviously, ignoring dropouts can lead to mistaken conclusions. In studies of chronically administered drugs, patients have a higher probability of dropping out of the study due to problems unrelated to either the drug or the disease. If the dropout rates for all reasons differ, then the treatment groups may not be comparable by the end of the study. Since it is important to determine the effect of a drug on long-term users, a repeat of the analysis of baseline variables should be done on the data of those patients who finish the study. If differences are noted, covariance analysis or other model modifications or analysis by patient subgroups can be performed.

5. *Hypotheses of No Difference*

Often the comparative drug in a clinical trial is the current standard, active drug and not placebo. The belief often is that the test drug is approximately equivalent in efficacy but may have fewer side effects.

Thus, the expected efficacy result may not have statistical significance. Hypothesis testing is not really designed to demonstrate equality. Since the null hypothesis of no difference can never be proved, equality studies are designed so as to ensure a sufficient (usually at least 0.08) power to detect small, not medically meaningful differences. Then, a statement can be made that the lack of statistical significance together with adequate power to detect a medically important difference is sufficient to support the conclusion that the drugs do not differ meaningfully. Another often used approach is to construct confidence intervals for the difference. The confidence interval should show whether the range of possible values for the difference does or does not include medically important differences.

6. *Problems with Multiplicity and Periodic Analysis*

Statistical problems arise when a large number of variables are analyzed, when multiple comparisons are made, and when repeated analyses on accumulating data are performed.

a. Multiplicity

When a large number of variables are analyzed, some of the comparisons are expected to produce statistically significant differences by chance alone. For example, if 100 independent variables are analyzed, it is expected that there will be five statistically significant differences due to chance alone when testing at $\alpha = 0.05$. The best way around this is to define two to four key endpoints (design variables) in the protocol. These are the principal indicators for interpreting the results of the study. When more than two treatments are included in a trial, several or all possible comparisons are often made. In this case a problem arises in selecting the appropriate p value. For discussion on problems of multiplicity see Tukey [28], Stoline [29] and Dunnett and Goldsmith [30].

b. Periodic Analysis

It can be demonstrated that repeated analysis on accumulating data at a fixed a level will lead to a declaration of statistically significant differences between two treatments even when in fact the treatments do not differ [31]. There are ethical, scientific, and economic justifications for periodic analyses of data. Therefore, to avoid the above problem a sequential design can be used [31] or the test statistic or p value can be adjusted using one of the methods described by Pocock [31,32], Canner [33,34], or DeMets and Ware [35].

7. *Inferences and Conclusions*

The statistician must ensure that the results of the study support the interpretation and inferences made. Unsupported inferences are speculation or hypotheses to be tested and should be so noted. If the present study conflicts with other studies, the statistician should help the physician (or clinical researcher) to reconcile the differences, raise hypotheses about the differences, or both.

Interpretations of clinical studies have been made in two ways: (1) the statistician interprets the results mathematically and then the physician using the statistical results, results of other studies, and his prior knowledge interprets the results medically; or (2) the statistician and physician collaboratively interpret the results.

The conclusions should present the findings of the study, tempered by a priori knowledge, constrained by the flaws in the study design, and compromised by the verified and suspected weaknesses of the data from the study.

X. REVIEW OF THE PROCESS

It is very helpful after an NDA has been filed to review the entire drug process with all team members and upper level management. The primary purpose is to review the strategy and tactics, to determine what went right and why and what went wrong and why and how the process might be improved. This is not a finger pointing or back patting exercise. It is an attempt to learn from past experiences, to interact further with other team members, to train new people, to record and maintain the invaluable experience gained, and to further improve one's skills. Managing clinical data as well as preclinical data in the pharmaceutical industry is a continuously challenging, dynamic process. As new regulatory requirements are established, as new drug types are developed, and as new indications for existing drugs are identified, management must be prepared to respond.

REFERENCES

1. C. R. Buncher and J. K. Tsay (Eds), *Statistics in the Pharmaceutical Industry*. Dekker, New York, 1981.
2. University Group Diabetes Program, A study of the effects of hypoglycemic agents on vascular complications in patients with adult-onset diabetes. Sections I and II. *Diabetes 19*(Suppl. 2): 747-830 (1970).
3. Hypertensive Detection and Follow-up Program Cooperative Group, Five-year findings of the Hypertensive Detection and Follow-up Program. I. Reduction in mortality of persons with high blood pressure, including mild hypertension. *JAMA 242:* 2562-2571 (1979).
4. Hypertensive Detection and Follow-up Program Cooperative Group, Five-year findings of the Hypertensive Detection and Follow-up Program. II. Mortality by race, sex and age. *JAMA 242:* 2572-2577 (1979).

5. A. E. Dorr, K. Gundersen, J. C. Schneider, Jr., T. W. Spencer, and W. B. Martin, Colestipol hydrochloride in hypercholesterolemic patients: Effect on serum cholesterol and mortality. *J. Chron, Dis. 31:* 5-14 (1978).
6. The Anturane Reinfarction Trial Research Group, Sulfinpyrazone in the prevention of cardiac death after myocardial infarction. *N. Engl. J. Med. 298:* 289-295 (1978).
7. The Anturane Reinfarction Trial Research Group, Sulfinpyrazone in the prevention of sudden death after myocardial infarction. *N. Engl. J. Med. 302:* 250-256 (1980).
8. The Norwegian Multicenter Study Group, Timolol-induced reduction in mortality and reinfarction in patients surviving acute myocardial infarction. *N. Engl. J. Med. 304:* 801-807 (1981).
9. H. S. Seltzer, A summary of criticisms of the findings of the University Group Diabetes Program (UGDP). *Diabetes 21:* 976-979 (1972).
10. C. Kilo, J. P. Miller, and J. R. Williamson, The Achilles heel of the University Group Diabetes Program. *JAMA 243:* 450-457 (1980).
11. R. Temple and G. W. Pledger, The FDA's critique of the Anturane reinfarction trial. *N. Engl. J. Med. 303:* 1488-1492 (1980).
12. U.S. Dept. of Health, Education and Welfare, Food and Drug Administration, *General Considerations for the Clinical Evaluation of Drugs.* FDA 77-3040 (1977).
13. U.S. Dept. of Health and Human Services, Food and Drug Administration, *General Statistical Documentation Guide for Protocol Development and NDA Submissions.* Docket No. 80D-0217 (1980).
14. Report of an international conference held at the Ditchley Foundations: The scientific and ethical basis of the clinical evaluation of medicines. *Eur. J. Clin. Pharmacol. 18:* 129-134 (1980).
15. E. A. Gehan and E. J. Freireich, Non-randomized controls in cancer clinical trials. *N. Engl. J. Med. 290:* 198-203 (1974).
16. S. J. Pocock, Allocation of patients to treatment in clinical trials. *Biometrics 35:* 183-197 (1979).
17. S. J. Pocock and R. Simon, Sequential treatment assignment with balancing for prognostic factors in the controlled clinical trial. *Biometrics 31:* 103-115 (1975).
18. R. Peto, M. C. Pike, P. Armitage, N. E. Breslow, D. R. Cox, S. V. Howard, N. Mantel, K. McPherson, J. Peto, and P. G. Smith, Design and analysis of randomized clinical trials requiring prolonged observations of each patient. I. Introduction and design. *Br. J. Cancer 34:* 585-612 (1976).
19. B. Iglewicz and C. B. Begg, A treatment allocation procedure for sequential clinical trials. *Biometrics 36:* 81-90 (1980).
20. D. P. Byar, R. M. Simon, W. T. Friedenwald, J. J. Schlesselman, D. L. DeMets, J. H. Ellenberg, M. H. Gail, and J. H. Ware,

Randomized clinical trials: Perspective on some recent ideas. *N. Engl. J. Med. 295:*74-80 (1976).
21. W. B. Brown, Jr., The crossover experiment for clinical trials. *Biometrics 36:*69-79 (1980).
22. G. W. Snedecor and W. G. Cochran, *Statistical Methods*, 7th ed. Iowa State University Press, Ames, Iowa, 1980.
23. R. G. D. Steel and J. H. Torrie, *Principles and Procedures of Statistics*, 1st ed. McGraw-Hill, New York, 1960.
24. B. J. Winer, *Statistical Principles in Experimental Design*, 1st ed. McGraw-Hill, New York, 1962.
25. J. G. Wagner, *Biopharmaceutics and Relevant Pharmacokinetics*, 1st ed. Drug Intelligence Publ., Hamilton, Ill., 1971.
26. K. S. Albert, (Ed.), *Drug Absorption and Disposition: Statistical Considerations*. American Pharmaceutical Association, Washington, 1980.
27. A. R. Feinstein, *Clinical Biostatistics*. Mosby, St. Louis, 1977.
28. J. W. Tukey, Some thoughts on clinical trials, especially problems of multiplicity. *Science 198:*679-684 (1977).
29. M. R. Stoline, The status of multiple comparisons in one-way ANOVA designs. *Am. Stat. 35:*134-141 (1981).
30. C. Dunnett and C. Goldsmith, When and how to do multiple comparisons. In *Statistics in the Pharmaceutical Industry* C. R. Buncher and J. Y. Tsay, Eds.). Dekker, New York, 1981 Chap. 16.
31. S. J. Pocock, Group sequential methods in the design and analysis of clinical trials. *Biometrika 35:*549-556 (1979).
32. S. J. Pocock, The size of cancer clinical trials and stopping rules. *Br. J. Cancer 38:*757 (1978).
33. P. L. Canner, Repeated analysis of clinical trial data. *Proc. 9th Int. Biometric Conf.* Vol. 1, pp. 261-275 (1976).
34. P. L. Canner, Monitoring treatment differences in long term clinical trials. *Biometrics 33:*603-615 (1977).
35. D. L. DeMets and J. H. Ware, Group sequential methods for clinical trials with a one-sided hypothesis. *Biometrika 67:*651-660 (1980).

9

DOCUMENT PREPARATION

Nelson H. Schimmel
Schering Corporation
Bloomfield, New Jersey

I. INTRODUCTION

Before the clinical research program can be initiated, most countries require the manufacturer to submit the results of their preclinical research. The submission in the United States is the IND, i.e., Notice of Claimed Investigational Exemption for a New Drug [1]. The content and format of the submission are dictated by the regulations of the country involved. The IND requirements are illustrative of a number of clinical trial permission requests.

Section 312.1 of the U.S. Code of Federal Regulations, 21.CFR.312.1, sets forth the requirements which must be fulfilled before beginning clinical trials in the United States. Form FD 1571 itemizes the 16 statements required from the sponsor.

Of these 16 statements, the following are of most significance to this discussion of document preparation:

> 6. A statement covering all information available to the sponsor derived from preclinical investigations and any clinical studies and experience with the drug as follows: a. Adequate information about the preclinical investigations, including studies made on laboratory animals, on the basis of which the sponsor has concluded that it is reasonably safe to initiate clinical investigations with the drug: Such information should include identification of the person who conducted each investigation; identification and qualifications of the individuals who evaluated the results and concluded that it is reasonably safe to initiate clinical investigations with the drug and a statement of where the investigations were conducted and where the records are available for inspection; and enough details about the investigations to permit scientific review. The preclinical investigations shall not be considered adequate to justify clinical testing unless they give proper attention to the conditions of the proposed clinical testing. When this information, the outline of the plan of clinical pharmacology, or any progress report on the clinical pharmacology, indicates a need for full review of the preclinical data before a clinical trial is undertaken, the Department will notify the sponsor to submit the complete preclinical data and to withhold clinical trials until the review is completed and the sponsor notified....
>
> 7. A copy (one in each of the three copies of the notice) of all information material, including label and labeling, which is to be supplied to each investigator: This shall include an accurate description of the prior investigations and experience and their results pertinent to the safety and possible usefulness of the drug under the conditions of the investigation. It shall not represent that the safety or usefulness of the drug has been established for the purposes to be investigated. It shall describe all relevant hazards, contraindications, side effects and precautions suggested by prior investigations and experience with the drug under investigations and related drugs for the information of clinical investigators....

In essence, the sponsor is asked to justify the exposure of citizens of a particular country to an unknown entity. In addition to the regulatory and public safety considerations (item 6 above), the company's management used this justification as the economic basis of going into clinical trials. The name of the summarizing preclinical document varies from company to company, but such a document is almost always prepared. Usually the preclinical data is divided into several major segments. Again, the manner of presentation varies widely but addresses the basic pharmacology, toxicology, and metabolism of the drug substance. This information is the basis for assuming the compound will have a therapeutic effect in man.

Usually, this preparation of the preclinical document becomes the responsibility of a science writer rather than the scientists who carried out the work. It is rarely possible for the pharmacologist for example to take "time out" to summarize all his data. In order for the writer to efficiently summarize the data from a specific discipline, it is necessary that he or she be familiar with the workings of the discipline. Generally, this requires that the writer has worked in the discipline being reported.

Another required part of the IND is the Investigator's Brochure (item 7 above). Ideally, this document undergoes a gradual series of revisions during the course of the clinical trials. While it is impossible to revise it with the acquisition of each bit of new data as the clinical program progresses and the characteristics of the drug emerge, this document evolves from a statement of preclinical information to a statement of clinical information covering dosage, duration, therapeutic effect, side effects, drug interaction, indications, usage, contraindications, etc. until in the final application, the licensing permit, it is the draft product information sheet (package insert).

The principles underlying the preparation of these documents are the same as those to be discussed for licensing applications. Although the submission permitting clinical trials to begin is a regulatory requirement, it should be recalled that the submission is also the first impression the regulatory agency has of the drug. In the United States, at least, this impression can result in the drug being assigned priority review status. The preclinical summary thus becomes an extremely important document. Its preparation should receive the utmost care.

If the development program is successful, the research conducted by a pharmaceutical company culminates in a registration document. In the United States, this document may be a New Drug Application (NDA), an International Health Registration, or a required report. The goal of such documents is usually to allow a product onto the market, i.e., to produce a return on investment. It is essential not only that it convey the necessary information but that it does so as efficiently as possible. Sound science can be discredited by a poor presentation.

The licensing application format varies widely from country to country, as does the amount and type of information required. In addition to these national variations, the requirements are constantly being revised. This discussion will therefore outline principles rather than provide a cookbook.

The NDA format and content also is prescribed in Title 21 of the Code of Federal Regulations Part 314.1 (21 CFR 314.1). Other agencies provide similar information in the equivalent document(s) of the country involved. For illustrative purposes, the FD 356H—New Drug Application—will be used as the basis of this presentation. The sponsor of an NDA is required to provide a total of 16 "attachments," 13 of which require significant preparation:

1. Table of contents
2. Summary
3. Evaluation of safety and effectiveness
4. Copies of the label and all other labeling to be used for the drug
5. A statement as to whether the drug is (or is not) limited in its labeling and by this application to use under the professional supervision of a practitioner licensed by law to administer it
6. A full list of the articles used as components of the drug
7. A full statement of the composition of the drug
8. A full description of the methods used in and the facilities and controls used for the manufacture, processing, and packaging of the drug
9. Samples of the drug and articles used as components
10. Full reports of preclinical investigations that have been made to show whether or not the drug is safe for use and effective in use
11. List of investigators
12. Full reports of clinical investigators that have been made to show whether or not the drug is safe for use and effective in use
13. If this is a supplemental application, full information on each proposed change concerning any statement in the approved application

II. OPTIONAL EXPANDED SUMMARY

The Federal Regulations allow the sponsor to combine Attachments 2 and 3 into an overall summary and evaluation which is to contain the following components:

A. Chemistry
B. Scientific rationale and purpose the drug is to serve
    1. Clinical purpose
    2. Highlights of preclinical studies
    3. Highlights of clinical studies
    4. Conclusions
C. IND number, etc.

D. Preclinical studies, primary and secondary pharmacology, metabolism, toxicology, etc.
E. Clinical studies, special studies, dose-ranging studies, controlled clinical studies, etc.

III. WORD PROCESSING

The more readily available word processing equipment in most centers will do much to modify the process of document preparation. It is now very easy to adjust spacing and location on the page so as to provide adequate margins for binding and pagination.

These versatile machines provide us with editing and retrieval capabilities never before possible on a practical basis. It is now possible to prepare submissions with standardized type. Italics and other fonts can be used where appropriate. Tables which are not computer generated can be prepared rapidly and be almost identical in appearance to those that come from the computer. This provides the reader with fewer distractions than if each section is prepared on different typewriters operated by different typists.

Since the goal of a submission is to obtain regulatory approval, all of these items take on significance because they have a favorable effect on the reviewer and increase the efficiency of the review.

Reports of individual development activities are the building blocks of a submission. Report writing is thus an underlying activity that ranks very close in importance to the scientific data being reported.

IV. ROLE OF THE SCIENCE/MEDICAL WRITER

It is a general principle that the summary and evaluation is the "essence" of the NDA. Thus, its clarity, ease of understanding, and overall impression are extremely important. An adequately presented summary and evaluation should make it possible for the reviewers to develop a full understanding of the product without the need to look further. Having achieved this understanding, it is necessary only to spot check the original data. The amount of checking required depends on a number of factors—not the least of which is the reviewer's personal preference.

The regulations covering the optional expanded summary provide for the presentation of highlights of the preclinical and clinical activities that make up the submission. These highlights can have a profound influence on the pharmacologist, chemist, or medical reviewer (or professionals with equivalent titles in other countries). It was stated earlier that the preclinical information should be written by an individual who is intimately familiar with the discipline involved in developing the data.

This individual must be able to communicate with his peers in the reviewing agency. The highlights of the summary and evaluation, which represent further distillation of the data, should logically be prepared by the same writer.

Individual study reports are usually written in great detail. The pharmacology experiments which characterize the compound are usually written as separate reports, as are the toxicology and clinical studies conducted on the drug. These individual reports may comprise several hundred pages. It is highly advantageous to the reviewer if the full report is preceded by a brief (1- or 2-page) synopsis.

This synopsis is in addition to the highlights called for by the regulation. The synopsis serves as an overview at the time the full report is being reviewed, but it also provides a ready source of reference in conjunction with reviews by other disciplines within the regulatory agency. For example, a clinical study report may wish to reference one of the pharmacological and toxicological reports as the basis for a particular protocol or the reason for carrying out a specific measurement. The medical reviewer need not read the entire preclinical study report if it is properly synopsized.

The full study report should contain all the data necessary to allow the reviewer to conduct his own reanalysis of the study without reference to the raw data. The location of the raw data within the submission, should it become necessary to review it, should be clearly identified.

To be an effective communicator, the writer should be intimately familiar with all stages of the research activity. If a medical writer wishes to report competently on a clinical study, for example, it is often helpful for him to be involved with the design of the study and to have access to interim reports on the progress of the study.

The role of the medical writer in the preparation of regulatory submissions has been a major concern of the American Medical Writers Association. The appreciation of the role of the writer by the U.S. Food and Drug Administration was recently expressed in a talk to this group by Lloyd G. Millstein, Ph.D., Acting Director of the FDA Division of Drug Advertising and Labeling [2].

V. ROLE OF THE COMPUTER

Another chapter presents an in-depth analysis of computer applications. This discussion deals only with document preparation. The computer is now ubiquitous. It is involved in the design of studies, preparation of dosage forms in some instances, and the analysis of the data generated. In addition, computers may be used to print labels, provide randomization, account for the test article shipped and returned, provide lists of investigators, etc.

Computer output, if properly programmed, can produce a final document for submission to the regulatory agency. In addition to creating

41999

TABLE
ONE MONTH ORAL TOXICITY STUDY OF SCH TEST IN RATS
CLINICAL SIGNS INCIDENCE: OBSERVATIONS/NUMBER EXAMINED

INTERVALS AND SIGNS	DOSE GROUPS (MG/KG) 0 (CONTRO	150 (LOW DO	300 (MID DO	600 (HIGH D
INTERVAL 1				
EXCRETA				
NO OBSERVABLE ABNORMALITIES				
NOT GRADED	1/1	1/1		
EYES/EARS/NOSE				
SALIVATION				
NOT GRADED			1/1	1/1
INTERVAL 2				
EXCRETA				
NO OBSERVABLE ABNORMALITIES				
NOT GRADED	1/1	1/1		
EYES/EARS/NOSE				
MYOSIS				
NOT GRADED				1/1
SALIVATION				
NOT GRADED			1/1	1/1
POSTURE/COAT/APPEARANCE				
URINE STAINED COAT				
NOT GRADED				1/1
INTERVAL 3				
EXCRETA				
NO OBSERVABLE ABNORMALITIES				
NOT GRADED	1/1	1/1		
EYES/EARS/NOSE				
MYOSIS				
NOT GRADED				1/1

04/14/82 - V 1.0

Figure 1 Computer-generated table.

tables of preclinical and clinical data, the computer can print out individual subject data (animal or human) to provide supporting evidence (as appendixes) for the full reports.

To avoid the need to "cut and paste" or to photoreduce computer printouts, it is necessary for the computer to be programmed to provide the output in the format required, centered on the page with adequate margins.

In an attempt to reduce the area of paper required for explanatory remarks on computer output, many computer operators tend to use symbols or codes to identify tables or lists. This practice requires translation in the form of footnotes or glossaries or the manual typing of titles, table numbers, etc. The need for all this additional activity can be avoided by proper planning and programming. Computer-generated tables (Fig. 1) can be produced in submissible form at significant cost savings over manually typed tables.

```
L 'R546.P2PRINTO.DATA' NONUM
 'R546.P2PRINTO.DATA'
1*ROUTE FILE 12                                                           00000160
                       0001
0PLOT WOR SCH SUM BY COD 1 BCOD 1234 EVE MON SEL ICO GT 1 $ KEY 29482116B00000170
 2029482116B           0002
1
 2029482116B           0003
0
 2029482116B           0004
0RUN DATE 26APR82 0814HRS          *****  SUMMARY  W O R K I N G  SCHEDULE  *****
 2029482116B           0005
0PROJECT  29482116
 2029482116B           0006
0                                                       SORT   CODES  1        BCODES 1234
 2029482116B           0007
 -----------------------------------------------------------------------------------------
 2029482116B           0008
                                                 MODE=0/FE     /..1982.../...1983..../...1984..../19
 2029482116B           0009
      NUMBER  D E S C R I P T I O N                             MAMJJASONDJFMAMJJASONDJFMAMJJASONDJFM
 2029482116B           0010
---------------------------------------------------------------I---------I---------I---------I------
 2029482116B           0011
                                                               I    I    I    I    I    I    I    I
 2029482116B           0012
      10207                                                    XXXXXXXXXXXXXXXXXXXXXXXXXXXXXXXXXXX
 2029482116BB071       0013
                                                               I    I    I    I    I    I    I    I
 2029482116B           0014
      20207                                                             XXXXXXXXXXXXXXXXXXXXXXXXXX  X
 2029482116BB072       0015
                                                               I    I    I    I    I    I    I    I
 2029482116B           0016
      30207                                                    XXXXXXXXXXXXXXXXXXXXXXXXXXXXXXXXXXX
 2029482116BB073       0017
                                                               I    I    I    I    I    I    I    I
 2029482116B           0018
      50207                                                              XXXXXXXXXXXXXXXXXXXXXXXXX
 2029482116BB075       0019
                                                               I    I    I    I    I    I    I    I
 2029482116B           0020
      70207                                                                                     XXX
 2029482116BB077       0021
                                                               I    I    I    I    I    I    I    I
 2029482116B           0022
---------------------------------------------------------------I---------I---------I---------I------
 2029482116B           00%!
READY
```

Figure 2 Dot matrix.

From the standpoint of clarity of computer-generated tables, attention must be given to the type of printer used. Very high-speed printers may produce a somewhat smudged print which becomes exaggerated when photocopied. As a general rule, formed character impact printers produce a sharper image, and dot matrix printers tend to produce a less-distinct image (Fig. 2).

VI. DATA ENTRY

There is no U.S. regulatory requirement governing data entry per se. Members of the Biometrics Division of the FDA have given talks and published papers which give their guidelines for presenting data. It is always prudent to review such information prior to assembling a submission. The FDA encourages presubmission conferences to discuss the type of data to be collected and the analysis of the data to be performed.

It is also prudent on the part of the sponsor to utilize these information sources when they are available. For U.S. submissions, the procedure employed in the data entry process (standard operating procedures) should be available for inspection. The conventions employed in the data processing of a specific study are useful to the statistical reviewer.

VII. THE DATA ENTRY SPECIALIST

In most situations, the transfer of information from the case report form (CRF) to the computer data base requires the intervention of data entry specialists. These individuals require knowledge of computer programming techniques and limitations and the clinical data to be entered. It is essential that they be involved in the design of the CRF if the transfer is to be accurate and efficient. For example, the units of laboratory test results must be established before the data are collected so as to provide space for it in the data base. The computer age has spawned another speciality.

Data entry carries with it an inherent error rate. Programs to call attention to missing data or to values outside the predetermined limits facilitate the location and correction of these errors. It is possible to provide the clinical monitor with a detailed list of information to be gathered at the next investigator contact.

When the submission contains a large number of data points, consideration should be given to presenting the findings from the quality control check done on the accuracy of transfer from case report to computer data base. Such sampling is usually done according to a probability standard such as Military Standard 105D and will establish an error rate of less than a predetermined amount. Such presentations will enable the reviewing statistician to check the statistical methods and the accuracy of calculations, and (if necessary) to recalculate them as a further check. The ready availability of these details develop a sense of confidence in the data and thus the overall submission.

A growing number of sponsors prepare a separate statistical submission to accompany the application in which the full statistical analyses of each study and the statistics of the total application are presented. These statistical documents usually contain the computer output showing statistical calculations and tables. The preparation of these documents is generally the responsibility of the statistician(s) who conducted the analysis—unlike the preclinical and clinical reports.

VIII. GRAPHICS

Graphics, especially computer-generated graphics, are relatively new in regulatory submissions. The capabilities are almost limitless but are dependent on the equipment available. Detailed discussion of them here is therefore impractical.

The old adage that a picture is worth a thousand words has not been disproven. The impact of graphic presentations of data, trends, results and comparisons should be weighed against the costs involved in producing them. In general, the cost is more than offset by the ease of understanding they provide the reviewer.

Most regulatory submissions are produced by photocopy machines which have limited capacity to reproduce color. Graphic presentations should, whenever possible, rely on shading, cross-hatching, and other devices which can easily be distinguished in black and white (Figs. 3 and 4).

IX. THE ANNOTATED PACKAGE INSERT

The labeling of a product, particularly the package insert or product information sheet, is a succinct summary of the whole development process. It distills the chemical, pharmacological, toxicological, metabolic, and clinical data into a few paragraphs. Negotiating the wording of this document with the regulatory agency can be one of the major events in the entire approval process.

There is considerable variation in the approach a given reviewer takes to a regulatory submission. Some leave the labeling until the review is complete and then attempt to decide if the application justifies the proposed labeling.

Other reviewers use the proposed labeling as an "orientation device," i.e., to learn quickly what the submission is intended to establish

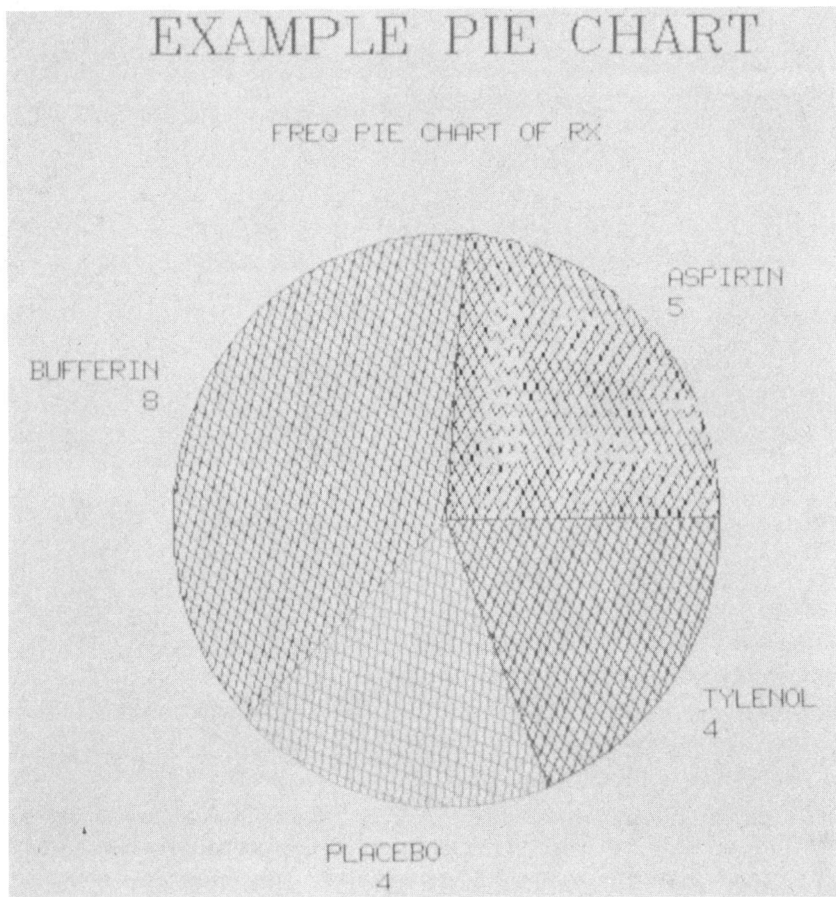

Figure 3 Computer-generated graphic.

about the drug in question. No matter which approach the reviewer takes, it is essential that he be able to quickly identify the basis of every statement made in the labeling.

This applies to internally generated documents and to published literature. It is prudent to provide copies of the pertinent references in close proximity to the proposed labeling for the convenience of the reviewer. Should this be impractical because of the volume of the references, the location elsewhere in the submission should be clearly identified with volume and page number.

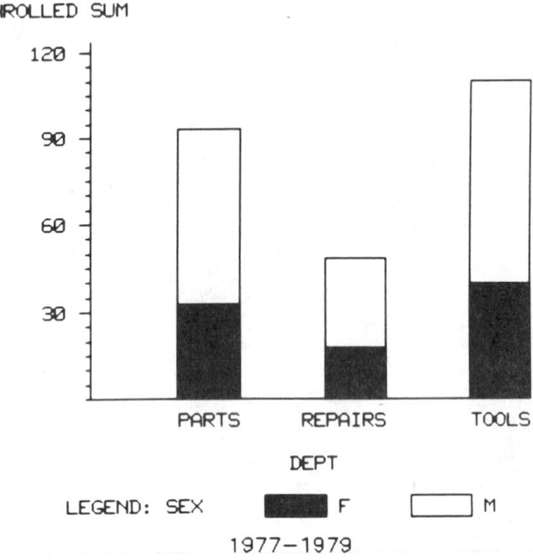

Figure 4 Computer-generated graphic.

X. LIBRARY/INFORMATION SCIENCES

The value of the published information in obtaining regulatory approval is influenced by a number of factors: quality, timing, regulations, etc. If in a particular situation the published literature is a major factor in presenting the data required for approval, it should be presented and analyzed with the same care as data generated directly by the sponsor.

Comprehensive listings of publications on a particular drug can be electronically generated by a number of fee-for-service organizations as well as the research library of many major companies. Some of these sources provide brief abstracts of the articles. When the published literature is the *basis* for approval, mere listing or abstracting does not provide the reviewer with the necessary synthesis of the data. A review of the literature presenting the sponsor's interpretation as well as the specific claim-supporting references should also be provided.

The reviewer should be able to accept the conclusions expressed by the labeling on the basis of the sponsor's comprehensive summary.

XI. DATA STORAGE

Although the computer age is upon us and the technology for electronic filing and retrieval of information exists, there is still a need to provide "hard copy." Regulatory agencies still require submissions on paper, and most sponsors will retain hard copy of the data. Regulations require that a sponsor be able to validate the data provided in support of a submission. This involves an audit trail which extends all the way back from the license application to the laboratory notebooks.

To be able to accomplish an audit in a timely fashion, the location of all of the raw data must be known. This is usually accomplished by the creation of a central file system. Some companies choose to have a clinical file and a nonclinical file; some combine all research files into a single unit. Rather than simply serve as physical storage, central files usually have a computerized data retrieval mechanism. The system may work in a number of ways, but one of its inherent characteristics is the ability to list all documents related to a particular topic. Depending on the sophistication of the system, the report can be a simple listing of documents by title or number or include an abstract of the document.

Clinical case records should be forwarded to the central file as soon as they are received. This will enable the creation of a log, copying for security purposes, and distribution to the data entry and medical groups for whatever additional processing is necessary.

The central files system allows the sponsor to have a running account of the progress of the clinical program. Since the clinical trials may run over several years, the log of case reports provides useful control information as well as fulfilling regulatory requirements to maintain up-to-date records. A number of non-U.S. agencies accept summary submissions as a matter of routine. The FDA has begun to accept some submissions or parts thereof on microfiche. Microfilming reduces file space but does not impact the amount of material for review. As indicated previously, a summary of the NDA is called for by the regulations, but it cannot suffice as the entire submission.

XII. TRENDS

A reduction in the size of regulatory submissions seems to be a trend of the future. The size of NDAs in the United States has grown to the point where they have become unmanageable. The amount of file space they require as well as the almost impossible task of reviewing the present-day submissions has triggered recent efforts to change the approach.

It therefore seems certain that some form of summary will eventually become acceptable worldwide. Just when this will occur is not predictable; but as the time approaches the clarity of presentation, the use of graphics, and computer output will assume even more importance than it has now. With present-day technology it is possible to foresee the time when "submissions" could become completely electronic.

The regulatory agency could be provided with the capability of directly accessing the sponsor's data base to spot check or completely reproduce items of interest. All word processing or computer-generated or -stored information could be electronically transferred. While this will eliminate the need for much of the paperwork now involved, it will not eliminate the need for summarization and presentation of the information in the most advantageous manner.

REFERENCES

1. Code of Federal Regulations, Title 21: Food and Drugs. Published by the Office of the Federal Register, National Archives and Records Service, Washington, D.C.
2. Lloyd G. Millstein, The role of the medical communicator in improving the quality and readability of product labeling and New Drug Application summaries. Paper presented at the American Medical Writers Association Meeting in Indianapolis, Ind., on June 3, 1981.

10
CLINICAL PHARMACOLOGY

William B. Abrams
Merck Sharp & Dohme Research Laboratories
Division of Merck & Co., Inc.
West Point, Pennsylvania

Keith H. Jones
Merck Sharp & Dohme Research Laboratories
Division of Merck & Co., Inc.
Rahway, New Jersey

I. INTRODUCTION

In the pharmaceutical industry, clinical pharmacologists, working either individually or in groups, bridge the gap between preclinical activities and broad-scale clinical investigations. Their principal role is the conduct of Phase I and Phase IIA programs. Phase I studies represent the first administration of new drug candidates to human beings. The objectives are to determine human tolerance, absorption, metabolism, elimination, pharmacodynamics, and dose range. In Phase IIA, new drugs are first used in patients with the disease states for which the drugs were developed. These studies provide initial evidences of efficacy and expand the safety data generated in Phase I. Phase I and IIA trials, although vigorously controlled, are small in scale to maximize efficiency and to minimize human exposure to early investigational drugs. In Phase IIB somewhat larger-scale trials are carried out to more precisely define dosage and safety in preparation for the very large-scale Phase III programs which precede the filing of a New Drug Application (NDA). Since the Phase I and II trials provide the principal data on which a decision is based as to whether to proceed to Phase III, the most expensive part of the drug development process, it is essential that the information generated be adequate in quality and quantity. Thus, the major mission of clinical pharmacologists in the drug industry is the timely generation of adequate efficacy and safety data permitting well-founded decisions to be made to proceed or not to proceed with new drug candidates.

Clinical pharmacologists are also often asked to conduct special studies on mechanism of action, drug-drug interactions, bioequivalence, disposition, and other such investigations. The nature of Phase I and II studies and special clinical pharmacology investigations determine the education, experience, and specific skills required by an industrial clinical pharmacology staff. Participants in the process include physicians, coordinators, statisticians, pharmacokineticists, and scientifically trained secretaries.

Physicians are expected to have specialty training in addition to a general medical background. Research experience is also essential, and many industrial clinical pharmacologists have a Ph.D. in a biological science as well as an M.D. degree. Coordinators are individuals with a science background who assist the physicians in planning, executing, monitoring, and reporting studies. Their backgrounds vary and may include laboratory research, nursing, medical secretarial skills, and computer sciences. Statisticians and pharmacokineticists may or may not be members of the clinical pharmacology staff but are integral members of the team. Since clinical pharmacology studies are characteristically small in size and intensive in design, it is essential that statisticians participate in the planning as well as the analysis to ensure that appropriate measurements and numbers of subjects are employed to give meaningful results. Bioavailability, bioequivalence, drug disposition, and drug interaction studies and those including drug blood and urine levels as a parameter require the participation of scientists skilled in drug metabolism and pharmacokinetics. Not only must the right samples be collected at the right times, but also in the right amounts, in the right tubes; and they must be stored, shipped, and analyzed under the proper conditions.

Dedicated, scientifically trained secretaries are especially important in a clinical pharmacology program. All of the functions of the professional members of the team are expressed through protocols, case report forms (CRFs) tabulations, and research reports, and the integrity and timeliness of these documents are in the hands of the secretaries. Needless to say, all members of the clinical pharmacology team must be familiar with governmental regulations pertaining to new drug studies in human beings.

The core operating group includes a physician investigator, one or more coordinators, and secretaries. Under the general supervision of a director, a representative department would comprise several such groups depending on the size of the company. Statisticians, pharmacokineticists, and other scientific specialists may be regular or ad hoc members of the group, depending on the specific project and organization of the company.

Although the industrial clinical pharmacologist's responsibility begins with initial clinical trials, research activity begins a year or two earlier when a drug candidate is identified as potentially interesting for human development. At this time, the clinical pharmacologist refreshes his knowledge of the field and interacts with preclinical scientists in planning the pharmacological, safety assessment, drug disposition, and biopharmaceutic studies which must be done before a drug can be given to humans. When sufficient information is available, the clinical pharmacologist prepares an overall plan for initial human studies. The important elements of this plan are the objectives of the program, the number and design of studies to be conducted, the nature and number of normal and patient volunteers to be involved, investigational supplies required, laboratory support needed, and the manpower and financial resources that will be required. This document is prepared collaboratively and thoroughly reviewed since it may determine the fate of the new drug brought forth at considerable expense and through the efforts of a large number of laboratory scientists. Furthermore, it commits actions by many groups in addition to the clinical pharmacology staff, as well as the financial resources of the company. Among the tests applied to the plan are the following: Will the proposed studies meet the stated objectives? Can the studies, as planned, be analyzed properly? Are the plans consistent with animal pharmacological and safety data? Can investigational drug supplies be prepared and packaged as described? Will the proposed timing ensure the smooth execution of the project, and does it fit with other programs of the company?

It is also important to consider whether the initial clinical trials will provide all the information necessary to conduct the large-scale Phase III programs which follow, if it is decided to proceed. Thus, in planning Phases I and II, the clinical pharmacologist consults with colleagues who will conduct the Phase III trials and who are often expert in the relevant disease states. Questions in this regard may deal with the firmness of dosage recommendations, the effects of renal and liver function on the drug, and the adequacy of safety information for expanded trials.

Finally, the plan for Phases I and II must meet governmental regulations and company policies; the latter are often more stringent than the former. In the United States, an Investigational New Drug (IND) application must be filed with the U.S. Food and Drug Administration (FDA) before such trials can begin. Schedule 9 of this document identifies the monitors and investigators who will be involved in the studies, including a summary of their training, experience, and facilities. Schedule 10 provides an outline of the clinical investigations to be performed and often includes the protocol for the first study. Each country where the drug may be studied has its own set of regulations which must be addressed. Company experts in regulatory affairs review the plan for Phases I and II carefully for its adequacy in this regard.

To summarize briefly, the principal role of clinical pharmacology in the pharmaceutical industry is to conduct Phase I and II investigations which are adequate in design and execution to permit a decision on the future course of new drug candidates. In planning for these trials, the clinical pharmacology staff must have expertise—or access to expertise—in medicine, including appropriate specialities, experimental design, biostatistics, data management, drug metabolism, and pharmacokinetics. Plans for the initial human studies require the collaboration and concurrence of the relevant preclinical laboratory scientists, the clinical groups who will use the information generated, and those responsible for compliance with company policy and governmental regulations. Once the plan has been approved, the process of implementation begins.

II. INDUSTRIAL CLINICAL PHARMACOLOGY UNITS

Many pharmaceutical companies sponsor clinical pharmacology activities in medical centers or community hospitals. The levels of involvement vary from single individuals to large departments. The nature of the relationship also varies. In some instances, the staffs of the clinical pharmacology units are employees of the companies, whereas in other situations they are staff or faculty members of the medical center. Clearly, each situation is governed by a specific understanding between the sponsor and the medical facility.

In 1927, Eli Lilly and Company initiated the first pharmaceutical company-sponsored research program in a community hospital. The Lilly Laboratory for Clinical Research at the Wishard Memorial Hospital, Indiana University Medical Center (formerly the Marion County General Hospital), still is one of the most prominent and successful enterprises of this type. This 67-bed facility is staffed by approximately 150 individuals led by a senior staff of 23 scientists with M.D. and/or Ph.D. degrees. Although they are all employees of the company, the seniors have full-time appointments at the Indiana University Medical School.

An alternative arrangement is exemplified by the Merck-sponsored Clinical Pharmacology Unit at the Jefferson Medical College in Philadelphia. In this instance, the staff of 4 seniors and 10 others are employees of the college and conduct studies offered by Merck on a priority basis. The unit is responsible for the clinical pharmacology teaching program at the college and is free to engage in research activities of its own choosing. Among other U.S. companies which sponsor clinical pharmacology units are Hoffmann-La Roche Inc., Schering-Plough Corporation, and The Upjohn Company.

The advantages to a pharmaceutical company of sponsoring a clinical pharmacology unit reside in the nature and importance of Phase I and IIA studies. As noted earlier, these involve initial trials in normal volunteers, bioavailability, bioequivalence, drug interaction studies, early efficacy trials, and investigations into mechanisms of action. A sponsored unit provides the company with rapid access to a facility capable of conducting such studies promptly and precisely. With experience, the unit staff, the local institutional review board, and the company get to know each other's requirements and administrative burdens are lessened. Monitoring is clearly easier when such studies are conducted near the monitor's home base. Indeed, in a sponsored unit, it is often possible for the monitor to directly participate in the research, enhancing his or her familiarity with the drug and confidence in the results. Such participation maintains medical and investigative skills, as well as credibility in the scientific community.

For the unit, pharmaceutical company sponsorship provides early access to new drugs, a base of scientific support, and financial stability. Since pharmaceutical companies are the predominant source of new drugs, the opportunity to conduct early and novel studies is an attractive advantage of such an arrangement to academic clinical pharmacologists. Similarly, pharmaceutical companies, more than academic institutions, have the requisite expertise in pharmacology, toxicology, and drug metabolism—fields of special importance to clinical pharmacology. The financial support does more than just pay the bills; it frees the senior staff from the need to generate funds, increasing productive time.

The essential components of a clinical pharmacology unit are a qualified professional staff, a well-trained group of nurses and technical coordinators, laboratory facilities for drug assay and clinical biochemical determinations, a properly constructed physical plant, and access to volunteers—both healthy subjects and patients. It is assumed the location in a medical center provides the necessary medical backup and a properly constituted investigational review board. The professional and coordinating staffs have the education and skills described for their industrial counterparts, but in the unit the skills are directly practiced. Furthermore, the physicians and scientists, through their teaching and patient care responsibilities, maintain a broad and practical knowledge of the practice of clinical pharmacology. This not only contributes to wise protocol construction but broadens the perspectives of the sponsor. Since the unit is dedicated to the conduct of Phase I and IIA studies, a high level of expertise is achieved—a property of great importance to the sponsor. The complementary contributions of the industrial and academic clinical pharmacology groups are the basis of the success of many sponsored programs throughout the world.

III. PHASE I STUDIES

The initial administration of an investigational new drug to human beings is usually accomplished in normal volunteers. Exceptions include anesthetic agents, anticancer drugs, and substances which may confer benefit to patients but carry a risk of adverse or toxic effects at doses close to those which may be therapeutic. Since healthy volunteers cannot derive benefit, risk must be minimal and the safety of the participating volunteer takes first consideration in planning Phase I studies. It is to be noted, however, that a review of Phase I study experience indicates safety to participants has indeed been accomplished [1]. Planning includes careful attention to animal pharmacology, toxicology, and drug disposition information, as well as to experimental design to minimize the study population and avoid the need for repetition of studies. It is likewise important to conduct a thorough prestudy clinical and laboratory screen to confirm the normality of the volunteers, thus reducing risk to the health of the subjects and to the safety profile of the new drug. In this regard, the use of placebo controls is essential from the earliest trials onward.

The objectives of Phase I studies include determination of pharmacodynamic and biochemical activity, safety, and tolerability, and the fate of the drug in the body. Pharmacokinetic analysis plus the biological response observed provide the basis for dosage regimens to be employed in future studies. To satisfy these objectives, studies of differing types and designs are required.

It is usual for these objectives to be achieved through an orderly sequence of studies of increasing dose exposure and complexity starting with initial single-dose administration. After the first dose is observed to be safe, incrementally increased single doses are given at intervals up to a predetermined maximum dose. The maximum is based on animal safety information. Blood or urinary drug concentrations may be determined to obtain preliminary data on drug absorption and elimination. Repeated dose studies are then conducted during which evidence may be obtained regarding tolerability and drug cumulation and elimination patterns. Observations are made of hematology, clinical biochemistry, and specific organ function tests, such as hepatic and renal function, to confirm safety. Specific tests of biochemical or organ function may also be conducted as indicators of drug activity, though separate pharmacodynamic studies in model systems are usually conducted separately from tolerability studies. These pharmacodynamic studies are designed to demonstrate the nature of drug action, its profile against time, and its relation to dose and concentration of drug in body fluids.

Pharmacokinetic studies to more precisely determine the rate and extent of drug absorption, its distribution in the body, and its elimination by various routes, including metabolism, are also conducted in normal volunteers and form the next logical series of Phase I studies. Comparative bioavailability or bioequivalence studies may also be conducted at this stage to determine the likely utility of various drug formulations. If it is important to determine the precise metabolic fate of the drug in man, then detailed studies often involving the use of radiolabeled tracer drug may be conducted at this stage in the Phase I investigations.

The study design will be decided by the clinical pharmacology monitor in association (as mentioned in Sec. I) with other clinical colleagues, preclinical pharmacologists, toxicologists, biochemists, formulation experts and biostatisticians. The investigator and other independent experts are also consulted at this stage. The design finally agreed on will be recorded in a protocol which represents the basic study document containing all essential details, including the way in which results are to be analyzed. The protocol must first be studied and agreed to by the investigator and then receive the acceptance of an ethical review committee or other independent body. It must also meet the requirements of Federal Regulations and the Declaration of Helsinki.

Details of the study design will depend on the immediate objective. Drugs being administered to humans for the first time may be given to only a small group or even one subject at a time to minimize risk. Early Phase I studies often rely more on in-depth clinical observations than on conventional statistical conclusions. Once the initial administration has been safely completed, group sizes are increased to permit study of tolerability at higher doses and intersubject variability in response. The application of pharmacological principles permits the rapid scale-up of doses to explore the range extrapolated from preclinical information. Repeated dose tolerability studies are usually conducted in larger populations, e.g., groups of 12 subjects. Bioavailability and bioequivalence studies usually involve groups of similar size, but occasionally up to 36 participants may be required to demonstrate a predetermined difference in the bioavailability between two formulations of the same drug.

Dosage is, of course, a most important consideration: starting dose, maximum dose, dosage increment, and dosage frequency. For early studies, guidance must be sought from the results of animal studies. The first dose to humans will be a low one, possibly one-tenth or even one-hundredth the pharmacologically active dose or a similar low fraction of the "no-toxic-effect level" in animals. The number of doses and frequency of administration depend upon the extent and duration of preclinical studies, the extent of absorption, pharmacokinetic data, and the slope of the dose-response curve.

It is customary to conduct investigations over a dose range adequate to determine a pharmacological effect and its dose relationship, or to achieve a fixed multiple of the proposed therapeutic dose, or to define a maximum tolerated dose. Predetermination of these dosage limits and the intervals between dosing are important aspects of study design. During early studies, dosage increments are frequently a matter of doubling, though numerous possible increment alternatives exist according to the characteristics of the drug substance, the objectives of the study, or the availability of the volunteers. If a maximum tolerated dose is established, then further studies proceed at a fraction (maybe one-half to two-thirds) of that dose to establish repeated dose tolerability and safety. Alternatively, if blood drug concentrations have been measured, subsequent dosing procedures may be determined, as noted above, on a pharmacokinetic basis. Occasionally, early pharmacokinetic or metabolic studies using single low-dose, high-specific-activity, radiolabeled drug may prove useful in establishing a further program of evaluation. This is especially so if such data have been found useful during preclinical and animal assessment.

The formulation used and the route of administration will be influenced by the results of studies in experimental animals. The oral route is the one usually chosen, and capsule formulations are most frequently employed as more sophisticated tablet formulations are neither available nor warranted at this stage of new drug development. To optimize absorption, single aqueous solutions may be used provided that the drug is sufficiently soluble and does not have a repulsive taste. In certain cases, it may be desirable to administer the first dose to humans by the intravenous route, thereby overcoming the problem of gastrointestinal absorption. When slow intravenous infusion is used, it also offers opportunity to more readily control the availability of drug to the systemic circulation and to react in the event of untoward effects. The formulations used for definitive bioavailability or bioequivalence studies are often more sophisticated and usually represent either advanced clinical trial or proposed marketing formulations.

Meticulous clinical and laboratory observations are crucial, and the importance of a good prestudy baseline has already been mentioned. During early Phase I studies, basic clinical and laboratory observations such as heart rate, blood pressure, respiration, electrocardiogram, and changes in mood and general well-being are made at frequent intervals throughout the expected duration of action of the drug. The subject may be linked to a continuous monitoring device, especially during the first administration. Equally intensive observation may be made at intervals during repeated dose studies. Clinical chemistry, hematology, and other appropriate laboratory observations are made at similar time intervals. As safety is of overriding importance, special attention is paid to laboratory tests of organ function, especially critical target organs such as the liver and kidney, as well as of the hematopoietic and reticuloendothelial systems. Special tests of function or change in function may be made depending on the known or expected mode of action of the drug substance. Blood and urine samples may also be taken at selected time points for assay of parent drug and/or metabolites. These data may prove useful in relating pharmacodynamic properties to drug concentration or in designing drug dosage regimens for subsequent studies. The data may also be valuable in determining compliance in repeated dose studies where supervision of dosing is not always possible.

It is important not only to make precise observations but to capture and record these observations in a permanent form at the time the study is conducted. Apart from the immediate need to work with these data, it may be necessary at a later stage to refer back to data on a mode of action which was not suspected at the time the study was conducted. Indeed, several important drug discoveries have been made as a result of careful observation during early human volunteer studies. To encourage and assist in the orderly maintenance of records, many companies follow the practice of issuing detailed CRFs and worksheets containing space for all relevant observations.

The responsibility for choice of a suitable investigator is that of the company clinical pharmacologist in association with clinical and preclinical colleagues. The investigator may be chosen for scientific background, clinical or clinical pharmacology experience, or prior association with the project team. The clinical facility and research team available to the investigator will also be crucially important. The investigator should be competent and experienced in clinical pharmacology or in an appropriate area of clinical investigation. He should be reliable in his selection and screening of volunteer subjects, be prepared to follow the agreed protocol, and have an efficient communication arrangement with the study director in the event of problems or need to change the protocol. Meticulous attention to the selection and prestudy screening of volunteers cannot be too heavily stressed. Inadequate preparation in this respect can be detrimental to the study, to the volunteers, and to the survival of a new drug. He, or an appropriate deputy, should be available for clinical supervision during the study and for a specified period after the expected duration of pharmacological effect of any new drug.

The implementation of Phase I studies may occur in any of several locations, including hospital or university departments of clinical pharmacology, internal medicine, or selected medical subspecialties. Such studies may also occur in clinical pharmacology units within the pharmaceutical industry itself or in qualified contract clinical research organizations. Wherever such studies are conducted, the staff, equipment, and facilities available should be commensurate with the demanding requirements of the work being conducted. With few exceptions, the first administration of a new drug to humans should be conducted in a well-equipped unit with immediate access to appropriate medical expertise and emergency resuscitation equipment. Exceptions include taste trials of substances which are not swallowed and which are administered in subpharmacologically active doses or topically administered drugs which do not become systemically available and hence do not cause systemic effects, including systemically mediated sensitivity reactions.

IV. PHASE IIA STUDIES

Many of the principles and practices just described for Phase I studies apply to Phase IIA and so will not be repeated here. The principal difference is a shift in emphasis from safety to efficacy. Patients as well as healthy volunteers may be involved. The basic question is: Does the drug behave in the clinic as it does in animal models? Since knowledge and management of disease states are involved, investigators may be clinical specialists as well as clinical pharmacologists. Study design will depend on preclinical animal data, information gathered in Phase I, and the nature of the disease under study. Some considerations involved in planning Phase IIA trials are listed in Table 1.

Design issues unique to Phase IIA relate to the specific therapeutic application and the limited information available on the new drug. If the efficacy parameters are quantitative and objective, e.g., blood pressure or natriuresis, the investigations may proceed from single to multiple doses as in Phase I but, nevertheless, should be double-blind and placebo or active drug controlled. A common approach is to employ a crossover design with the control agent(s) used as one or two treatment alternatives versus several doses of the test drug.

Efficacy information may be obtained in healthy volunteers when the primary pharmacological activity of the test drug is the same as the therapeutic intent [2]. For example, a series of studies conducted at the Jefferson Clinical Pharmacology Unit [3,4] and elsewhere [5] in normal male volunteers defined the diuretic, natriuretic, and uricosuric properties of the racemic loop diuretic indacrinone and its individual enantiomers. The key observations were the greater natriuretic potency of the (−) enantiomer (Fig. 1) and the greater uricosuric effect of the (+) enantiomer (Fig. 2). A double-blind, multicenter, repeat dose trial conducted in Europe and also involving normal volunteers demonstrated the graded net effects of various ratios of the enantiomers on serum uric acid levels (Fig. 3) [6]. The control agents were hydrochlorothiazide and ticrynafen. A final product candidate could then be

Table 1 Considerations in the Design of Phase IIA Trials

I. *Patients:*
- Criteria for inclusion or exclusion
- Number required for significance
- Allocation to the treatments
- Frequency of interview and examination
- Standardization of environmental conditions
- Informed consent

II. *Drugs:*
- Dosage strengths and route of administration
- Frequency and time of administration
- Fixed or flexible dosage regimen
- Pharmaceutical form
- Use or exclusion of concomitant drugs
- Identification and matching of drugs to be compared
- Mode of distribution to patients
- Control of ingestion and absorption
- Duration of treatment
- Control of deviations from protocol

III. *Effects:*
- Parameters to be measured; subjective or objective
- Quantification of responses
- Frequency of observations
- Recording of observations
- Laboratory procedures
- Terms of comparison (pre- and post-drug, placebo, standard drug)
- Use of double-blind technique or not
- Experimental design
- Compilation of data
- Statistical methods to be used

selected. The translation to clinical therapeutics was assured by demonstrating the efficacy of indacrinone in hypertension [7] and edema [8].

Design considerations change when the therapeutic application involves a disorder with great natural variability or subjective efficacy parameters. An example of the former is arthritis and of the latter, mental health. In these situations, one might elect to conduct single- or multiple-dose, rising-dose trials in patients with an equal group of patients on placebo carried in parallel. Assignment to treatments is randomized and the studies are conducted double-blind. Quantitation is achieved by the use of appropriate rating scales. For example, a

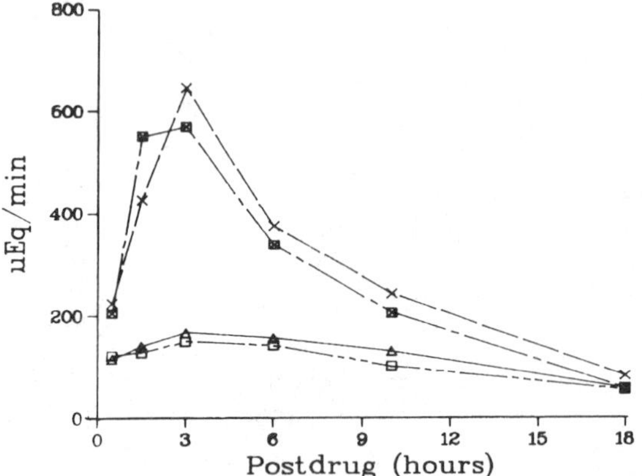

Figure 1 Time course of mean urinary sodium excretion in study A after single doses of indacrinone (×) 20 mg (n = 10), its (+) enantiomer (□) 10 mg (n = 5), its (−) enantiomer (⊠) 10 mg (n − 5), and (△) placebo (n − 10). [From P. H. Vlasses et al., Pharmacology of enantiomers and (−) p-OH metabolite of indacrinone. *Clin. Pharmacol. Ther.* *29:*798-807 (1981).]

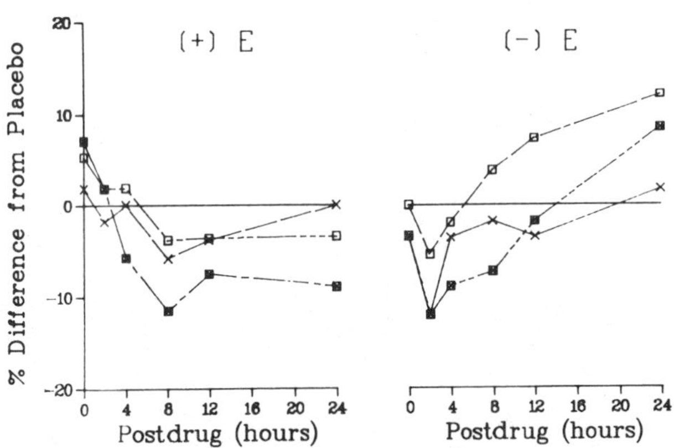

Figure 2 Time course of mean serum uric acid concentration (percent difference from placebo) after 10 (×), 20 (□), and 40 (⊠) mg of either the (+) enantiomer (+E) or (−) enantiomer (−E) of indacrinone (study A). For the 10 and 40 mb values, n = 5; for the 20 mg values, n = 10. [From P. H. Vlasses et al., Pharmacology of enantiomers and (−) p-OH metabolite of indacrinone. *Clin. Pharmacol. Ther.* *29:*798-807 (1981).]

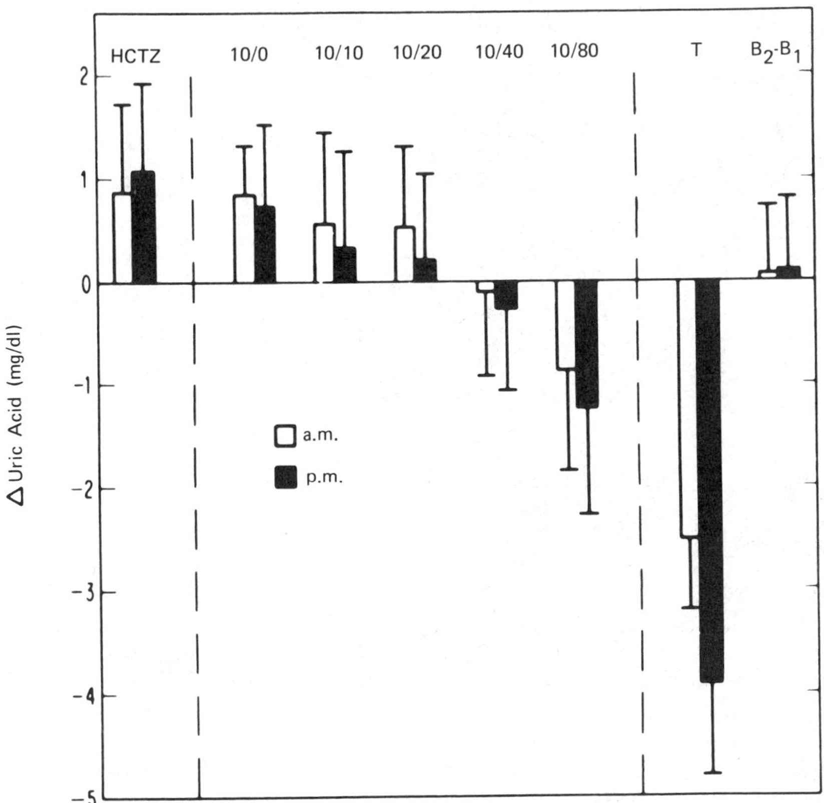

Figure 3 Change in plasma uric acid ($\bar{x}$ ± SD) at day 7. HCTZ 50 mg (n = 18, morning; n = 17, evening), I(−)/(+) 10/0 (n =18), I(−)/(+) 10/10 (n = 18), I(−)/(+) 10/20 (n = 16), I(−)/(+) 10/40 (n = 18), I(−)/(+) 10/80 (n = 18), and T 250 mg (n = 18). $B_2 - B_1$ (n = 63) is the change in pretreatment plasma uric acid from period 1 to period 2 and is an indication of plasma uric acid changes due to random fluctuation. [From J. A. Tobert et al., Enhancement of uricosuric properties of indacrinone by manipulation of the enantiomer ratio. *Clin. Pharmacol. Ther. 29:* 344-350 (1981).]

rating scale designed by Norris [9] for the estimation of subjective central nervous system reponses can be used in normal volunteer as well as patient studies. In this procedure the subject marks an ungraduated 10-cm line to express his feelings on each of the 16 items described in Table 2. The 16 items represent four categories of effects; thus, the changes in the subject's ratings during investigation under double-blind conditions provides a qualitative and quantitative estimate of the

Table 2 Rating Battery for Estimating Subjects' Feelings

1.	Alert	—Drowsy	9.	Happy	—Sad
2.	Calm	—Excited	10.	Antagonistic	—Amicable
3.	Strong	—Feeble	11.	Mentally slow	—Quick-witted
4.	Muzzy	—Clear headed	12.	Tense	—Relaxed
5.	Well-coordinated	—Clumsy	13.	Attentive	—Dreamy
6.	Lethargic	—Energetic	14.	Interested	—Bored
7.	Contented	—Discontented	15.	Withdrawn	—Gregarious
8.	Troubled	—Tranquil	16.	Incompetent	—Proficient

Source: Reprinted with permission from H. Norris, the action of sedatives on brainstem oculomotor systems in man. *Neuropharmacology 10:*181-191 (1971). Copyright 1971 Pergamon Press, Ltd.

central nervous system effects of a new drug during early clinical trials. The analog scale system can be applied to many similar situations. In all types of efficacy trials, methods should be as standard as possible so that comparisons can be made over time and with other related drugs.

Although estimation of efficacy has been emphasized thus far, Phase IIA studies also extend the safety and tolerability information gained in Phase I. In addition to clinical observations, laboratory safety data are collected at regular intervals before, during and after drug administration. Testing of the drug is now made more rigorous with its administration to ill persons. Proper controls must be utilized to avoid an unwarranted issue of toxicity. A special aspect of safety is the determination of drug disposition in individuals with compromised renal and liver function. Since drugs and their metabolites are eliminated by the liver and/or kidneys, it is important to identify the potential for cumulation and toxicity in such patients. Proper precautions may then be exercised in future studies and in clinical use.

Although the initial phases of human drug trials were discussed here separately, they are usually integrated in planning and carried out sequentially. A Phase I-IIA program recently conducted by the Merck Clinical Pharmacology group with a new antihypertensive agent serves as an example of this. The preclinical research laboratories had identified three related compounds which were potent inhibitors of the enzyme that converts the inactive decapeptide angiotensin I to the vasoactive, aldosterone-releasing octapeptide angiotensin II [10]. Another converting enzyme inhibitor developed by Squibb & Sons, Inc., captopril, had been shown to be an effective antihypertensive agent [11] but subject to potentially serious side effects [12]. The Merck drug candidates were devoid of the mercapto group present in the captopril molecule thought to be responsible for much of the side effect profile. The assignment to the clinical pharmacology staff was to identify the most promising of the three candidates and define its activity and safety.

A study was arranged to compare single doses of each drug in 20 normal volunteers [13]. The initial dose of 1.25 mg was substantially lower than the safe doses in two species of animals and less than that expected to be fully active. The basic test was the pressor response to exogenously administered angiotensin I. In addition, changes in plasma renin activity, aldosterone, angiotensin II, and converting enzyme activity, as well as safety and tolerability, were evaluated. Proceeding carefully, it was possible to explore four doses of each agent, permitting the construction of dose-response curves (Fig. 4). As information became available, the relative bioavailabilities of the two best drugs were compared in the same center [14]. Twelve volunteers received a single dose of each drug at weekly intervals. Serum and urine samples were obtained and sent to the Merck Sharp & Dohme Research Laboratories for drug level assay. Biochemical measurements related to the renin angiotensin system were also made. The results of these studies made it possible to select a candidate for full development which was designated as MK-421. The first study in the United States was a dose-ranging investigation in hypertensive patients conducted at the Merck-sponsored Jefferson Clinical Pharmacology Unit [15]. The study provided information on the relationships among dose, blood pressure lowering (Fig. 5), plasma levels of active drug (Fig. 6), and the biochemical consequences of converting enzyme inhibition (Fig. 7). Subsequent studies were reported on such issues as intravenous (i.v.) activity, i.v./oral pharmacokinetics, multiple-dose efficacy, combined therapy with a diuretic dosage regimen, bioequivalence of formulations, and mechanism of action. Simultaneously, the Merck clinical pharmacology group placed studies elsewhere to determine material balance with radiolabeled drug, multiple-dose pharmacokinetics, pharmacokinetic and pharmacodynamic interaction with a diuretic, effect of and on renal and hepatic function, hemodynamic and metabolic responses, and interactions with a number of drugs. Small-scale, multicenter trials in hypertensive patients in both the United States and International centers were arranged to confirm the relevance of the clinical pharmacology data to the clinical setting. Sufficient information was generated to permit the initiation of the Phase IIB-III program approximately 1 year after the primary candidate was selected. Clinical pharmacology involvement with this and other successful drugs continues through Phase III and long after marketing.

It is not always possible to employ placebo controls in Phase IIA studies. For example, it would generally be considered unethical to design placebo-controlled studies with antibiotics. Although preclinical microbiology and Phase I bioavailability data provide a reasonable indication of probable efficacy, factors such as the effect of disease on

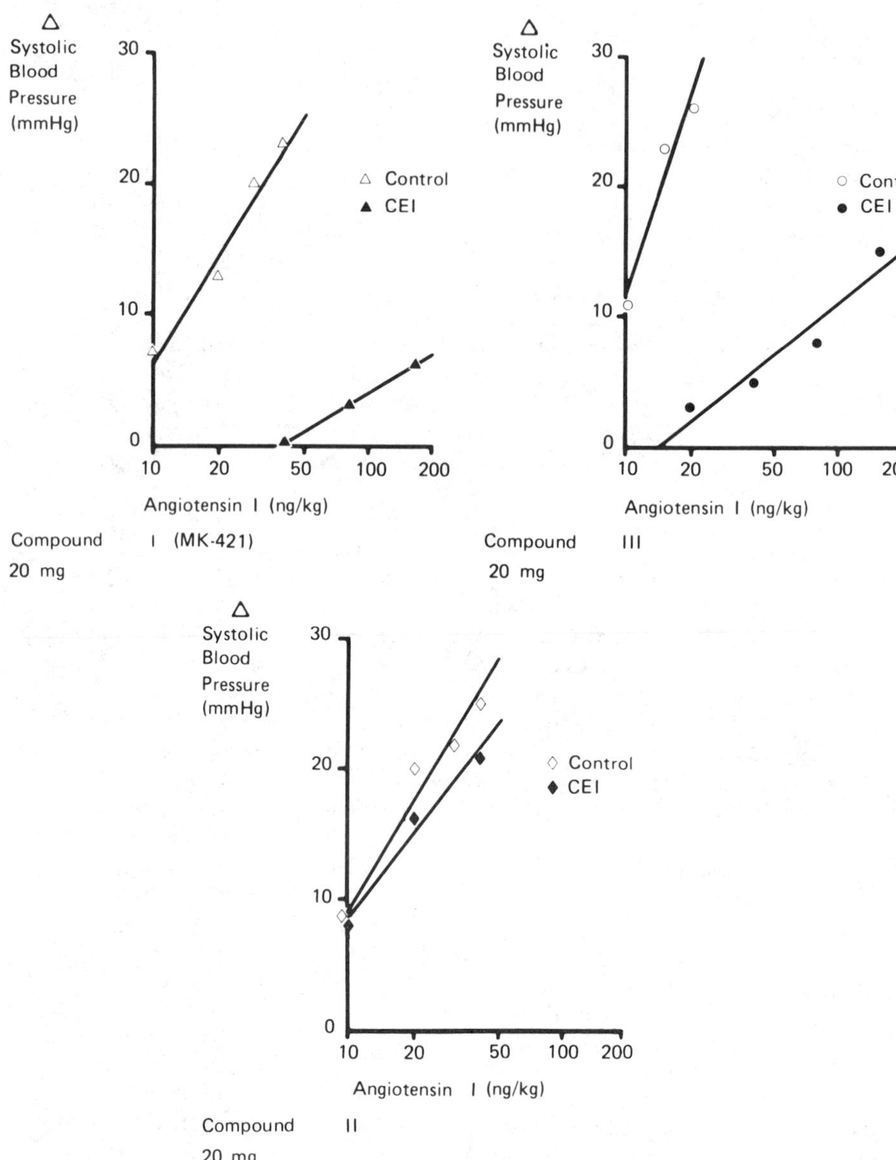

Figure 4 Dose-systolic blood pressure response curves to angiotensin I before and 3 to 4 h after 20 mg of one of the three converting enzyme inhibitors to one subject by mouth. [From J. Biollaz et al., Three new long-acting converting enzyme inhibitors: Relationship between plasma converting enzyme activity and response to angiotensin I. *Clin. Pharmacol. Ther. 29*:665-670 (1981).]

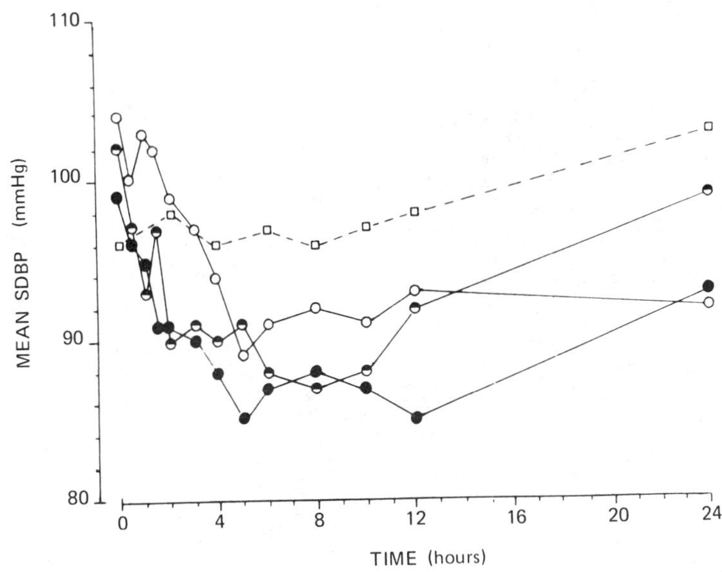

Figure 5 Mean SDBP at various times after placebo (□), and on alternate days, after the 2.5(○), 5(●), and 10(◒) mg doses of enalapril maleate (MK-421) during the inpatient study. (Source: Ref. 15.)

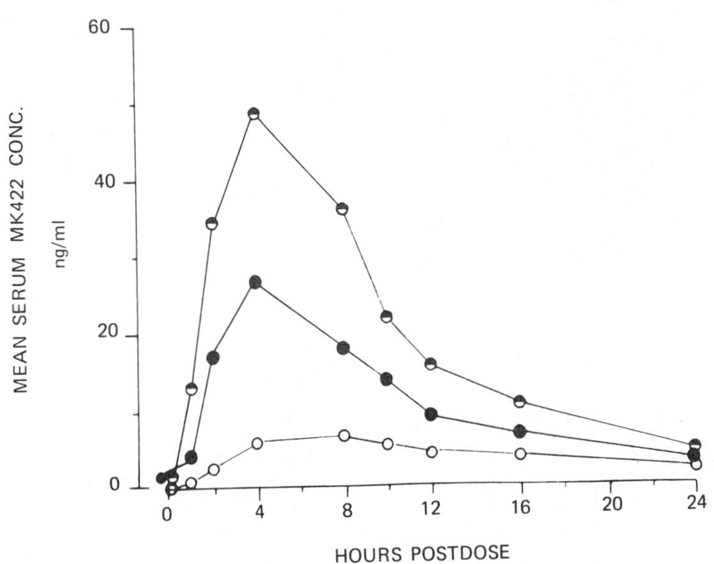

Figure 6 Mean serum MK-422 concentrations at various times after the 2.5(○), 5(●), and 10(◒) mg doses of enalapril maleate (MK-421) during the inpatient study. Mean AUC^{0-24} and peak concentrations were significantly ($p < 0.01$) different among doses. (Source: Ref. 15.)

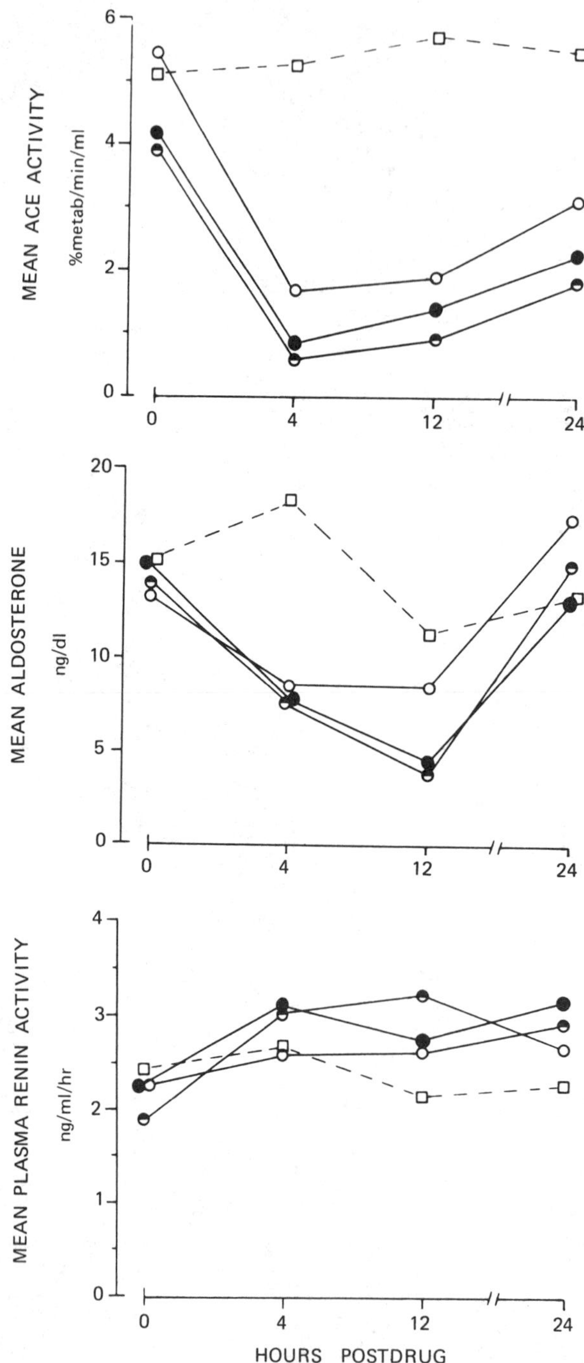

Figure 7 Mean ACE, PAC, and PRA at selected times after placebo (□), the 2.5(○), 5(●), and 10(◓) mg doses of enalapril maleate (MK-421) during the inpatient study. (Source: Ref. 15.)

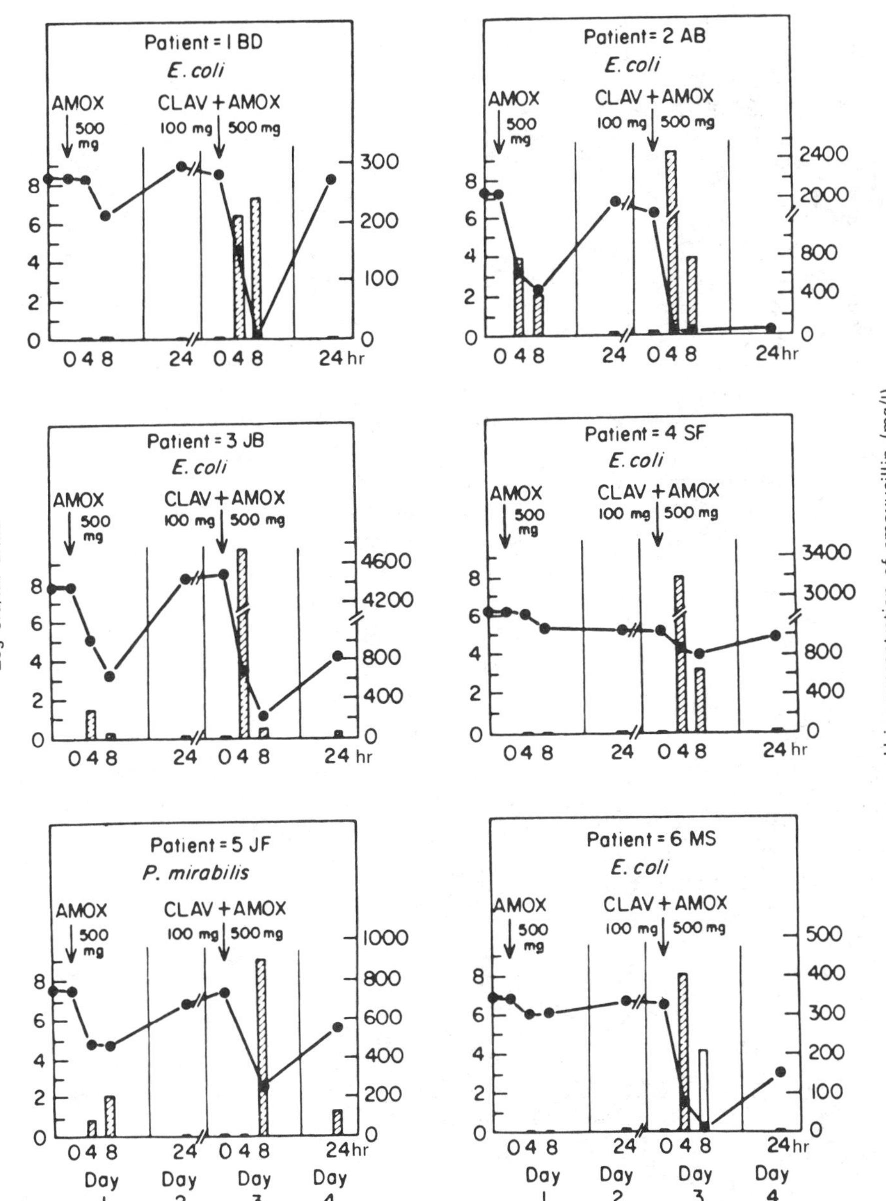

Figure 8 Changes in the bacterial count (line) and amoxicillin concentration (hatched bars) in the urine of six patients, treated consecutively with amoxicillin and amoxicillin plus clavulanic acid. (Source: Ref. 16.)

absorption and elimination processes and penetration of antibiotics to the site of infection are important and can be investigated only in patients. The nature of the infection to be treated will be determined by the antibacterial spectrum of the antibiotic. A useful early assessment may be achieved with quantitative bacteriological colony counts in treated subjects with asymptomatic bacteriuria. This may be followed by similar studies in patients with more severe urinary tract infections. The results of such a quantitative study are shown in Fig. 8 [16]. Careful sequential urinary colony counts with simultaneous measurement of urinary and serum concentration of antibiotic provide a useful means of determining the dose-response and true profile of antibiotic activity in the diseased state. With prudent choice of sample timing, which should extend beyond the expected duration of antibiotic effectiveness, useful information can be obtained by which to propose treatment regimens in different types of infection. Infections at other sites are seldom amenable to such investigation, though similar designs using fluid from tissue cages or from superficial skin blisters have been used to study penetration of antibiotics. These are usually conducted in normal volunteers, however, and early evidence of effectiveness in tissue infections is best determined by carefully monitored therapeutic studies using dosage regimens derived from microbiology and bioavailability data.

V. CONCLUSION

In summary, clinical pharmacology in the pharmaceutical industry is responsible for the conduct of Phase I and IIA investigations with two principal objectives: (1) the generation of adequate information to permit a decision on the future course of new drug candidates, and (2) the generation of adequate information to proceed into Phase IIB-III programs if the decision is affirmative. These obligations require expertise in medicine, including appropriate specialties, experimental design, biostatistics, data management, drug metabolism, pharmacokinetics, and governmental regulations. Studies are conducted in-house, in a sponsored academic unit, and/or in a qualified medical center. In the modern drug development environment, clinical pharmacology issues are extensive and complex; therefore, an integrated worldwide approach seems most efficient.

REFERENCES

1. W. L. Pines (Ed.), A primer on new drug development. *FDA Consumer 7:* 13 (Dec. 1973/Jan. 1974).
2. W. B. Abrams, The generation of efficacy data during initial clinical trials. In *Principles and Techniques of Human Research and*

Therapeutics (F. G. McMahon, Ed.), Vol. 1. Futura, Mount Kisco, N.Y., 1974, p. 203.

3. J. D. Irvin, P. H. Vlasses, P. B. Huber, R. K. Ferguson, J. J. Schrogie, and R. O. Davies, Comparison of oral indacrinone with furosemide. *Clin. Pharmacol. Ther. 28:*376-383 (1980).
4. P. H. Vlasses, J. D. Irvin, P. B. Huber, R. B. Lee, R. K. Ferguson, J. J. Schrogie, A. G. Zacchei, R. O. Davies, and W. B. Abrams, Pharmacology of enantiomers and (−) p-OH metabolite of indacrinone. *Clin. Pharmacol. Ther. 29:*798-807 (1981).
5. B. T. Emmerson, L. Thompson, and K. Mitchell, Uricosuric response to increasing doses of the indanone diuretic MK-196 in healthy volunteers. *Nephron 23*(Suppl. 1):38-40 (1979).
6. J. A. Tobert, V. J. Cirillo, G. Hitzenberger, I. James, J. Pryor, T. Cook, A. Buntinx, I. B. Holmes, and P. M. Lutterbeck, Enhancement of uricosuric properties of indacrinone by manipulation of the enantiomer ratio. *Clin. Pharmacol. Ther. 29:*344-350 (1981).
7. J. A. Vedin, C. E. Wilhelmsson, C. Moerlin, P. Lund-Johansen, C. Vorburger, W. Enenkel, P. M. Lutterbeck, V. J. Cirillo, J. Bolognese, and K. F. Tempero, A double-blind comparison of a novel indacrinone diuretic (MK-196) with hydrochlorothiazide in the treatment of essential hypertension. *Br. J. Clin. Pharmacol. 8:*261-266 (1979).
8. W. LaCorte, J. D. Irvin, A. K. Jain, P. B. Huber, J. R. Ryan, J. J. Schrogie, R. O. Davies, and F. G. McMahon, MK-196 in congestive heart failure. *Clin. Pharmacol. Ther. 25:*233 (1979).
9. H. Norris, The action of sedatives on brain stem oculomotor systems in man. *Neuropharmacology 10:*181-191 (1971).
10. A. A. Patchett, E. Harris, E. W. Tristram, M. J. Wyvratt, M. T. Wu, D. Taub, E. R. Peterson, T. J. Ikeler, J. tenBroeke, L. G. Payne, D. L. Ondeyka, E. D. Thorsett, W. J. Greenlee, N. W. Lohr, R. D. Hoffsommer, H. Joshua, W. V. Ruyle, J. W. Rothrock, S. D. Aster, A. L. Maycock, F. M. Robinson, R. Hirschmann, C. S. Sweet, E. H. Ulm, D. M. Gross, T. C. Vassil, and C. A. Stone, A new class of angiotensin converting enzyme inhibitors. *Nature 288:*280-283 (1980).
11. R. K. Ferguson, H. R. Brunner, G. A. Turini, H. Gavras, and D. N. McKinstry, A specific orally active inhibitor of angiotensin converting enzyme in man. *Lancet i:*775-778 (1977).
12. Editorial: Inhibitors of angiotensin I converting enzyme for treating hypertension. *Br. Med. J. 281:*630-631 (1980).
13. J. Biollaz, M. Burnier, G. A. Turini, D. B. Brunner, M. Prochet, H. J. Gomez, K. H. Jones, F. Ferber, W. B. Abrams, H. Gavras, and H. R. Brunner, Three new long-acting converting enzyme inhibitors: Relationship between plasma converting enzyme activity and response to angiotensin I. *Clin. Pharmacol. Ther. 29:* 655-670 (1981).

14. E. H. Ulm, M. Hichens, H. J. Gomez, A. E. Till, E. Hand, T. C. Vassil, J. Biollaz, H. R. Brunner, and J. L. Schelling, Enalapril maleate and a lysine analog (MK-521): Disposition in man. *Br. J. Clin. Pharmacol. 14:*357-362 (1982).
15. R. K. Ferguson, P. H. Vlasses, B. N. Swanson, P. Mojaverian, J. Hichens, J. D. Irvin, and P. B. Huber, effect of enalapril, a new converting enzyme inhibitor, in hypertension. *Clin. Pharmacol. Ther. 32:*48-53 (1982).
16. F. W. Goldstein, M. D. Kitzis, and J. F. Acar, Effect of clavulanic acid and amoxycillin formulation against β-lactamase producing Gram-negative bacteria in urinary tract infections. *J. Antimicrob. Chemother. 5:*705-709 (1979).

11
CLINICAL RESEARCH

Allen E. Cato and Linda Cook
Burroughs Wellcome Co.
Research Triangle Park, North Carolina

I. STAFFING A DEPARTMENT OF CLINICAL RESEARCH

Medical departments in pharmaceutical companies have traditionally been comprised mostly of physicians. Prior to 1962, the primary function of the medical department in industry was answering inquiries from practioners regarding clinical questions about a firm's products. Research, such as it existed, mainly involved clinical experience by investigators outside the company, utilizing simple research designs. Company physicians had rarely received or needed research training, as their involvement in research was minimal.

In the 1960s, the role of the pharmaceutical M.D. necessarily changed, due to at least two factors: (1) the 1962 Kefauver-Harris Amendment of the U.S. Food and Drug Administration (FDA) regulations requiring proof of efficacy as well as safety, and (2) the increasing complexity and rigor of clinical research itself. In the ensuing years, clinical research has become increasingly more rigorous while the practice of medicine has become increasingly complex, largely due to scientific advances. Hence, physicians in industry now must be not only thoroughly schooled in medicine but must have significant scientific training as well.

Currently, most medical research departments consist primarily (in numbers) of M.D.'s, with supporting help from Ph.D.'s and/or other scientifically trained non-doctorate-level individuals. Groups or sections within the department are almost always headed by physicians, who then must ultimately handle scientific, medical, and administrative decisions.

The complexity of research and of medicine today is such that one individual can hardly be expected to know thoroughly either of the two subjects, much less both areas. Furthermore, in most companies, the physicians's role involves no direct contact with patients. The physician is at risk then, of progressively losing the very skills for which he or she was hired.

Recognizing this dilemma some years ago, the decision was made at Burroughs Wellcome to restructure the Department of Clinical Research. It was realized that, of the total work involved in proceeding from an Investigational New Drug (IND) to a New Drug Application (NDA), a relatively small amount *requires* an M.D. The rest of the work can be done equally well, and in many cases better, by individuals with training other than that required for clinicians. A simple reading test demonstrates the varying orientation of individuals with different educational backgrounds:

> *Instructions:* Read this sentence only once, and count the number of f's.
>
> "Fewer Fatal Failures are the Result of Years of Scientific Study Combined With the Experience of Years."

The test has been administered to a number of potential job candidates—M.D.'s, Ph.D.'s, secretaries, and others. Without fail, scientists and clinicians record 3 or 4 F's in the sentence. Secretaries generally score at or near the correct answer: 7. It is nearly impossible for an individual highly trained in science or medicine to look at a sentence and fail to read for content. In so doing, the f in "of" cannot be seen.

By forming groups of individuals with varying backgrounds, the opportunity exists for combining multiple but different strengths into the complex task of drug research. A key member of the team is the physician. That part of the job which requires medical training is crucial and demands input from well-trained, up-to-date clinicians.

In order to attract highly qualified physicians and to allow physicians to maintain the clinical and/or laboratory skills for which they were hired, Burroughs Wellcome Co. encourages each physician to spend up to 40% of the week at one of the medical schools nearby (Duke or the University of North Carolina). The physicians's activities at the university are entirely dependent upon his own training, the needs of the university, and the department head to which he reports. There is no requirement made by Burroughs Wellcome in terms of defining activities at the university, except that remuneration may not be received. It is expected, and indeed required at the time of hiring, that training be sufficient to enable each company physician to secure an academic appointment at one of the universities. For example, each week one physician spends a day at Duke University seeing patients with pediatric pulmonary disease and another day at the University of North Carolina teaching a course in the School of Pharmacy.

Since physicians spend 40% of the work week at the university, day-to-day administrative chores must be performed by other personnel. Also, clinical research from the medical department of a pharmaceutical company involves as much or more science than medicine. Thus, the staff comprises primarily Ph.D.'s, Pharm.D.'s, and individuals with M.S. or B.S. degrees in pharmacy or one of the allied sciences. The interdisciplinary backgrounds of these individuals allow each person to contribute something special to the complex subjects of study design, execution, analysis, and final reports. Scientists combine with clinicians to forge a symbiotic relationship in conducting clinical drug research.

A. Structure of a Section

The structure of a section is illustrated in Fig. 1. Each section comprises a section head and other levels including senior clinical research scientist (SCRS) I and II, clinical research scientist (CRS) I and II, clinical research associate (CRA) I and II, clinical associate (CA) I and

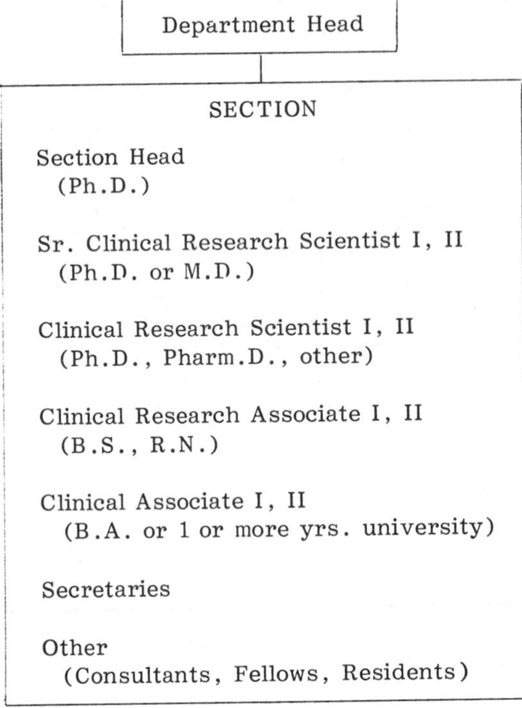

Figure 1 Structure of a section.

II, and secretaries. Also included may be physician consultants and a pharmacy fellow or resident.

B. Description of Positions

The section head is usually a Ph.D., and in most instances he serves as project leader of an investigational drug. In addition, he supervises other personnel and projects, attends to administrative matters, and serves as liaison between the section and the department head.

Senior clinical research scientists are usually M.D.'s or Ph.D.'s, and in many instances are project leaders of developmental drugs. The SCRS performs at the primary level of direct responsibility under the general direction of the section head. He is the senior person in the overall developmental plans for a new drug and has responsibility for the conceptualization of clinical study designs. Also, the SCRS establishes and maintains liaison with the clinical investigators with respect to clinical study planning.

The clinical research scientist may have a Ph.D., Pharm.D., or other scientific degree. The CRS performs at the secondary level, under the general direction of the SCRS, the backup functions of clinical study initiation and execution. He also directs support of marketed products. Responsibilities include review and critique of clinical study protocols, design of data collection forms, and other functions relating to study execution.

The clinical research associate usually is an inexperienced person with a B.S., pharmacy degree, R.N., or other special scientific medical training. The CRA is involved in the technical aspects of the design and execution of clinical studies and associated data analysis functions. He also assists in support of marketed products. The CRA usually assists, at different levels, in all general functions of drug development. The scope of activities includes initial drafting of protocols and medical reports, and monitoring of studies once they are in progress.

The clinical associate position is an entry level job for an inexperienced individual with a B.A. or other nonscientific degree. Alternatively, it may be an advanced level position for an experienced but non-degreed person. However, one or more years of college are required. The CA is engaged in multifacted activities and serves as technical and administrative assistant.

Entry level for any of the various positions is dependent on a combination of factors such as amount of experience and the degree held. Also, at any given position, there is usually an overlap of individuals with varying degrees (Ph.D., M.S., B.S.) since at all levels similar functions are performed, but the degree of responsibility varies for different positions. However, individuals holding higher degrees will usually advance to upper levels more quickly.

The department is categorized according to body function rather than by drug function, e.g., "Respiratory" rather than "Anti'infective." Obviously, an anti-infective compound would usually have efficacy in other than respiratory diseases. Compounds to be studied in more than one area are assigned to the section in which the drug's primary use is thought to lie. Studies for secondary indications jointly involve: (1) individuals with general knowledge of the drug from the section to which the drug is primarily assigned, and (2) individuals with knowledge of the disease to be studied from the section involving the secondary indication.

C. Clinical Pharmacodynamics Section

One section, the Clinical Pharmacodynamics (CP) Section, has three special functions which distinguish it from other sections such as Cardiovascular, Respiratory, Psychiatry, Neurology, and Anesthesiology:

1. The CP Section is responsible for the planning of all pharmacokinetic and metabolism studies required for the development of

a drug. All pharmacokinetic study protocols either originate or are reviewed in this section. The actual execution of the studies may occur solely in the section or jointly with another section. depending on the interests, ability, and availability of personnel in other sections.

2. Drugs which are likely to have primary indications involving more than one section would be assigned to the CP Section. For example, prostacyclin, having vasodilator and platelet aggregation inhibiting properties, might involve the Cardiovascular Section (hypertension), Respiratory Section (pulmonary embolism or pulmonary hypertension), and Neurology/Anesthesiology Section (stroke). Potential drugs targeted for uses which fit none of the other sections might also be assigned here (e.g., antidiarrheal).
3. The section also serves as a manpower resource pool for the other sections. Pharmaceutical clinical drug research suffers from the "inverse funnel" phenomenon. Ordinarily, during Phase I and early Phase II, studies are few in number and must be conducted sequentially. Hence, usually only a few individuals are assigned to a drug, since increasing the number of workers does little to speed the progress. Once efficacy is defined, however, multiple studies usually begin, and by Phase III the number of clinical research personnel assigned may become rate limiting to drug development. Even if more workers are hired, it usually takes months or years to train them. Additionally, the investigation of a drug may be suddenly terminated, for example, by uncovering unexpected toxicity, possibly resulting in highly trained excess manpower. Thus, the CP Section supplies flexible generalists to work with the specialists in other sections where they are most needed at any given time.

All three functions of the CP Section, then, involve individuals capable of working in other sections as members of a team.

D. Pharmacists and Pharmacy Training

Because of combined pharmaceutical and clinical training, pharmacists with postgraduate training have a background particularly applicable to industrial clinical research. Such individuals are particularly valuable in the CP Section with its multidisciplinary approach, but at Burroughs Wellcome pharmacists are employed in every section. At present there are nine pharmacists with advanced degrees in the Department of Clinical Research, two of whom currently are section heads.

The similarity of interests and activities of clinical pharmacists at Burroughs Wellcome to those at the University of North Carolina School of Pharmacy has led to close ties between the two institutions. Most of the pharmacists (and some nonpharmacists) at Burroughs Wellcome are on the School of Pharmacy faculty and are involved in teaching activities. A graduate level course entitled "Clinical Development of Drugs" is taught entirely by Burroughs Wellcome staff.

Further interaction with the School of Pharmacy has occurred through the residency rotations in conjunction with the Pharmacy Master of Science degree program [1]. Residents spend 3 months in the Department of Clinical Research gaining knowledge and experience in the methodology and execution of clinical trials. This additional research exposure during training has aided clinical pharmacists in choosing career pathways, and a number of residents have subsequently joined industrial clinical research teams.

As an expanded endeavor of the residency rotations, a 2-year novel fellowship program has begun, jointly sponsored by the School of Pharmacy and Burroughs Wellcome. The program is being offered to provide specialized training both in drug development and clinical research, with training time shared between the two institutions. Candidates are selected for the program on a nationally competitive basis. Applicants must have completed advanced professional studies and a residency in clinical pharmacy.

The general objective of the new 2-year fellowship is to provide to the postgraduate student (M.S., Pharm.D., or Ph.D.) specialty training in (1) academic clinical drug research from the perspective of an investigator, and (2) drug development research from a company's viewpoint as sponsor of clinical investigations.

At the School of Pharmacy, the Fellow works with faculty members from the Division of Pharmacy Practice in current clinical research projects. During the 2-year period the Fellow is expected to conceive, design, and complete an individual research project of adequate scope that the work is publishable in an appropriate journal.

At Burroughs Wellcome, the Fellow can select rotation from a variety of specialty areas of drug development as well as pharmacokinetics, pharmacodynamics, and bioavailability. The Fellow is assigned "real" tasks such as coauthoring an investigator's brochure for a compound beginning clinical trials, writing a protocol for Phase I-IV studies, and coauthoring final medical study reports. Along with clinical scientists from the appropriate section, the Fellow monitors clinical studies at the study site.

The Fellow is enabled to see both sides of clinical research and its commitments and obligations through the simultaneous but contrasting experiences provided (1) at the industry site (where the Fellow learns of the obligations of a company sponsoring a drug's development), and (2) through the work at the School of Pharmacy and the clinical research unit (where the Fellow performs as the investigator).

II. OVERALL CLINICAL DEVELOPMENT PLANS

When it has been demonstrated for a new compound that potential benefits outweigh risks, based on data from preclinical studies in toxicology, pharmacology, and metabolism, clinical trials can begin. Only then is exposure of the compound to humans warranted. The subsequent clinical development of a potential new therapeutic entity is arbitrarily divided into three phases, each representing a stage of progress toward filing an NDA.

A. Phase I

Phase I studies are designed to determine the safety and tolerance of a new compound and to gather pharmacokinetic and bioavailability data upon which to plan future studies. The initial Phase I is designed usually for single doses to be given in increasing amounts until a pharmacological effect or toxicity is reached. The results are subsequently used to plan a multiple-dose trial [2].

Generally, normal volunteers (some 20-40 on drug) are utilized. However, for some indications and when drugs are known to be toxic (as in anticancer agents), patients are used in the initial trials. In either case, volunteers are usually incarcerated to allow maximum observation for safety.

In the report, discussion of Phase I trials is included under the heading "Clinical Pharmacology," although some firms (such as Burroughs Wellcome) include all phases of clinical trials under the heading "Clinical Research."

B. Phase II

Once an acceptable margin of safety is established for a new compound from Phase I trials, Phase II begins. The initial Phase II study is defined as the first study in which measurements will be made in patients to determine efficacy. Clearly, then, the dividing line between Phase I and Phase II is often nebulous and arbitrarily delineated.

The major objectives of development at this stage are (1) to determine if the drug is effective for one or more indications, and (2) to determine if the drug is safe by close monitoring of the incidence and severity of adverse reactions. An appropriate dosing schedule is determined, and further data are gathered in patients instead of normal volunteers on pharmacological, pharmacokinetic, and metabolic profiles.

Phase II studies are usually conducted in closely monitored patients (some 100-200 on drug). Open, uncontrolled trials may be utilized in early Phase II. Open trials are employed in order to make necessary changes in optimum dosing range, patient selection, and other factors which are not easily varied in more rigorous study designs.

Inpatients or outpatients may be used, depending on the drug under investigation and the targeted disease. Patients should ordinarily be free of hematological, renal, hepatic, cardiac, or other serious diseases and receive no concomitant therapy if possible.

At this stage, many problems are faced for the first time: criteria for patient selection; qualified and suitable investigator selection; sample size; sampling procedures; data analysis (dropouts, withdrawals, incomplete data); concomitant drugs; data collection and transmission; and multicenter trials.

Later in Phase II consideration of all of these factors is important in designing the pivotal clinical trials, i.e., those trials which determine and define the therapeutic value of a new drug. Double-blind, placebo-controlled trials are usually required unless ethical or practical considerations dictate otherwise. At the end of Phase II, sufficient data should have been generated to allow a decision regarding further large-scale testing of the new drug. Increasingly, the decision is made in conjunction with the FDA at an "end of Phase II" meeting.

Thus, Phase II studies are the crucial stage in a drug's development. It is wise to contact experts in the disease categories for which the drug is being developed. Their advice in planning the studies and in evaluation of the drug's effects on the disease process should be incorporated at an early stage.

C. Phase III

Once the overall safety and effectiveness of a drug has been clearly established from early clinical trials, Phase III begins with expansion

Table 1 Regulations on Foreign Clinical Trials

1. Well-qualified investigators; adequate facilities
2. Ethical conduct according to the Declaration of Helsinki *or* the regulations of the country, whichever offers greater protection
3. Institutional Review Board (IRB) approval, *or* study conducted according to standards of the country
4. Although desirable, IRB need not contain nonscientists
5. Adequate foreign clinical data may be substituted for Phase I and Phase II studies in the United States

Source: M. J. Finkel, Acceptance of non-US data by the FDA: current position. Presented at American Society for Clinical Pharmacology and Therapeutics, New Orleans, March 1981.

Table 2 Analysis of Certain[a] NDAs Approved in 1974 Through 1980 for the Presence and Significance of Foreign Clinical Studies

Number of NDAs analyzed	NDAs containing foreign studies (%)	NDAs containing foreign studies considered significant and/or pivotal for approval (%)
166	84 (51%)	29 (35%)

[a]Included are all new molecular entities, salts or esters of new molecular entities and major new indications.
Source: M. J. Finkel, Acceptance of non-US data by the FDA: Current position. Presented at American Society for Clinical Pharmacology and Therapeutics, New Orleans, March 1981.

of controlled and uncontrolled trials (usually 600 to 1,500 patients on drug). Information to be gathered during this stage includes additional evidence of safety and efficacy, long-term tolerance, drug interactions, comparisons with existing drugs already marketed, abuse potential, and variations in dosing regimens.

D. Foreign Clinical Trials

In developing an overall study plan, contribution of data from foreign clinical trials should be considered. As long as some specific criteria are met (Table 1) and studies are well designed, data from foreign clinical trials are increasingly being used in the United States to support NDAs (Table 2). In a significant number of cases, the foreign clinical data were crucial to subsequent approval. Utilizing foreign investigators might allow a time savings for some projects in reaching NDA, as well as saving money, as foreign trials cost considerably less than comparable U.S. trials.

At the same time, other aspects of the development of the drug must be considered. Some of these aspects include bulk manufacturing for market, formulation problems, stability studies, packaging, labeling, storage, package inserts, advertising, and preparations for NDA submission.

III. SELECTION OF AN INVESTIGATOR

Choosing the correct investigator(s) for a given study would seem to be a relatively simple part of clinical research. However, a mistake made during selection of an investigator sets a course for ultimate disaster before a study is completed.

Too often the investigator is selected for the wrong reason(s). Not infrequently, management may force an investigator upon a clinical monitor because of past favors or obligations, marketing considerations, desire for a "name in the speciality," or budgetary factors. In fact, specific requisites concerning selection of an investigator should be thoroughly reviewed before a clinical study is mutually agreed upon by the sponsor and the investigator (Table 3). Some of the factors listed are required by the U.S. Food and Drug Administration (FDA) [2], some involve common sense, and some have evolved through trial and considerable error. At Burroughs Wellcome, the general research policies with respect to company-investigator interactions are prospectively explained to potential investigators both verbally and in writing. This is particularly important because in many cases the final decision relative to selection of investigational sites is competitive and potential investigators need to know the rules.

A. Qualifications of the Primary Investigator

The primary investigator must be qualified in the field in which the study is to be conducted [3]. FDA inspections have occurred when a psychiatrist was routinely doing studies of low back pain and a pediatrician was studying oral contraceptives [4]. It is particularly helpful if an investigator has a known track record in performing clinical drug research and if he is known to be reliable in delivering prompt and accurate results. However, experience has repeatedly demonstrated that experts in a given disease field are not necessarily the best investigators. Many authorities have experience in clinical research but little knowledge of experimental designs and statistical considerations that are necessary for applied pharmaceutical clinical research.

B. Qualifications of the Support Staff

For a study to be properly conducted, it is equally important that the colleagues and support staff of the primary investigator meet certain qualifications. The best investigative team may consist of a primary physician, a committed research Fellow (see Sec. I.D), and a study coordinator. The coordinator should have the authority and a sufficiently aggressive personality to coordinate the interactions among the various study personnel and the monitors from the sponsoring company. Generally, senior level medical staff have neither the time nor the inclination to execute the routine and mundane tasks necessary for a thorough and accurate study. A nurse or technician is usually required to see that the study continues on the allocated time schedule and according to protocol.

In addition, someone must be responsible to maintain up-to-date and accurate patient data collection records. Inaccurate records (i.e.,

Table 3 Factors for Selection of Investigator(s)

1. Qualifications of primary investigator
   a. Medical training
   b. Scientific expertise for disease/drug investigated
   c. Experience
2. Qualifications of colleagues and support staff
   a. Communications between monitors and study personnel (study coordinator)
   b. Protocol adherence (nurse, technician)
   c. Recordkeeping and data collection forms (study coordinator)
   d. Clinical trial material accountability (pharmacist)
3. Cooperativeness
   a. Monitor's interactions and communications with investigator
   b. Team concept
   c. Sponsoring company's philosophies
4. Patient population
   a. Numbers sufficient
   b. Availability to meet protocol criteria
5. Motivation
   a. Scientific interest
   b. Pertinent or appropriate publications
   c. Constructive criticism of protocol
6. Personality
   a. Cooperativeness
   b. Integrity
7. Time commitments
   a. Degree of delegation of authority
   b. Delineation of areas of responsibility
8. Physical facilities
   a. Ability to perform specialized studies
   b. Laboratories licensed
   c. Accessibility to transportation
9. Budget proposals
   a. Competitive
   b. Basis calculated per completed patient

containing omissions and errors) submitted to the sponsoring company can be time consuming and costly to correct and may jeopardize the study. Federal (FDA) inspectors have stated that they feel free to inspect not only records in the physician's office but hospital records, nursing home records, and various lab reports. Indeed, they may look at records reasonably prior to and subsequent to the study, as well as during study execution [4]. For instance, if a patient is said to be drug free for a month prior to the study, the validity may be checked by looking at the records of prior clinic visits. For a patient who is purported to have had a disease for 2 years, there should be data to substantiate the time of the diagnosis. Federal inspectors may also check to see whether symptomatology occurring during the study that was not present at baseline is followed until it is no longer present, returns to baseline, or is no longer clinically significant.

Usually a pharmacist should be appointed at the study site to control properly the storage, dispensation, and accountability of clinical trial material. Failure to do so can result in problems relative to study medication since few physicians are oriented toward or interested in the nuances of clinical trial material. Failure to assign an individual to be responsible for clinical trial material has resulted in numerous and serious problems. In one study of a cardiovascular medication, an investigator changed hospital affiliations within the same city. During the move, half of the clinical trial material was reported as "lost." In another example, a monitor made a site visit to an investigator not known for neatness or organization but who was cooperative and thorough in his patient care. The investigator had "stored" the clinical trial medication on top of the radiator during the winter in Boston!

It is up to the clinical monitor, then, to train the investigator in the many facets of documentation and to explain the requirements for good clinical practices, management of clinical trial material, and study design necessary for performing quality research. Conversely, no matter how thoroughly the clinical monitor has acquainted himself with the disease process under study, the input of the practicing physician who is expert in the treatment of the disease is critical. Cooperation between investigators and monitors facilitates a most rewarding experience for both based on a common interest in accurate and ethical drug development.

C. Cooperativeness

Cooperativeness is best achieved with investigators when expectations are delineated very early during interaction. There are some investigators who are not willing to enter into a cooperative interaction with clinical monitors, who do not wish to share their ideas with others whom they see as competitors, who do not wish to give up some autonomy in order to be part of a team, or who do not consider clinical monitors as experts in applied clinical research and therefore wish to dictate the terms of the study [5].

Such negative attitudes occur less frequently if the contributions and philosophies of the sponsoring company are explained during initial contacts with a potential investigator. At Burroughs Wellcome, all potential investigators are sent a letter which includes the following paragraphs:

> Because of our interest and input into good scientific principles as well as good clinical practices, Burroughs Wellcome encourages publication of accurate materials of general scientific interest. We strongly support co-authorship with our investigators. However, all abstracts and manuscripts must be reviewed and approved by Burroughs Wellcome before submission for publication. Our staff of statisticians, chemists, physicians, clinical pharmacologists and pharmacokineticists will make recommendations relative to their field of expertise and their comments should be given careful consideration by the participating authors prior to drafting a final manuscript. Our purpose is not to withhold or censor either positive or adverse data on potentially useful compounds from studies sponsored by Burroughs Wellcome, but to assure the accuracy of these reports and to maintain and not jeopardize patent rights.
>
> As required by the Food and Drug Administration, representatives at Burroughs Wellcome will visit your site periodically to aid you in adhering to these guidelines [3,6] as well as to update you and your staff on the general development of [the drug being studied]. Burroughs Wellcome and all investigators associated with a particular study will have confidential exchanges of clinical and/or scientific information when it relates to either patient safety and/or drug efficacy.

These statements allow an early understanding to be reached between the clinical monitor and the potential investigator concerning subjects which, if not dealt with early, can create considerable disagreement and disharmony. For instance, agreement on publication and authorship is best settled before the study begins. If a company representative(s) is to be included as coauthor of a publication, he must know the subject matter thoroughly and be able to make substantial donations to the design and conduct of the study.

D. Patient Population

Another factor involved in selection of investigators is the availability of an appropriate patient population in numbers sufficient to allow the study to progress at a reasonable rate [7]. It is almost axiomatic that the investigator, after reviewing his records in comparison to the inclusion and exclusion criteria of a study, will determine that his patient population is more than adequate [8]. If the clinical monitor has not checked selected patient records to determine that patient numbers are indeed sufficient to achieve the objectives set forth in the protocol in the allotted time period, it is almost certain that patient enrollment will be less than hoped for or anticipated. Experience has shown that when clinical monitors review patient records which physicians judge as meeting entry criteria, the monitor will find many instances in which these patients fail to satisfy criteria specified in the protocol. One reason for initiation of FDA inspections occurs when an investigator seems to have too many patients with a given disease for the locale or setting in which he practices.

Even when eligible patient numbers seem sufficient, there are still many reasons why the rate of acquisition of patients may not be adequate. For example, once the study begins, some patients will not agree to participate, others will be unavailable for various administrative reasons, and some may drop out of the study for different reasons (moving, not enough time, etc.). For others the disease changes, and hence some patients eligible at a given point in time may not later meet entrance or exclusion criteria.

E. Motivation

Motivation is another important factor in choosing a potential investigator. Is he mainly concerned with science or money? Scientific interest is usually reflected by significant and worthwhile contributions in the clinical literature. Constructive criticisms and suggestions of the protocol under discussion also indicate an interest in the ultimate results. Often the most motivated investigators are the younger ones not yet established in a high-powered medical school environment but who are well trained, enthusiastic, and ambitious.

F. Investigator Personality

Similarly, personality of an investigator often contributes heavily to the selection process. "Prima donna" investigators may change opinions frequently, both before and during the study. Such investigators are rarely willing to recognize the team approach—in which contributions are made by the clinical monitor as well as the investigator. The likelihood of such mutual cooperation and understanding can usually be determined from early interactions.

The integrity of the investigator must be above question. In one case a clinical monitor was treated to lunch by a potential investigator. After lunch the investigator offered the receipt to the monitor, suggesting that the monitor could then charge both meals to his company and keep the money. Such a questionable proposition might well foreshadow worse things to come—practices which could not only result in invalid data but could also trigger an FDA audit. For example, an FDA inspection may occur when the efficacy for a drug appears too good or the toxicity or adverse reactions appear too few when compared with the results of other physicians studying the same drug. In one factual case, even though the protocol required physician supervision of all experiments, complex and potentially dangerous cardiovascular tests were performed by a research nurse in the absence of any clinician. Due to the investigator's negligence and lack of judgment, the study was terminated by the sponsoring company and the investigator was not used in any subsequent trials.

G. Time Commitment

The time commitment and availability of the physician and his support staff must be such that projected dates for initiation and completion of the study not be unduly delayed. For many physicians, clinical research is a minimal part of their professional activity, most of which is spent in the management of patient care or in basic research settings. "Experts" are often overextended in their time commitments. If the investigator is conducting other studies, they may interfere with the conduct of the planned study.

The degree of delegation of authority should be carefully inspected. In one study of an antidepressant, an investigator was questioned about ratings he had assigned to patients recently evaluated. The investigator was unable to elucidate the reasons for his own ratings or even to recall the patients. Additional inquiries revealed that many of the ratings were those of the investigator's junior staff members. Further complicating the situation was an expression by the junior staff members that they frequently disagreed with the principal investigator's ratings of the patients' status.

It is not uncommon for experts to delegate primary study responsibility to junior staff members who may not have the expertise, capability, or the interest required. Furthermore, even when junior staff members are capable and interested, frequent staff changes often occur, resulting in study delays, protocol violations, and inaccurate and incomplete data collection forms.

H. Physical Facilities

A determination of the adequacy of any proposed facility to be used during the clinical investigation is also a part of selecting an investigator. Physical facilities should reflect adequate hospital and emergency equipment and any other specialized equipment required by specific techniques called for in the protocol. Laboratory facilities should preferably be licensed. Accessibility of the site to public and private means of transportation may also be a factor.

I. Budget Proposals

The proposed budget for a potential study may also play a role in the selection of an investigator. Competitive budgetary considerations may be a significant factor in deciding on a potential study site, particularly for Phase I and bioavailability studies. For one proposed study the same protocol submitted to different investigators (sites) resulted in proposed budgets differing by more than threefold and a total of $200,000.

It is best to agree upon a study budget calculated on a per completed patient basis. This method discourages physicians from loosely entering patients who ultimately are unlikely to fit protocol criteria. It may also be a motivator for the investigator to enter the maximum number of patients to complete the study. Additionally, the studies may be discontinued for a variety of reasons (e.g., the company no longer being interested in the drug; FDA intercession; inadequate rate of enrollment of patients; or information from other studies making the study design invalid). Therefore, a completed patient basis allows a fair compensation for work accomplished to the point of discontinuation.

In the final analysis the qualifications governing the selection of an investigator will be determined by the strategic and operating goals of the desired study. The responsibilities of the potential investigator and of the sponsoring company must be specifically delineated and then carefully discussed and agreed upon before an investigator is ultimately selected.

IV. INVESTIGATOR-SELECTION SITE VISIT

When the choice of investigator(s) is still under consideration, an investigator-selection site visit to each potential investigator must be made. The ensuing discussions will likely be of a general nature to permit evaluation of the factors listed in Table 3 (see Sec. III). Once these factors have been carefully considered, the investigator can be chosen. For those investigators who are turned down, it is crucial that an early understanding be reached that "many are called though few are chosen" and that reasons for refusal may be beyond the control of the clinical monitor or the investigator (e.g., company budget resources may be insufficient to meet the needs of the investigator). Otherwise, an investigator turned down for one study may not be available for the next.

V. PRESTUDY SITE VISIT

For an investigator whose credibility is established from past experience with the sponsoring company, the selection process may already

Table 4 Items to Be Reviewed with Investigator(s)

A. Investigator Manual
   1. Review of preclinical data (pharmacological activity, toxicity, etc.)
   2. Review of clinical data (efficacy/safety)

B. Obligations of investigators and monitors
   1. *Federal Register 42*(187), Sept. 27, 1977
   2. *Federal Register 43*(153), Aug. 1978

C. Clinical protocol
   1. Primary objectives of study
   2. Study design (blinding conditions, basic test groups, sample sizes, randomization procedure, duration of study, etc.)
   3. Test articles and dosage
      a. Names, dosage forms, unit dose sizes
      b. Dosage instructions (unit dose, frequency, daily dose, duration of treatment, etc.)
      c. Drug packaging and labeling as related to study design
      d. Decoding procedure for emergency conditions
      e. Accounting procedures and records for test articles
         (1) Identification of shippers/recipients
         (2) Dates of shipping/receiving—spot check of invoices and similar documents on contents of packages
         (3) Subject usage—amount dispensed, used, returned
         (4) Test article identification
      f. Instructions for return of unused/reusable test articles
      g. Storage instructions for test articles
      h. Other disposition of drug supplies by investigator/sponsor (general)
   4. Experimental procedures
      a. Review of overall time/events schedule
      b. Screening procedures—inclusion/exclusion and other admission criteria
      c. Measurement/evaluations—screening, baseline, treatment and posttreatment phases
   5. Adverse reactions
      a. Definition of serious (alarming) adverse reactions
      b. Immediate reporting of serious adverse reactions
         (1) Procedure for immediate contact between investigator and clinical monitor
         (2) Recordkeeping and adverse reaction reporting—FD 1639's
         (3) Procedure for follow-up as agreed upon by the investigator and clinical monitor

Table 4 (Continued)

---

6. Protocol modifications—review of plans and records for:
   a. Addendums to the clinical protocol made in writing
   b. Authorization of modifications (sponsor)
   c. IRB approval (if required)
7. Statistical analysis (description, plans, reporting, and schedule for interim and final analyses)
8. Consent and clearance provisions
   a. Institutional review obligations
   b. Subject consent obligations

D. Data collection forms (DCFs)
   1. Review of plans/schedule for collection/review
   2. Field edit—legibility (black ink), completeness, missing dates, documentation of study dropouts and reasons, etc.
   3. Review necessity of signature of investigator on appropriate DCFs
   4. Necessity for monitor to periodically spot check DCF entries and compare with corresponding entries in the investigator's records when possible.

E. Facilities check
   1. Availability of specialized equipment, if required
   2. Specifications of backup facilities, if required
   3. Licensure of laboratories

F. Other (general)
   1. Review general records retention obligations
   2. Review of financial agreement relative to cost of study, time and amount of partial payments, etc.
   3. Assignment of personnel at the site for form review, control, etc.
   4. Review provisions for a "Continuation Study Protocol" and patient entry to it, since some patients may want to remain on drug after completion of the study

---

be completed without an investigator-selection site visit. By whichever means the investigator is selected, a prestudy site visit is then made. The prestudy site visit is primarily for purposes of discussing the proposed study. At this point, a number of procedures relative to clinical study initiation usually will have been executed [2].

Before the visit with the investigator and his support staff, the following documents should be in the possession of the monitor or investigator so that they can be discussed at the time of the visit.

Table 5 The Most Common Deficiencies Discovered During Routine FDA Audits

Common deficiencies	Number	Percentage of 137 audits[a]
1. Failure to establish adequacy of laboratory facilities used by clinical investigator	33	24
2. Failure to maintain adequate records of drug accountability	32	23
3. Absence of standard monitoring procedures	30	22
4. Failure to review patient records	28	20
5. Failure to ensure IRB approval	24	18
6. Failure to document monitoring visits	19	14
7. Failure to visit study site before and during clinical study	13	9
8. Insufficient number of monitors who have indicated that they have compared clinical records with case reports	44	32

[a]This sample includes a small number of noncommercial sponsors.

Regarding responsibilities of the monitor:

1. FD 1572/73—signed by the clinical investigator
2. CV/publications—investigator
3. Grant proposal
4. Clinical study protocol/data collection forms (DCFs)—protocol signed by clinical investigator
5. Institutional Review Approval (document signed by the appropriate individual(s) representing the committee)
6. Patient Consent Form (sample)
7. Clinical laboratory data—normal ranges
8. Documentation that supplies of test articles were sent and received by the investigator

Regarding responsibilities of the investigator:

1. Investigator Manual and other informational material

2. A checklist review of obligations of investigators and monitors
3. Clinical study protocol/data collection forms (DCFs)
4. Method and schedule of grant payment

If any of the above procedures and documents resulting therefrom are not completed prior to the prestudy visit, they should be completed at the time of the visit.

Arrangements for the prestudy site visit should usually include prior communication with the investigator in order to ensure that all of the appropriate individuals involved in conducting the study will be present at the meeting. The preparation and forwarding of an agenda to the proposed attendees is also recommended.

A detailed list of potential items to be discussed/reviewed is given in Table 4. Table 5 lists the most common deficiencies (relative to the proposed regulations [6]) discovered during routine FDA audits of companies sponsoring clinical drug research [4]. It should be obvious from a comparison of the two tables that a thorough prestudy site visit with proper documentation can obviate most potential deficiencies.

Although a prestudy visit is not absolutely mandatory prior to the start of every clinical study, it is highly recommended in the majority of cases. It becomes mandatory for those studies of a complex nature requiring detailed preparation and direct communication between the clinical monitor and clinical investigator.

REFERENCES

1. J. H. Hull, A. W. Pittman, A. E. Cato, F. M. Eckel, and G. Cloutier, Training the "clinical scientist" through a combined industrial/academic fellowship. *Am. J. Pharm. Ed. 47:*30-34 (Spring 1983).
2. General considerations for the clinical evaluation of drugs. FDA 77-3040, Bureau of Drugs Clinical Guidelines (Sept. 1977).
3. Obligations of clinical investigators of regulated articles. *Federal Register 43:*No. 153 (Aug. 8, 1978).
4. A. B. Lisook, FDA audit of investigators and sponsors. *Drug Information J.* 97-101 (Jan./June 1982).
5. R. J. Crossley, Project planning, investigator selection and data harmonization. *Drug Information J.* 35-43 (Jan./June 1982).
6. Proposed establishment of regulations on obligations of sponsors and monitors. *Federal Register 42:*No. 187 (Sept. 27, 1977).
7. R. Peto, M. C. Pike, P. Armitage, N. E. Breslow, D. R. Cox, S. V. Howard, N. Mantel, K. McPherson, J. Peto, and P. G. Smith, Design and analysis of randomized clinical trials requiring prolonged observation of each patient. *Br. J. Cancer 34:*585-612 (1976).

8. J. A. Gerringe, Initial preparation for clinical trials. In *The Principles and Practice of Clinical Trials* (E. L. Harris and J. D. Fitzgerald, Eds.). Williams & Wilkins, Baltimore, 1970, pp. 47-65.

12
THE MONITORING PROCESS

C. L. Bendush* and F. J. Novello
Clinical Investigations Division
Lilly Research Laboratories
Eli Lilly and Company
Indianapolis, Indiana

**Present Affiliation:* E. R. Squibb and Sons, Inc., Princeton, New Jersey

One must go seek more facts, paying less attention to technics of handling the data and far more to the development and perfection of methods of obtaining them.

—From Hill [1]

I. INTRODUCTION

Modern drug development is dependent upon purposeful pursuit of scientific information which characterizes the drug and documents its usefulness. The monitoring process, i.e., project management, is a fairly recent, and now essential, addition to this pursuit. Earlier in the history of drug development clinical studies were supported by the sponsor in a haphazard fashion with very little planning or direction. The result was a data base that by today's standards was deficient and unscientific. Essentially today's monitoring process is a discipline that has resulted from the search for more effective techniques and procedures with which to manage the activities and resources necessary for efficient generation of scientifically sound drug information.

In practice, the monitoring process is a specialized management method or system by which the monitoring team regulates, adjusts, and controls the sequence of drug studies for the purpose of developing a data base that will scientifically document the basic properties and clinical usefulness of the drug under study. The monitoring process begins during the preclinical phase of drug development and continues into the postmarketing phase of the drug's life cycle, as long as the sponsor continues to collect drug-related experiences and information.

For the most part, drug development is concerned with the interactions between the drug developer or sponsor (usually a pharmaceutical company), the outside investigators, and the governing regulatory agencies. Each of the parties has a scientific as well as a societal role in the process of drug development and use. Since each tends to see his role differently, conflicting views are inevitable. While such conflicts are often counterproductive and therefore lengthen the process, little doubt remains that constructive interactions produce drugs that are scientifically supportable and useful to society.

To some extent the drug development process is complicated by the fact that each of the parties in the research effort has his own objectives and biases. The pharmaceutical innovator needs business growth and has the natural desire to demonstrate the uniqueness of any new drug. The investigators want new and interesting findings to publish while offering a new therapy to their patients. Also, both the investigators and their patients frequently expect useful results, thus producing a placebo effect. Finally, regulators, remembering unscientific studies and practices, have doubts and suspicions regarding the motives of those they are required to regulate.

Nevertheless, all three concerned parties are united by a common purpose: to produce sound ethical scientific research and present it to society in a manner consistent with good medical standards. The monitoring process is essential to fulfilling that objective; it is the glue that bonds the many elements of the scientific drug research process together.

II. BACKGROUND AND OBJECTIVES

Humanity's interest in medicine, therapy, and illness dates back to antiquity, when bias, suggestion, and religious beliefs were at the center of medical treatments. Against this background, medicines were empirically chosen and dogmatically accepted by medical practitioners for centuries. Overtones of this rich history are still evident in current approaches to illness and therapy. Indeed, the "art of medicine" which may be too empirical, frequently conflicts with a strictly scientific approach to therapy, which may be "lacking" in compassion.

Recognition of the need for a scientifically based approach to drug development dates back to the beginning of the twentieth century when most drugs were considered to be worthless. This situation was best described by William Osler and Oliver Wendell Holmes, two of the greatest contemporary physicians: Osler said, "The young physician starts off in life with twenty drugs for each disease, and the old physician ends up by having a single drug for twenty diseases!" And Holmes said, "If the whole *materia medica*, as now used, could be sunk to the bottom of the sea it would be all the better for mankind—and all the worse for the fishes." (Both quotations are from Ref. 2.)

Rapid strides have been made in the process of drug development. The complex process of today is a far cry from the unplanned efforts of yesterday. The modern monitoring process, dedicated to sound scientific research principles, requires the interaction of the sponsor, investigator, and regulator. They all contribute significantly to the four basic steps in the drug research process. The four steps are (1) planning, (2) implementation, (3) documentation, and (4) interpretation.

The initial step in the drug research process is strategic, i.e., definition of specific study objectives. The second step is tactical, which involves planning the specific steps with time estimates needed to achieve the objectives. Included in the tactical phase is organization of the monitoring team and estimates of resources necessary, i.e., drug requirements, financial support, and personnel needs. The third step, i.e., documentation, specifies the precise means and methods by which the data will be generated, collected, and organized. This involves (1) development of protocols for individual or multiclinic studies ("core protocols"); (2) design of the case report form (CRFs); and (3) organization of the data handling systems. Although members of the monitoring team are usually involved in the planning step, the monitoring process is viewed as starting at the implementation step. Finally, as the studies are completed, the data are collected and organized so that the last step consisting of data interpretation begins.

Each of these four steps in the research process must be carried out with a clear understanding of the regulatory requirements or must follow discussions at appropriate intervals with regulatory officials. Advice from outside investigators or other outside experts is also very useful in the initial planning. It is essential that the outside investigators who will conduct the clinical trials be involved in the development of the protocol and CRF.

Unfortunately, the outside investigators and academicians are usually not involved in the interpretation of the data, information, and results. Even in the case of outside clinical studies, the investigators frequently do not receive final study reports from the sponsor. A clinical study should not be considered closed until the investigator has been furnished with a final report of his study.

In essence, the objective of the drug development process is therapeutic progress, produced by asking the right questions, documenting the conduct of the study, collecting the data, and interpreting the results with sound scientific methods. The result will be credible scientific conclusions to which others can agree.

III. RESPONSIBILITIES

The monitoring process is a highly specialized system of controlled research activities designed to systematically define the basic properties and clinical usefulness of a new chemical entity. Such activities take place over time and produce progressive advances in the state of knowledge regarding the drug under study.

The monitoring process is responsible for facilitating the development of a drug through five basic stages in its life cycle·

1. *Preclinical*—Animal and laboratory studies leading to the filing of an Investigational New Drug (IND) application.
2. *Phase I (clinical pharmacology)*—Initial in-patient introduction of a drug into humans (usually volunteers) for early study of dose ranging, toxicity, drug dynamics, and metabolism, including occasionally early efficacy, e.g., oncolytic drugs. Studies frequently include comparative drug or placebo controls.
3. *Phase II (clinical investigation)*—Includes early controlled studies (usually hospitalized patients) to demonstrate effectiveness and relative safety and to further define dosage.
4. *Phase III (clinical trials)*—Greatly expanded well-controlled and uncontrolled trials to further define effectiveness, safety, statistical significance, and general usefulness. Later, following approval of the New Drug Application (NDA), additional Phase III IND studies are frequently undertaken as part of the Phase IV program. Such studies typically explore "unapproved" uses, dosages, and indications.
5. *Phase IV (postmarketing)*—Surveillance of medical practice experiences following approval of the NDA, with emphasis on rare untoward effects: sometimes prospective multiclinic trials will be initiated to seek out rare events, while other studies may explore comparative features with competitive drugs.

The above stages in drug development are shown schematically in Fig. 1. It is useful to divide them among five interrelated plans (i.e., Preclinical and Plans A, B, C, and D), with a team leader assigned the responsibility of monitoring the progress for each plan (see Table 1).

The NDA decision point (see Fig. 1 and Table 1) is greatly dependent upon the monitoring process because it is at this time that a decision is made by sponsor management to proceed toward the filing of an NDA. This decision initiates Plan B (Figs. 1 and 2) with a greatly expanded clinical trials program and a substantial increase in financial and manpower requirements.

The monitoring team ensures that the individual clinical studies are conducted according to the time and events schedule of the various plans. Accordingly, a determination must be made that:

1. Each clinical study conforms to regulatory requirements and accepted medical standards for the protection of subjects and patients.
2. The investigator and his associates are thoroughly instructed in their study responsibilities, including reporting procedures and study drug accountability.
3. All relevant data and information are accurately recorded and reflect the source documents (physician records or hospital records).

4. The investigator receives periodic progress reports as the study progresses, and a final report when the study has been completed.
5. Any side effects are promptly evaluated and communicated as appropriate to other investigators and regulatory agencies.
6. The investigator is reimbursed for expenses according to the terms of the study grant.

The monitoring process, if not born in a climate of increased regulatory requirements, at least accelerated its development under those conditions. Prior to 1962, when the Harris-Kefauver amendment to the Federal Pure Food and Drug Act was adopted, clinical testing of drugs was carried out in a very unscientific and haphazard manner [3]. Proof of efficacy was not required; consequently, very little information was generated regarding benefit vs. risk of new drugs. This changed considerably when proof of efficacy was mandated by the amendment. In such a regulatory climate, it became all too apparent that "approvable" NDAs would depend upon a sound scientific plan and a disciplined monitoring process.

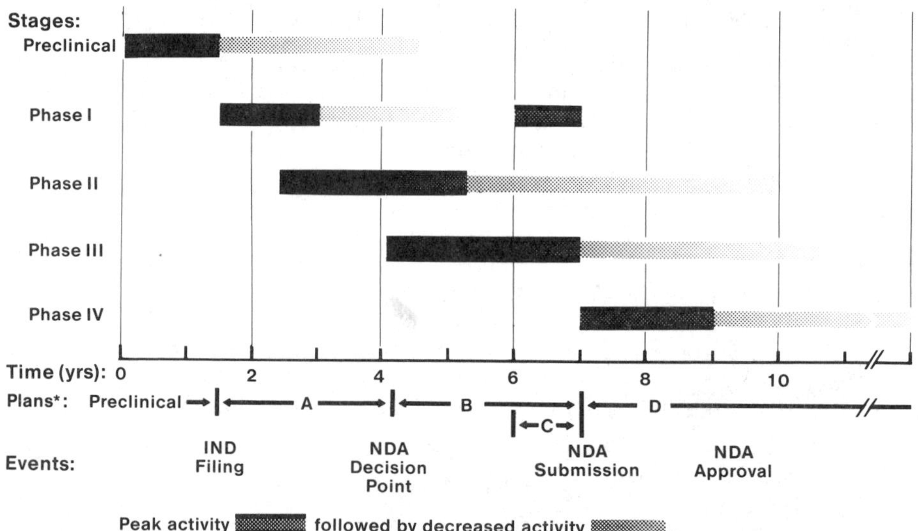

Figure 1 Schematic representation of the time and events for the drug development process by stages.

Table 1 Research Planning and Monitoring

Planning	Team leader	Team members	Functions
Preclinical plan	Research scientist	Research staff, clinical pharmacologist, new products planner	Preclinical studies leading up to IND; initial assessment of marketing potential
Plan A	Clinical pharmacologist	Above members plus clinical investigations monitor and paramedical associates [i.e., clinical research associates (CRAs)]	Phase I and early Phase II studies; NDA decision point
Plan B	Clinical investigations monitor	Research staff (minimal), clinical pharmacologist, CRAs, new product planner	Definition of desired claims in package insert; Phase III culminating in an NDA
Plan C	Clinical pharmacologist or clinical investigations monitor	Same as for Plan A, plus clinical trials monitor	Bioavailability and physical characterization of dosage form to be sold; special pharmacology studies, i.e., drug interaction, tissue distribution, and metabolism in pregnant patients, renal impaired, etc.
Plan D	Clinical trials monitor	Clinical investigations monitor, CRAs, international and domestic marketing plans managers	Postmarketing surveillance

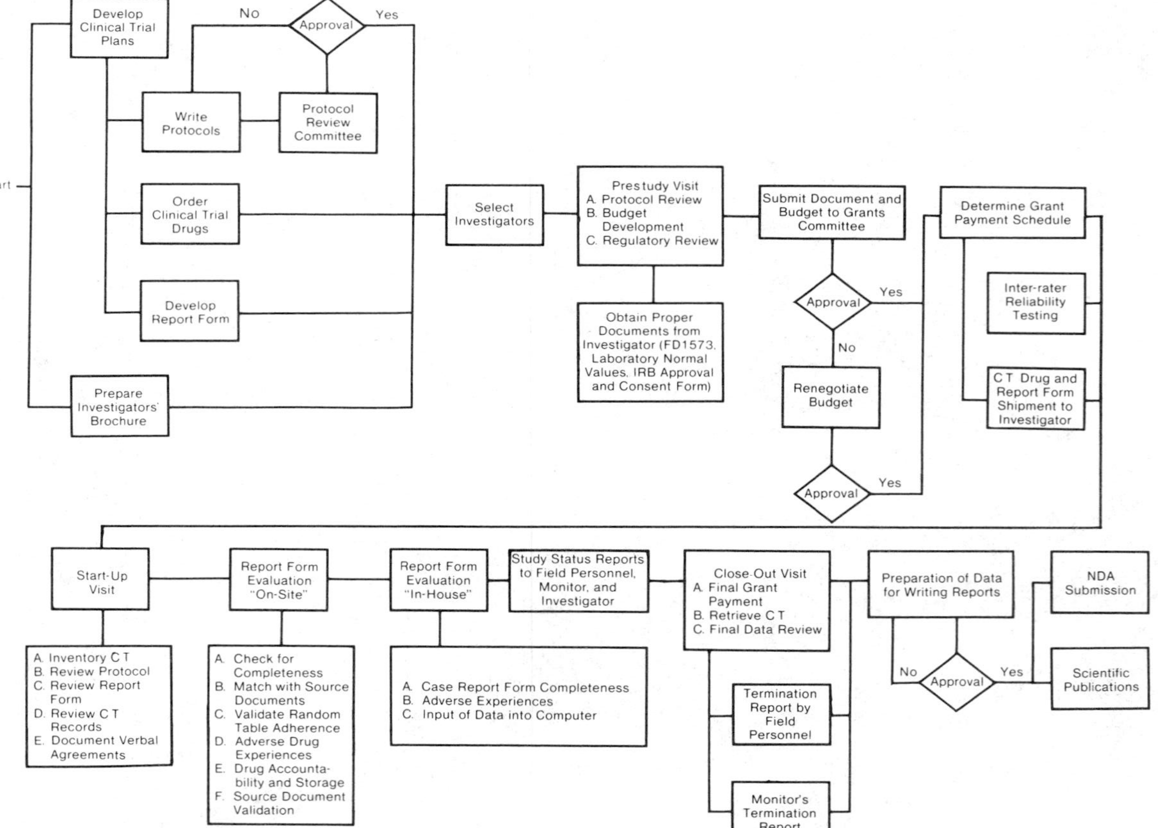

Figure 2 Flow diagram of events for a typical Plan B.

IV. PROJECT TEAM ORGANIZATION

Organization and staffing of the project team is a critical step in the drug development process. Staffing of the project team is the overall responsibility of the sponsor's top management group. The members of the project team come from several functional units within the corporation. The fact that top management selects the members ensures that the strategic or planning phase of the drug research program will address the desires of management. Once the staffing is completed, then it is the job of the project team to proceed with the tactical or implementation phase of the drug development process. As the drug moves through the various phases in its life cycle (Fig. 1), the membership of the project team changes, with some members being added while others are leaving. Effective operation of the project team depends upon:

1. Integration of the diverse talents of its members
2. A high level of commitment to team interaction and research plan objectives
3. Providing visibility for the accomplishments of its members
4. A clear understanding by each member of his role in the implementation of the research plans
5. Communication to management regarding the status of the research projects along with ongoing resource needs of the team

There are three basic organizational styles that a project-monitoring team can follow: (1) functional, (2) project, and (3) matrix.

The functional style, with the members remaining in their permanent organizational positions, is the oldest of the three styles and probably today the least effective for complex drug development projects due to communication problems.

The project style is one in which all members on the project team operate as one organizational unit, with members reporting both functionally and organizationally to the project team leader. Such a style is probably not well suited to drug development, which takes place over several years. The project style is probably best suited to dealing with specific, well-defined problems that require intensive input by a variety of people over a relatively short period of time, i.e., several months to a year or two.

The matrix style is probably overall the most effective of the three styles; the members have a close working relationship with the project team leader while maintaining traditional organizational ties to their line management.

It is important for management to carefully organize, staff, and monitor the project team so that the strategic needs can be clearly identified and implemented through the tactical operations of an effective project team.

DEPARTMENT OF HEALTH AND HUMAN SERVICES
PUBLIC HEALTH SERVICE
FOOD AND DRUG ADMINISTRATION

NOTICE OF CLAIMED INVESTIGATIONAL EXEMPTION FOR A NEW DRUG

Form Approved OMB No. 57-R0030
Use of this form is prohibited after 6/30/82.

NOTE: No drug may be shipped or study initiated unless a complete statement has been received. (21 CFR 312.1(a)(2)).

Name of Sponsor ______________________

Address ______________________

Date ______________________

Name of Investigational Drug ______________________

To the Secretary of Health, Education and Welfare
For the Commissioner of Food and Drugs
Bureau of Drugs (HFD-106)
5600 Fishers Lane
Rockville, Maryland 20857

Dear Sir:

The sponsor, ______________________, submits this notice of claimed investigational exemption for a new drug under the provisions of section 505(i) of the Federal Food, Drug, and Cosmetic Act and §312.1 of Title 21 of the Code of Federal Regulations.

Attached hereto in triplicate are:

1. The best available descriptive name of the drug, including to the extent known the chemical name and structure of any new-drug substance, and a statement of how it is to be administered. (If the drug has only a code name, enough information should be supplied to identify the drug.)

2. Complete list of components of the drug, including any reasonable alternates for inactive components.

3. Complete statemen tof quantitative composition of drug, including reasonable variations that may be expected during the investigational stage.

4. Description of source and preparation of, any new-drug substances used as components, including the name and address of each supplier or processor, other than the sponsor, or each new-drug substance.

5. A statement of the methods, facilities, and controls used for the manufacturing, processing, and packing of the new drug to establish and maintain appropriate standards of identity, strength, quality, and purity as needed for safety and to give significance to clinical investigations made with the drug.

6. A statement covering all information available to the sponsor derived from preclinical investigations and any clinical studies and experience with the drug as follows:

a. Adequate information about the preclinical investigations, including studies made on laboratory animals, on the basis of which the sponsor has concluded that it is reasonably safe to initiate clinical investigations with the drug: Such information should include identification of the person who conducted each investigation; identification and qualifications of the individuals who evaluated the results and concluded that it is reasonably safe to initiate clinical investigations with the drug and a statement of where the investigations were conducted and where the records are available for inspection; and enough details about the investigations to permit scientific review. The preclinical investigations shall not be considered adequate to justify clinical testing unless they give proper attention to the conditions of the proposed clinical testing. When this information, the outline of the plan of clinical pharmacology, or any progress report on the clinical pharmacology, indicates a need for full review of the preclinical data before a clinical trial is undertaken, the Department will notify the sponsor to submit the complete preclinical data and to withhold clinical trials until the review is completed and the sponsor notified. The Food and Drug Administration will be prepared to confer with the sponsor concerning this action.

b. If the drug has been marketed commercially or investigated (e.g. outside the United States), complete information about such distribution or investigation shall be submitted, along with a complete bibliography of any publications about the drug.

c. If the drug is a combination of previously investigated or marketed drugs, an adequate summary of preexisting information from preclinical and cliical investigations and experience with its components, including all reports available to the sponsor suggesting side-effects, contraindications, and ineffectiveness in use of such components: Such summary should include an adequate bibliography of publications about the components and may incorporate by reference any information concerning such components previously submitted by the sponsor to the Food and Drug Administration. Include a statement of the expected pharmacological effects of the combination.

d. If the drug is a radioactive drug, sufficient data must be available from animal studies or previous human studies to allow a reasonable calculation of radiation absorbed dose upon administration to a human being.

7. A total (one in each of the three copies of the notice) of all informational material, including label and labeling, which is to be supplied to each investigator: This shall include an accurate description of the prior investigations and experience and their results pertinent to the safety and possible usefulness of the drug under the conditions of the investigation. It shall not represent that the safety or usefulness of the drug has been established for the purposes to be investigated. It shall describe all relevant hazards, contra-indications, side-effects, and precautions suggested by prior investigations and experience with the drug under investigation and related drugs for the information of clinical investigators.

8. The scientific training and experience considered appropriate by the sponsor to qualify the investigators as suitable experts to investigate the safety of the drug, bearing in mind what is known about the pharmacological action of the drug and the phase of the investigational program that is to be undertaken.

FORM FDA 1571 (4/80) PREVIOUS EDITIONS ARE OBSOLETE.

Figure 3 Notice of claimed investigational exemption for a new drug (Form FDA 1571). (Courtesy of Food and Drug Administration, U.S. Department of Health and Human Services, Rockville, Md.)

9. The names and a summary of the training and experience of each investigator and of the individual charged with monitoring the progress of the investigation and evaluating the evidence of safety and effectiveness of the drug as it is received from the investigators, together with a statement that the sponsor has obtained from each investigator a completed and signed form, as provided in subparagraph (12) or (13) of this paragraph, and that the investigator is qualified by scientific training and experience as an appropriate expert to undertake the phase of the investigation outlined in section 10 of the "Notice of Claimed Investigational Exemption for a New Drug." (In crucial situations, phase 3 investigators may be added and this form supplemented by rapid communication methods, and the signed form FD-1573 shall be obtained promptly thereafter.)

10. An outline of any phase or phases of the planned investigations and a description of the institutional review committee, as follows:

a. Clinical pharmacology. This is ordinarily divided into two phases: Phase 1 starts when the new drug is first introduced into man only animal and in vitro data are available with the purpose of determining human toxicity, metabolism, absorption, elimination, and other pharmacological actoin, preferred route of administration, and safe dosage range; phase 2 covers the initial trials on a limited number of patients for specific disease control or prophylaxis purposes. A general outline of these phases shall be submitted, identifying the investigator or investigators, the hospitals or research facilities where the clinical pharmacology will be undertaken, any expert committees or panels to be utilized, the maximum number of subjects to be involved, and the estimated duration of these early phases of investigation. Modification of the experimental design on the basis of experience gained need be reported only in the progress reports on these early phases, or in the development of the plan for the clinical trial, phase 3. The first two phases may overlap and, when indicated, may require additional animal data before these phases can be completed or phase 3 can be undertaken. Such animal tests shall be designed to take into account the expected duration of administration of the drug to human beings, the age groups and physical status, as for example, infants, pregnant women, premenopausal women, of those human beings to whom the drug may be administered, unless this has already been done in the original animal studies. If a drug is a radioactive drug, the clinical pharmacology phase must include studies which will obtain sufficient data for dosimetry calculations. These studies should evaluate the excretion, whole body retention, and organ distribution of the radioactive material.

b. Clinical trial. This phase 3 provides the assessment of the drug's safety and effectiveness and optimum dosage schedules in the diagnosis, treatment, or prophylaxis of groups of subjects involving a given disease or condition. A reasonable protocol is developed on the basis of the facts accumulated in the earlier phases, including completed and submitted animal studies. This phase is conducted by separate groups following the same protocol (with reasonable variations and alternatives permitted by the plan) to produce well-controlled clinical data. For this phase, the following data shall be submitted:

i. The names and addresses of the investigators. (Additional investigators may be added.)

ii. The specific nature of the investigations to be conducted, together with information or case report forms to show the scope and detail of the planned clinical observations and the clinical laboratory tests to be made and reported.

iii. The aproximate number of subjects (a reasonable range of subjects is permissible and additions may be made), and criteria proposed for subject selection by age, sex, and condition.

iv. The estimated duration of the clinical trial and the intervals, not exceeding 1 year, at which progress reports showing the results of the investigations will be submitted to the Food and Drug Administration.

c. Institutional review committee. If the phases of clinical study as described under 10a and b above are conducted on institutionalized subjects or are conducted by an individual affiliated with an institution which agrees to assume responsibility for the study, assurance must be given that an institutional review committee is responsible for initial and continuing review and approval of the proposed clinical study. The membership must be comprised of sufficient members of varying background, that is, lawyers, clergymen, or laymen as well as scientists, to assure complete and adequate review of the research project. The membership must possess not only broad competence to comprehend the nature of the project, but also other competencies necessary to judge the acceptability of the project or activity in terms of institutional regulations, relevant law, standards of professional practice, and community acceptance. Assurance must be presented that neither the sponsor nor the investigator has participated in selection of committee members; that the review committee does not allow participation in its review and conclusions by any individual involved in the conduct of the research activity under review (except to provide information to the committee); that the investigator will report to the committee for review any emergent problems, serious adverse reactions, or proposed procedural changes which may affect the status of the investigation and that no such change will be made without committee approval except where necessary to eliminate apparent immediate hazards; the reviews of the study will be conducted by the review committee at intervals appropriate to the degree of risk, but not exceeding 1 year, to assure that the research project is being conducted in compliance with the committee's understanding and recommendations; that the review committee is provided all the information on the research project necessary for its complete review of the project; and that the review committee maintains adequate documentation of its activities and develops adequate procedures for reporting its findings to the institution. The documents maintained by the committee are to include the names and qualifications of committee members, records of information provided to subjects in obtaining informed consent, committee discussion on substantive issues and their resolution, committee recommendations, and dated reports of successive reviews as they are performed. Copies of all documents are to be retained for a period of 3 years past the completion or discontinuance of the study and are to be made available upon request to duly authorized representatives of the Food and Drug Administration. (Favorable recommendations by the committee are subject to further appropriate review and rejection by institution officials. Unfavorable recommendations, restrictions, or conditions may not be overruled by the institution officials.) Procedures for the organization and operation of institutional review committees are contained in guidelines issued pursuant to Chapter 1-40 of the Grants Administration Manual of the U.S. Department of Health, Education, and Welfare, available from the U.S. Government Printing Office. It is recommended that these guidelines be followed in establishing institutional review committees and that the committees function according to the procedures described therein. A signing of the Form FD-1571 will be regarded as providing the above necessary assurances. If the institution, however, has on file with the Department of Health, Education, and Welfare, Division of Research Grants, National Institutes of Health, an "accepted general assurance," and the same committee is to review the proposed study using the same procedures, this is acceptable in lieu of the above assurances and a statement to this effect should be provided with the signed FD-1571. (In addition to sponsor's continuing responsibility to monitor the study, the Food and Drug Administration will undertake investigations in institutions periodically to determine whether the committees are operating the accord with the assurances given by the sponsor.)

(The notice of claimed investigational exemption may be limited to any one or more phases, provided the outline of the additional phase or phases is submitted before such additional phases begin. This does not preclude continuing a subject on the drug from phase 2 to phase 3 without interruption while the plan for phase 3 is being developed.)

Ordinarily, a plan for clinical trial will not be regarded as reasonable unless, among other things, it provides for more than one independent competent investigator to maintain adequate case histories of an adequate number of subjects, designed to record observations and permit evaluation of any and all discernible effects attributable to the drug in each individual treated, and comparable records on any individuals employed as controls. These records shall be individual records for each subject maintained to include adequate information pertaining to each, including age, sex, conditions treated, dosage, frequency of administration of the drug, results of all relevant clinical observations and laboratory examinations made, adequate information concerning any other treatment given and a full statement of any adverse effects and useful results observed, together with an opinion as to whether such effects or results are attributable to the drug under investigation.

11. A statement that the sponsor will notify the Food and Drug Administration if the investigation is discontinued, and the reason therefor.

12. A statement that the sponsor will notify each investigator if a new-drug application is approved, or if the investigation is discontinued.

13. If the drug is to be sold, a full explanation why sale is required and

Figure 3 (Continued)

should not be regarded as the commercialization of a new drug for which an application is not approved.

14. A statement that the sponsor assures that clinical studies in humans will not be initiated prior to 30 days after the date of receipt of the notice by the Food and Drug Administration and that he will continue to withhold or to restrict clinical studies if requested to do so by the Food and Drug Administration prior to the expiration of such 30 days. If such request is made, the sponsor will be provided specific information as to the deficiencies and will be afforded a conference on request. The 30-day delay may be waived by the Food and Drug Administration upon a showing of good reason for such waiver; and for investigations subject to institutional review committee approval as described in item 10c above, and additional statement assuring that the investigation will not be initiated prior to approval of the study by such committee.

15. When requested by the agency, an environmental impact analysis report pursuant to §25.1 of this chapter.

16. A statement that all nonclinical laboratory studies have been, or will be, conducted in compliance with the good laboratory practice regulations set forth in Part 58 of this chapter, or, if such studies have not been conducted in compliance with such regulations, a statement that describes in detail all differences between the practices used in conducting the study and those required in the regulations.

Very truly yours,

SPONSOR	PER
	INDICATE AUTHORITY

(This notice may be amended or supplemented from time to time on the basis of the experience gained with the new drug. Progress reports may be used to update the notice.)

ALL NOTICES AND CORRESPONDENCE SHOULD BE SUBMITTED IN TRIPLICATE.

Figure 3 (Continued)

V. DRUG DEVELOPMENT PARTICIPANTS

A. The Drug Sponsor

When clinical studies are initiated by the drug manufacturer, that company acts as the sponsor by filing an IND, i.e., "Notice of Claimed Investigational Exemption for a New Drug" [4] (Form FDA 1571, shown in Fig. 3). The term "sponsor" also is used to refer to a company's duly authorized representative(s), such as the monitor and paramedical associates. (In general, the term refers to any person or institution that initiates a clinical investigation by providing financial resources or investigational drugs but does not actually perform the clinical trials.)

B. The Monitor

The principal monitor for a study drug may be a physician or a nonphysician health professional working under the supervision of a physician. The monitor may be assisted by paramedical personnel, i.e., clinical research associates (CRAs)—sometimes referred to as medical information associates (MIAs) or clinical information administrators (CIAs)—who work in close association with the monitor. These personnel are usually responsible for data management and overall supervision of the day-to-day activities of studies. Frequently, other paramedicals such as the clinical research coordinators (CRCs) and the clinical information representatives (CIRs) are located in the field and are involved in investigator contacts and data acquisition. The field force of CRCs represent the research division to investigators through personal contact with them: they review protocols with established investigators to determine their interest in the project and current ability to engage in a study; they negotiate budget requests; and they frequently review some early case report forms (CRFs) for completeness and protocol compliance. Additionally, they provide the research group with contacts with potential new investigators and assess their capability as study participants.

The CIRs act as data collectors through routine visits at key study centers. They review CRFs and audit representative samples of cases against the source documents.

Many of the paramedical personnel have B.S. degrees in pharmacy or another field of science; many will also have had prior experience in marketing or sales.

Figure 4 (Overleaf) Statement of investigator (clinical pharmacology, Form FD 1572). (Courtesy of Food and Drug Administration, U.S. Department of Health and Human Services, Rockville, Md.)

FORM FD 1572
DEPARTMENT OF HEALTH, EDUCATION, AND WELFARE
FOOD AND DRUG ADMINISTRATION

STATEMENT OF INVESTIGATOR

(Clinical Pharmacology)

TO: SUPPLIER OF DRUG *(Name and Address, include Zip Code)*	NAME OF INVESTIGATOR *(Print or Type)*
Lilly Research Laboratories A Division of Eli Lilly and Company 307 East McCarty Street Indianapolis, Indiana 46285	John Smith, M.D.
	DATE January 2, 1982
	NAME OF DRUG Moxalactam

Dear Sir:

The undersigned, John Smith, M.D., submits this statement as required by section 505(i) of the Federal Food, Drug, and Cosmetic Act and §312.1 of Title 21 of the Code of Federal Regulations as a condition for receiving and conducting clinical pharmacology with a new drug limited by Federal *(or United States)* law to investigational use.

1. A Statement of the education and training that qualifies me for clinical pharmacology

INVESTIGATOR — ATTACH CV OR PROVIDE INFORMATION HERE

☒ SEE CURRICULUM VITAE ☐ SEE FOLLOWING

2. The name and address of the medical school, hospital, or other research facility where the clinical pharmacology will be conducted

University of America School of Medicine
City Hospital
New York, N.Y.

3. If the experimental project is to be conducted on institutionalized subjects or is conducted by an individual affiliated with an institution which agrees to assume responsibility for the study, assurance must be given that an institutional review committee is responsible for initial and continuing review and approval of the proposed clinical study. The membership must be comprised of sufficient members of varying background, that is, lawyers, clergymen, or laymen as well as scientists, to assure complete and adequate review of the research project. The membership must possess not only broad competence to comprehend the nature of the project, but also other competencies necessary to judge the acceptability of the project or activity in terms of institutional regulations, relevant law, standards of professional practice, and community acceptance. Assurance must be presented that the investigator has not participated in the selection of committee members; that the review committee does not allow participation in its review and conclusions by any individual involved in the conduct of the research activity under review (except to provide information to the committee); that the investigator will report to the committee for review any emergent problems, serious adverse reactions, or proposed procedural changes which may affect the status of the investigation and that no such change will be made without committee approval except where necessary to eliminate apparent immediate hazards; that reviews of the study will be conducted by the review committee at intervals appropriate to the degree of risk, but not exceeding 1 year, to assure that the research project is being conducted in compliance with the committee's understanding and recommendations; that the review committee is provided all the information on the research project necessary for its complete review of the project; and that the review committee maintains adequate documentation of its activities and develops adequate procedures for reporting its findings to the institution. The documents maintained by the committee are to include the names and qualifications of committee members, records of information provided to subjects in obtaining informed consent, committee discussion on substantive issues and their resolution, committee recommendations, and dated reports of successive reviews as they are performed. Copies of all documents are to be retained for a period of 3 years past the completion or discontinuance of the study and are to be made available upon request to duly authorized representatives of the Food and Drug Administration. (Favorable recommendations by the committee are subject to further appropriate review and rejection by institution officials. Unfavorable recommendations, restrictions, or conditions may not be overruled by the institution officials.) Procedures for the organization and operation of institutional review committees are contained in guidelines issued pursuant to Chapter 1-40 of the Grants Administration Manual of the U.S. Department of Health, Education, and Welfare, available from the U.S. Government Printing Office. It is recommended that these guidelines be followed in establishing institutional review committees and that the committees function according to the procedures described therein. A signing of the Form FD-1572 will be regarded as providing the above necessary assurance; however, if the institution has on file with the Department of Health, Education, and Welfare, Division of Research Grants, National Institutes of Health, an "accepted general assurance," and the same committee is to review the proposed study using the same procedures, this is acceptable in lieu of the above assurances and a statement to this effect should be provided with the signed FD-1572. (In addition to sponsor's continuing responsibility to monitor the study, the Food and Drug Administration will undertake investigations in institutions periodically to determine whether the committees are operating in accord with the assurances given by the sponsor.)

00 NS 9102(1) PRINTED IN USA JUN 80 **RETURN TO INDIANAPOLIS — FOR SUBMISSION TO REGULATORY AFFAIRS**

Figure 4

4. The estimated duration of the project and the maximum number of subjects that will be involved

6 Mo. Maximum of 20 subjects

5. A general outline of the project to be undertaken *(Modification is permitted on the basis of experience gained without advance submission of amendments to the general outline, but with the approval of the review committee and upon notification of the sponsor.)*

Compound X will be administered IV to a maximum of 20 subjects in single doses of 250 mg. Blood and urine samples will be collected at 0, 1/4, 1/2, 1, 2, 4, 6, 8, and 10 hourly intervals to determine serum and urine levels of the drug.

6. THE UNDERSIGNED UNDERSTANDS THAT THE FOLLOWING CONDITIONS GENERALLY APPLICABLE TO NEW DRUGS FOR INVESTIGATIONAL USE GOVERN HIS RECEIPT AND USE OF THIS INVESTIGATIONAL DRUG

a. The sponsor is required to supply the investigator with full information concerning the preclinical investigation that justifies clinical pharmacology.

b. The investigator is required to maintain adequate records of the disposition of all receipts of the drug, including dates, quantity, and use by subjects, and if the clinical pharmacology is suspended, terminated, discontinued, or completed, to return to the sponsor any unused supply of the drug. If the investigational drug is subject to the comprehensive Drug Abuse Prevention and Control Act of 1970, adequate precautions must be taken, including storage of the investigational drug in a securely locked, substantially constructed cabinet, or other securely locked, substantially constructed enclosure access to which is limited, to prevent theft or diversion of the substance into illegal channels of distribution.

c. The investigator is required to prepare and maintain adequate case histories designed to record all observations and other data pertinent to the clinical pharmacology.

d. The investigator is required to furnish his reports to the sponsor who is responsible for collecting and evaluating the results, and presenting progress reports to the Food and Drug Administration at appropriate intervals, not exceeding 1 year. Any adverse effect which may reasonably be regarded as caused by, or is probably caused by, the new-drug shall be reported to the sponsor promptly; and if the adverse effect is alarming it shall be reported immediately. An adequate report of the clinical pharmacology should be furnished to the sponsor shortly after completion.

e. The investigator shall maintain the records of disposition of the drug and the case reports described above for a period of 2 years following the date the new-drug application is approved for the drug, or if no application is to be filed or is approved until 2 years after the investigation is discontinued and the Food and Drug Administration so notified. Upon the request of a scientifically trained and specifically authorized employee of the Department, at reasonable times, the investigator will make such records available for inspection and copying. The names of the subjects need not be divulged unless the records of the particular subjects require a more detailed study of the cases, or unless there is reason to believe that the records do not represent actual studies or do not represent actual results obtained.

f. The investigator certifies that the drug will be administered only to subjects under his personal supervision or under the supervision of the following investigators responsible to him.

Fred J. West, M.D.

and that the drug will not be supplied to any other investigator or to any clinic for administration to subjects.

g. The investigator certifies that he will inform any patients or any persons used as controls, or their representatives, that drugs are being used for investigational purposes, and will obtain the consent of the subjects, or their representatives, except where this is not feasible or, in the investigator's professional judgment, is contrary to the best interests of the subjects.

h. The investigator is required to assure the sponsor that for investigations involving institutionalized subjects the studies will not be initiated until the institutional review committee has reviewed and approved the study. (The organization and procedure requirements for such a committee should be explained to the investigator by the sponsor as set forth in Form FD-1571, division 10, unit c.)

Very truly yours,

John Smith M.D.
(Signature of Investigator)

City Hospital
(Address)

New York, N.Y.

00 NS 9102(2) PRINTED IN USA JUN 80 RETURN TO INDIANAPOLIS – FOR SUBMISSION TO REGULATORY AFFAIRS

Figure 4 (Continued)

C. The Investigator

The term "investigator" refers to an individual who is an expert in his field of study and is qualified by scientific training and experience to clinically evaluate investigational new drugs. On the basis of his evaluation, other experts should be confident that the drug will have the effects it is represented to have under the conditions of use prescribed, recommended, or suggested in the proposed labeling [5]. Others involved in assisting the clinical investigator may include research nurses, study coordinators, resident physicians, and pharmacy and laboratory personnel. The principal investigator must file a statement of investigator form, either Form 1572 (Clinical Pharmacology) or Form 1573 [4] (See Figs. 4 and 5, respectively), as well as a curriculum vitae under the sponsor's IND.

D. Institutional Review Board

Before any new drug can be administered to patients or subjects by an investigator, the protocol for the trial must be approved by an Institutional Review Board (IRB). The investigator must give his assurance that an IRB has approved the protocol and is responsible for continuing to review the conduct and results of the proposed clinical trial. The membership must be composed of sufficient members of varying background, i.e., lawyers, clergymen, or laymen, as well as scientists, to ensure complete and adequate review of the research project. The investigator presenting the research protocol cannot be a member nor select any members of the IRB committee [6].

E. Regulatory Agencies

The Food and Drug Administration (FDA), a division of the U.S. Department of Health and Human Services, is the principal agency concerned with the regulation of drug research and labeling in the United States. Most other countries have their own governing regulations for drug development and registration; however, some will register a drug after it has been approved in the country of origin. Preclinical and clinical research data may be used for international registration purposes; however, requirements vary. Complete familiarity with the regulations in those countries where registration will be sought is essential.

Figure 5 Statement of investigator (Form FD 1573). (Courtesy of Food and Drug Administration, U.S. Department of Health and Human Services, Rockville, Md.)

FORM FD 1573
DEPARTMENT OF HEALTH, EDUCATION, AND WELFARE
FOOD AND DRUG ADMINISTRATION

STATEMENT OF INVESTIGATOR

TO: SUPPLIER OF DRUG *(Name and Address, include Zip Code)*	NAME OF INVESTIGATOR *(Print or Type)*	
Lilly Research Laboratories A Division of Eli Lilly and Company 307 East McCarty Street Indianapolis, Indiana 46285	John Smith, M.D.	
	DATE 1/2/82	SOC. SEC. NO. 123-45-6789
	NAME OF DRUG Moxalactam	

Dear Sir:
The undersigned, John Smith, M.D., submits this statement as required by section 505(i) of the Federal Food, Drug, and Cosmetic Act and §312.1 of Title 21 of the Code of Federal Regulations as a condition for receiving and conducting clinical investigations with a new drug limited by Federal (*or United States*) law to investigational use.

1. STATEMENT OF EDUCATION AND EXPERIENCE

a. Colleges, universities, and medical or other professional schools attended, with dates of attendance, degrees, and dates degrees were awarded

INVESTIGATOR — ATTACH CV OR PROVIDE INFORMATION HERE

☒ SEE CURRICULUM VITAE ☐ SEE FOLLOWING

b. Postgraduate medical or other professional training *(Indicate dates, names of institutions, and nature of training)*

INVESTIGATOR — ATTACH CV OR PROVIDE INFORMATION HERE

☒ SEE CURRICULUM VITAE ☐ SEE FOLLOWING

c. Teaching or research experience *(Indicate dates, institutions, and brief description of experience)*

INVESTIGATOR — ATTACH CV OR PROVIDE INFORMATION HERE

☒ SEE CURRICULUM VITAE ☐ SEE FOLLOWING

d. Experience in medical practice or other professional experience *(Indicate dates, institutional affiliations, nature of practice, or other professional experience)*

INVESTIGATOR — ATTACH CV OR PROVIDE INFORMATION HERE

☒ SEE CURRICULUM VITAE ☐ SEE FOLLOWING

e. Representive list of pertinent medical or other scientific publications *(Indicate titles of articles, names of publications and volume, page number, and date)*

INVESTIGATOR — ATTACH CV OR PROVIDE INFORMATION HERE

☒ SEE CURRICULUM VITAE ☐ SEE FOLLOWING

00 NS 9103(1) PRINTED IN USA OCT 81

RETURN TO INDIANAPOLIS — FOR SUBMISSION TO REGULATORY AFFAIRS

Figure 5

2a. If the investigation is to be conducted on institutionalized subjects or is conducted by an individual affiliated with an institution which agrees to assume responsibility for the study, assurance must be given that an institutional review committee is responsible for initial and continuing review and approval of the proposed clinical study. The membership must be comprised of sufficient members of varying background, that is, lawyers, clergymen, or laymen as well as scientists, to assure complete and adequate review of the research project. The membership must possess not only broad competence to comprehend the nature of the project, but also other competencies necessary to judge the acceptability of the project or activity in terms of institutional regulations, relevant law, standards of professional practice, and community acceptance. Assurance must be presented that the investigator has not participated in the selection of committee members; that the review committee does not allow participation in its review and conclusions by any individual involved in the conduct of the research activity under review (except to provide information to the committee); that the investigator will report to the committee for review any emergent problems, serious adverse reactions, or proposed procedural changes which may affect the status of the investigation and that no such change will be made without committee approval except where necessary to eliminate apparent immediate hazards; that reviews of the study will be conducted by the review committee at intervals appropriate to the degree of risk, but not exceeding 1 year, to assure that the research project is being conducted in compliance with the committee's understanding and recommendations; that the review committee is provided all the information on the research project necessary for its complete review of the project; and that the review committee maintains adequate documentation of its activities and develops adequate procedure for reporting its findings to the institution. The documents maintained by the committee are to include the names and qualifications of committee members, records of information provided to subjects in obtaining informed consent, committee discussion on substantive issues and their resolution, committee recommendations, and dated reports of successive reviews as they are performed. Copies of all documents are to be retained for a period of 3 years past the completion or discontinuance of the study and are to be made available upon request to duly authorized representatives of the Food and Drug Administration. (Favorable recommendations by the committee are subject to further appropriate review and rejection by institution officials. Unfavorable recommendations, restrictions, or conditions may not be overruled by the institution officials.) Procedures for the organization and operation of institutional review committees are contained in guidelines issued pursuant to Chapter 1-40 of the Grants Administration Manual of the U.S. Department of Health, Education, and Welfare, available from the U.S. Government Printing Office. It is recommended that these guidelines be followed in establishing institutional review committees and that the committees function according to the procedures described therein. A signing of the Form FD 1573 will be regarded as providing the above necessary assurances; however, if the institution has on file with the Department of Health, Education, and Welfare, Division of Research Grants, National Institutes of Health, an "accepted general assurance," and the same committee is to review the proposed study using the same procedures, this is acceptable in lieu of the above assurances and a statement to this effect should be provided with the signed FD 1573. (In addition to sponsor's continuing responsibility to monitor the study, the Food and Drug Administration will undertake investigations in institutions periodically to determine whether the committees are operating in accord with the assurances given by the sponsor.)

2b. A description of any clinical laboratory facilities that will be used. (If this information has been submitted to the sponsor and reported by him on Form FD 1571, reference to the previous submission will be adequate).

City Hospital

3. The investigational drug will be used by the undersigned or under his supervision in accordance with the plan of investigation described as follows: (Outline the plan of investigation including approximation of the number of subjects to be treated with the drug and the number to be employed as controls, if any; clinical uses to be investigated; characteristics of subjects by age, sex and condition; the kind of clinical observations and laboratory tests to be undertaken prior to, during, and after administration of the drug; the estimated duration of the investigation; and a description or copies of report forms to be used to maintain an adequate record of the observations and test results obtained. This plan may include reasonable alternates and variations and should be supplemented or amended when any significant change in direction or scope of the investigation is undertaken.)

IND Protocol No. 20 Estimated Number of Patients 50

Estimated Completion Date 10/83

Is Publication Intended? ☒ Yes ☐ No

00 NS 9103(2) PRINTED IN USA OCT 81

RETURN TO INDIANAPOLIS — FOR SUBMISSION TO REGULATORY AFFAIRS

Figure 5 (Continued)

4. THE UNDERSIGNED UNDERSTANDS THAT THE FOLLOWING CONDITIONS, GENERALLY APPLICABLE TO NEW DRUGS FOR INVESTIGATIONAL USE, GOVERN HIS RECEIPTS AND USE OF THIS INVESTIGATIONAL DRUG:

a. The sponsor is required to supply the investigator with full information concerning the preclinical investigations that justify clinical trials, together with fully informative material describing any prior investigations and experience and any possible hazards, contraindications, side-effects, and precautions to be taken into account in the course of the investigation.

b. The investigator is required to maintain adequate records of the disposition of all receipts of the drug, including dates, quantities, and use by subjects, and if the investigation is terminated, suspended, discontinued, or completed, to return to the sponsor any unused supply of the drug. If the investigational drug is subject to the Comprehensive Drug Abuse Prevention and Control Act of 1970, adequate precautions must be taken including storage of the investigational drug in a securely locked, substantially constructed cabinet, or other securely locked substantially constructed enclosure, access to which is limited, to prevent theft or diversion of the substance into illegal channels of distribution.

c. The investigator is required to prepare and maintain adequate and accurate case histories designed to record all observations and other data pertinent to the investigation on each individual treated with the drug or employed as a control in the investigation.

d. The investigator is required to furnish his reports to the sponsor of the drug who is responsible for collecting and evaluating the results obtained by various investigators. The sponsor is required to present progress reports to the Food and Drug Administration at appropriate intervals not exceeding 1 year. Any adverse effect that may reasonably be regarded as caused by, or probably caused by, the new drug shall be reported to the sponsor promptly, and if the adverse effect is alarming, it shall be reported immediately. An adequate report of the investigation should be furnished to the sponsor shortly after completion of the investigation.

e. The investigator shall maintain the records of disposition of the drug and the case histories described above for a period of 2 years following the date a new-drug application is approved for the drug; or if the application is not approved, until 2 years after the investigation is discontinued. Upon the request of a scientifically trained and properly authorized employee of the Department, at reasonable times, the investigator will make such records available for inspection and copying. The subjects' names need not be divulged unless the records of particular individuals require a more detailed study of the cases, or unless there is reason to believe that the records do not represent actual cases studied, or do not represent actual results obtained.

f. The investigator certifies that the drug will be administered only to subjects under his personal supervision or under the supervision of the following investigators responsible to him,

Fred J. West, M.D.

and that the drug will not be supplied to any other investigator or to any clinic for administration to subjects.

g. The investigator certifies that he will inform any subjects including subjects used as controls, or their representatives, that drugs are being used for investigational purposes, and will obtain the consent of the subjects, or their representatives, except where this is not feasible or, in the investigator's professional judgment, is contrary to the best interests of the subjects.

h. The investigator is required to assure the sponsor that for investigations involving institutionalized subjects, the studies will not be initiated until the institutional review committee has reviewed and approved the study. (The organization and procedure requirements for such a committee should be explained to the investigator by the sponsor as set forth in Form FD 1571, division 10, unit c.)

Very truly yours,

John Smith, M.D.
(Signature of Investigator)

City Hospital
(Address)

New York, N.Y.

(This form should be supplemented or amended from time to time if new subjects are added or if significant changes are made in the plan of investigation.)

00 NS 9103(3) PRINTED IN USA OCT 81

RETURN TO INDIANAPOLIS – FOR SUBMISSION TO REGULATORY AFFAIRS

Figure 5 (Continued)

VI. BEFORE THE STUDY

A. Planning the Monitoring Process

Once the drug research plan has been finalized, it is necessary to carefully lay out the steps of the monitoring process. Management of the numerous clinical activities in the drug development process, as shown for example in Fig. 2 for Plan B, requires careful planning of the sequence of activities and time scheduling for each of the resultant events. The project-monitoring team must see that the activities necessary for completion of each event, e.g., protocol and CRF, start-up visit, close-out visit, and finally NDA, are completed in proper sequence and according to the study-plan timetable.

Using a flow diagram (Fig. 2) it is possible to develop a network of activities with time estimates for achieving each of these events. However, this usually is not done in such detail because it is too time consuming and too difficult to make such estimates with precision. Today's practice involves defining the start date, then estimating the NDA decision point and NDA submission date. Once this is done the monitoring team attempts to keep the project on schedule.

Two techniques that may be applied, at least in part, to the management of drug development, are PERT (Project Evaluation Review Technique) and CPM (Critical Path Management):

1. PERT is applied to research activities where exact time estimates are difficult to determine. This method utilizes the calculation of probability distribution for projected activity duration using three time estimates—optimistic, most likely, and pessimistic.
2. CPM is a technique that involves identification of the critical path through the research activities network, i.e., events schedule, then managing the project accordingly. The critical path is a path through the activities network that takes the longest time to complete and cannot be shortened. Thus, these critical activities can be identified for special attention in terms of resources, since they are the activities that are "rate limiting."

Both of these techniques require organization of the various activities in the drug development process into an activities (clinical studies) network. The activities are listed in the network in the order, according to time sequence, in which they are to be performed. Time estimates then are assigned to each one of the activities leading to desired events, e.g., IND filing date, NDA decision point, or NDA submission date. Thus, it becomes apparent that some activities can be accomplished concurrently whereas others must be done sequentially. Early in the development of a drug, activities usually are sequential; however, after the NDA decision point is reached, they usually are concurrent. Therefore, it is desirable to develop a critical path of activities with time estimates for each of the phases of drug development.

Research planning involves definition of various activities with sequencing of those activities and, therefore, preparation of an activities network, which in turn leads to time and events scheduling. Such techniques are useful for the effective management of the drug development process. Interested readers should refer to standard texts on project management for more information on CPM and PERT.

Although the ultimate responsibility for the conduct of the clinical trials rests with the clinical monitor, it is incumbent upon the monitoring team to devise an overall tactical plan for implementation of the trials. Medical research management, with the help of others, such as a protocol committee, a grants committee, and marketing representatives, can define the scope and direction of the trials. With a plan in place, the monitoring team can now set about the task of preparing the materials required for the implementation of the trial.

B. The Investigator's Brochure

A manual to guide the outside investigator must be prepared. This document is a summary of the research information prepared for the express use of clinical trial investigators. Sections are usually provided to reflect available data on the chemistry, preclinical pharmacology, toxicology, human pharmacology and toxicity, dosage and route of administration information, and formulation ("how supplied"). The manual must be comprehensive and up to date. The monitor and his associates are usually responsible for its preparation.

C. Clinical Trial Drugs

Clinical trial drugs must be produced and packaged well in advance of the anticipated starting date of the clinical trial. The package design is coordinated by the monitor and his staff with the production and packaging staff.

When the study is a blinded comparative trial, it is necessary to provide enough lead time to obtain supplies of the control drug and to reformulate it to look identical to the study drug. Finally, bioavailability must be determined for the control drug. Obviously, in the case of nonblinded, comparative, multiclinic studies it will be necessary to order adequate supplies of the control drug and repackage them as dictated by the protocol.

C. Protocols and Case Report Forms

Another step in the preparatory stages of the clinical trials program involves the preparation of the protocols and the CRFs. Since these subjects are discussed in detail elsewhere in this book, it is sufficient to say that the protocol must be explicit in detail and that the CRF must be integrated with the protocol, be easy to follow, and function as a data collection tool. The writing of the protocol is also a responsibility of the monitor and the staff, but it should always be subject to final review and criticism by other parties, e.g., the sponsor's Protocol Review Committee, outside investigators, and IRBs.

E. Investigator Selection

When all of the preliminary planning and preparation are completed, the process of investigator selection is initiated. The selection of investigators is a joint responsibility shared by the members of the monitoring team with the final selection resting with the monitor. Some companies employ an outside field staff to seek out reputable and qualified investigators. Others rely on the monitor to identify investigators.

The primary objectives of the investigator selection process are to ensure that the potential investigator is qualified to do the study and can complete it on schedule according to protocol. It is necessary, therefore, to establish that the investigator has an appropriate patient population as well as the time and ancillary personnel to fulfill all of the requirements set forth in the protocol. This selection process is the key to a smooth-flowing clinical trial, for it is the one activity over which the monitoring staff has limited control once the study starts. Careful attention should be given to the investigator's team members. It is important that they understand their role in performing the study. Frequently, the nurse coordinator, pharmacist, or resident physician can ensure the smooth running of a study.

F. Regulatory Requirements

The investigator must thoroughly understand the protocol and legal obligations as set forth in the FD 1572 or FD 1573 (Statement of Investigator) (Figs. 4 and 5) [4]. These regulatory documents must be completed by the investigator and sent to the sponsor for filing as part of the IND. The sponsor's representatives should conduct at least one prestudy visit to review the study protocol and CRF, along with the FDA requirements, as described in the FD 1572/1573 and other federal regulations. The materials necessary for submission to the investigator's IRB can be supplied at this meeting, i.e., investigator's brochure, CRF, protocol, and informed consent form. At the same time the funding requirements can be finalized and the terms of the study grant agreed upon.

G. Interrater Reliability Testing

Multiclinic trials usually necessitate interrater reliability testing of critical clinical and laboratory parameters. This process involves the training of observers in the techniques of scoring certain variables to be measured repeatedly by each center. The intent of this training is to develop a degree of consistency of scoring among observers in several clinics. Multiclinic trials of antidepressants, for example, are amenable to interrater reliability testing since a large number of subjective data points are collected and the data are pooled. When laboratory values are critical to assessing safety or efficacy, steps must be taken to ensure that inter- and intralaboratory variations will allow pooling of the laboratory data.

VII. BEGINNING THE STUDY: START-UP VISIT

When clinical trial drugs and CRFs have arrived at the study site, the clinical monitor and his associates usually schedule a start-up visit. During this meeting details of the study and requirements for regulatory compliance can be reviewed (see Fig. 6 for an outline of subjects to be discussed). At this time arrangements are made for periodic site visits by the sponsor's representatives. These follow-up visits will facilitate the review of CRFs and the verification of these data against the source documents, as well as provide assurance that the study is being properly conducted. (See Sec. VIII.B, below.)

The start-up visit will set the tone of the clinical trial by giving the investigator and his staff confidence in the sponsor as a partner in the project.

VIII. DURING THE STUDY

A. Contacts with the Investigator

The monitor should carefully follow the progress of the study for compliance with the protocol and governing regulations. At yearly intervals, the IRB must review the status of the study and issue a renewal, if needed [6]. The monitor should obtain a copy of this renewal for his files.

In the course of data evaluation, the monitor will be made aware of any departure from the protocol. The reasons for protocol violations should be investigated. Many times practical necessity may force an investigator to deviate from the protocol. If changes in the protocol are required, then the protocol must be amended and the IRB and FDA notified.

INFORMATION FOR CLINICAL INVESTIGATORS

Investigators who conduct clinical trial studies may be inspected by the FDA. We are asked from time to time what such an inspection would involve. Obviously, it is not possible to predict every type of inquiry, but we believe the outline below covers most of the issues you should be prepared to handle.

It is also important to remember you may be asked to supply documentary evidence to support your replies.

A. INSTITUTIONAL REVIEW BOARD (IRB)

1. If the study required **Institutional Review Board** approval, did the investigator submit the following materials and obtain approval from the IRB before any human subjects were allowed to participate in the investigation:
   a. Protocol or investigational plan
   b. Report of prior investigations
   c. Materials to be used in obtaining informed consent
2. Has the protocol remained unchanged with respect to:
   a. Patient selection
   b. Number of patients
   c. Frequency of patient observations
   d. Dosage
   e. Route of administration
   f. Frequency of dosage
   g. Blinding procedures (if applicable)
3. If the investigator modified his protocol, were the modifications **submitted to** and **approved by** the IRB prior to implementation?
4. Are all approvals of modifications and other required documents by the IRB in writing?
5. Are records maintained of all **submissions to** and all **actions by** the IRB regarding the study?

B. TEST ARTICLE ACCOUNTABILITY

1. **Who**, in addition to the investigator, **is authorized** to administer the test article?
2. **For human studies**, are additional investigators listed in the Form FD-1572, or FD-1573?
3. Are accounting procedures maintained for all usage of the test article, including records showing the following:
   a. Receipt date(s) and quantity received
   b. Date(s) and quantity dispensed, including identification of patients who received it
   c. Date(s) and quantity returned to sponsor or alternate disposition
4. Are records sufficient to allow a comparison of the investigator's test article usage against the amount shipped to him, minus the amount returned to the sponsor?
5. Were all unused supplies of the test article returned to the sponsor when either:
   a. The investigator discontinued or completed his participation in the clinical investigation, **or**
   b. The investigation was discontinued or terminated by the sponsor, **or**
   c. The investigation was discontinued or terminated by FDA
6. **If no above**, did alternate disposition of the test article expose humans or food-producing animals to experimental risk?
7. If a test article is a substance listed in any schedule of the Controlled Substances Act (21 U.S.C. 802; 21 CRF Part 1308), is it **securely locked** in a substantially constructed enclosure?
8. Is **access** to the controlled substance **restricted** by the investigator?
9. Did the investigator refrain from disseminating any promotional material representing that the test article is safe and effective for the purposes for which it is under investigation or otherwise commercialize the article?

C. PROTOCOL

1. Does the investigator have a **written protocol?**
2. Are all **changes** to the protocol, and **reasons therefor:**
   a. Documented by the investigator?
   b. Dated?
   c. Maintained with the protocol?
   d. Reported to the sponsor?

Figure 6 Information for clinical investigators. (Courtesy of Eli Lilly Research Laboratories, Indianapolis, Ind.)

D. CONSENT OF HUMAN SUBJECTS

1. Was **informed consent** obtained from **all subjects prior** to their entry into the study?
2. Was **written consent** obtained?

E. RECORDS REGARDING SUBJECTS

1. Does the investigator **maintain records** on each subject, which include:
   a. Observations, information, and data on the condition of the subject at the time the subject **entered** into the investigation
   b. Documentation regarding the **consent** of the subjects
   c. Records of **exposure** of the subject to the test article
   d. Records of **exposure** to any concomitantly or concurrently administered drugs, including date (and time, if relevant)
   e. Observations and data on the **condition** of the subjects throughout their participation in the investigation (This includes results of all laboratory tests performed and the appearance of factors which might alter the effects of the test article, e.g., development of unrelated illness.)

F. REPORTING

1. If the study is subject to the Institutional Review Board requirement, has the investigator submitted **periodic reports** to the IRB on the progress of the investigation at intervals **not** exceeding one year?
2. If the investigation has been **completed, terminated,** or **discontinued,** did the investigator submit a **final report within 3 months** of such action to the:
   a. Institutional Review Board
   b. Sponsor
3. If **adverse experiences** occurred was it determined if they were regarded as caused by or associated with the test article and if they were previously anticipated (in nature, severity, or incidence) in any written information regarding the article?
4. If the **adverse experiences** were regarded as caused by or associated with the test article, did the investigator submit a report of this to the:
   a. Institutional Review Board
   b. Sponsor
5. How are adverse experiences reported to the sponsor?
6. If deaths occured among study subjects, was it determined if they were regarded as **caused by** or **associated with** the test article and if they were previously anticipated in any written information regarding the article?
7. If deaths were regarded as **caused by** or **associated with** the test article, did the investigator submit a report of this to the:
   a. Institutional Review Board
   b. Sponsor
8. Were the reports concerning any events in Questions 4 and 7 submitted **within 10 days** after the investigator discovered the effects?
9. Does the investigator **maintain copies** of all reports submitted to the:
   a. Institutional Review Board
   b. Sponsor

G. RECORDS RETENTION

1. Does the investigator **maintain custody** of his required records?
2. If no above, did the investigator **notify the sponsor** in writing?
3. Are records retained for the specified timeframe as follows:
   a. A period of **two (2) years** following the date on which the test article is approved by FDA for marketing for the purposes which were the subject of the clinical investigation; **or**
   b. A period of **five (5) years** following the date on which the results of the study are submitted to FDA in support of an application or petition to approve the test article for marketing for the purposes which were the subject of the clinical investigation; **or**
   c. A period of **two (2) years** following the date on which the entire clinical investigation is terminated or discontinued

THE LILLY RESEARCH LABORATORIES

Figure 6 (Continued)

B. Quality Assurance of Case Report Forms

Quality assurance of clinical trial CRFs is a vital step in good monitoring practice. It not only serves to ensure proper data collection from source documents but also provides definitive feedback as to protocol compliance.

It is impossible for the sponsor to check all data on each CRF in every clinical trial. For that reason, quality assurance starts with instructing the investigators and their associates as to their responsibilities, as well as determining that each investigator has the time to do the study and has ample administrative help.

For the most part, it is prudent to check all CRFs from the pivotal controlled studies—at the beginning and as the studies progress. These are the "core" or multiclinic comparative studies that are the cornerstone of the NDA data base.

In the remaining studies the data from a specified number of CRFs (usually the first six to ten) should be checked against the source documents very early in the course of the trial. This review provides the clinical monitor with a decision point in terms of how well the study is being conducted. If these initial CRFs are accurately completed and in good agreement with the source documents, the investigator should proceed with enrollment of additional patients. If the contrary is true, the study can be terminated or corrective measures can be instituted. As the study progresses, further audits of the CRFs against the source documents can be done at intervals determined by the quality of the records.

Although the audit procedure may seem to be a time-consuming and expensive way of ensuring good study practices, experience has shown that it improves the overall quality of the study, thereby saving time and money. When CRFs are evaluated away from the study site, the data may appear to be in good order but in fact may not accurately reflect what is actually in the source document. Missing data are much more difficult and expensive to retrieve months after the patient has been treated. Furthermore, the investigator is operating at a disadvantage if he does not know the status of his study early in its course.

The audit procedure serves an added function of maintaining the investigator's files in proper order in case they are inspected by the FDA. Studies should be organized to facilitate such an inspection (see Table 2) [4].

C. Internal Quality Control

Quality assurance of study-generated data is but one step in good clinical trial management. All completed CRFs, including those that are evaluated thoroughly at the study site, are forwarded to the spon-

Table 2 Clinical Investigator Inspection Report (Items to be checked by FDA inspector)

A. Visits by the monitor to the clinical investigator:
Were visits made prior to the trial and was determination made of the investigators' qualification and work load? Did the monitor visit with the investigator during the study?

B. Institutional Review Board (IRB):
Is the study subject to IRB review and has it been submitted for consideration before human subjects were allowed to participate in the investigation? Has a copy of the informed consent been submitted to the IRB? Has the protocol remained unchanged and if not has the IRB been informed? Are all IRB records maintained?

C. Test article accountability:
Who is authorized to administer the test drug? Are they listed in the FD 1572/3? Are test drug accounting records maintained? Is storage of the test drug adequate?

D. Protocol:
Is there a written protocol and has it remained unchanged? Are all changes documented, dated, and reported to the sponsor and IRB?

E. Consent of human subjects:
Was informed consent obtained from all subjects prior to their entry with the study?

F. Records regarding subjects:
Did the investigator maintain records on which case reports are based on each subject, including: condition of the subject; consent; records of administration of the test drug; exposure to concomitantly administered drugs; records of condition of subject throughout their participation in the study; and identification of all persons involved in generating raw data?

G. Reporting:
Did the investigator send periodic reports to the IRB and to the sponsor? Did the study terminate before completion? Did the investigator submit a final report to the sponsor and IRB within 3 months of completion? Have adverse reactions occurred, and were they related or unrelated to the test drug? Were they reported to the sponsor and IRB? Have deaths occurred; if so, were they reported to the sponsor and IRB?

H. Records retention:
Does the investigator have custody of his records? Are records retained for 2 years following the date on which the test drug is approved by FDA for marketing or 5 years following the date on which the results of the study are submitted to FDA or 2 years after the study is terminated?

Source: Modified from the *Food and Drug Administration Compliance Program Guidance Manual*, Chap. 48.

sor. They are first "logged in" to the medical department, then evaluated by members of the data management team and the clinical monitor. The logging system consists of a computer program which tracks the progress of each CRF through the data management system. The system can supply status reports to the monitoring team at any point in the study and tell them the condition of the data (degree of completeness or accuracy) for each visit, each patient, each investigator, or for the entire project. During the CRF review the clinical monitor must pay particular attention to reports of adverse drug experiences.

One of the primary goals of the clinical monitoring team is to establish and maintain a "clean" data base and, through this data base, track the progress of the clinical trial. If the data base is "dirty," e.g., missing or obviously incorrect data, the resulting conclusions will not be credible. Any flaws found in a study should be corrected and reviewed carefully with the investigator and members of the investigator's team.

D. Adverse Drug Experiences

Federal (FDA) regulations [4] and good monitoring practices dictate that all serious adverse drug experiences be brought promptly to the attention of the sponsor by the clinical investigator. The sponsor must in turn notify other clinical investigators (thus their respective IRBs) and the FDA. The CRF logging system provides an early warning for all adverse drug experience information. Such reports allow the clinical monitor and research management to make decisions about the fate of the drug or the design of the trial.

Adverse drug experiences which occur after a drug has been marketed must also be reported to the FDA. This is done by completing a Drug Experience Report Form (FDA 1639) (see Fig. 7) [7]. The investigator or practicing physician may fill out the form; however, the frequent practice is that they notify the sponsor of a suspected side effect and the sponsor completes the form. Such reports should be carefully documented, and the occurrence of suspected adverse drug experiences carefully evaluated by the monitor. The occurrence of suspected adverse drug experiences may dictate that prospective clinical trials be undertaken to further define their incidence and importance. If such reports are not reflected in the package literature,

---

Figure 7 Drug experience report. (Courtesy of Food and Drug Administration, U.S. Department of Health and Human Services, Rockville, Md.)

DEPARTMENT OF HEALTH AND HUMAN SERVICES
PUBLIC HEALTH SERVICE
FOOD AND DRUG ADMINISTRATION
ROCKVILLE, MD 20857

FORMAT APPROVED 5/11/82 by FDA as required by 21 CFR 310.301 (a)(1).

FDA CONTROL NO.

ACCESSION NO

DRUG EXPERIENCE REPORT

I. REACTION INFORMATION

1. Patient ID/Initials (In Confidence)	2. Age	3. Sex	4. Wgt (lbs.)	5. Ht. (in.)	6. Reporting Date	7. Reaction Onset Date
R. R.	65	F	125	5'5"	1/2/82	11/25/81

8. Describe Suspected Reaction(s)

Tinnitus and Nausea after 2 days of therapy

11/25/81

9. Outcome of Reaction to Date
- [ ] Alive with sequelae
- [x] Recovered
- [ ] Still under treatment for reaction
- [ ] Died (Give cause/date)

10. Tests/Laboratory Data Confirming Reaction (Include biopsy and/or autopsy results)

None

11. Was Outpatient Treatment for Reaction Required? [ ] Yes [x] No

12. Was Hospital Treatment for Reaction Required? [ ] Yes [x] No

II. SUSPECT DRUG(S) INFORMATION

13. Suspect Drug(s) - Trade/Generic Name(s), Manufacturer, IND/NDA No

Aspirin (325 mg)

14. Total Daily Dose

3 grams/day

15. Route of Administration

Oral

16. Indication(s) for Use

Pain due to arthritis

Therapy Dates (*From/To*)

11/25/81 | 11/27/81

18. Therapy Duration

1/25/81 to 11/27/81

19a. Was Treatment with Suspected Drug Reduced in Dosage? [ ] Yes [ ] No OR [x] Discontinued

19b. Did Reaction Abate? [x] Yes [ ] No

20a. Was Drug Reintroduced or Dose Increased? [ ] Yes [x] No

20b. Did Reaction Reappear? [ ] Yes [x] No

III. RECENT/CONCOMITANT DRUGS AND MEDICAL PROBLEMS

21. Other Drug(s)	Total Daily Dose	Route	Dates/Duration of Administration (*From/To*)	Indications
Milk of Magnesia	1 Tbsp	PO	11/25/81 - 11/27/81	Constipation

22. Describe Other Relevant Medical History (i.e. allergies, environmental or occupational exposure, previous drug reactions, pregnancy with gravidity/parity)

None reported

Your cooperation is needed to insure comprehensive, accurate, and timely use and interpretation of these data.

23. Mfr. Name and Address

Eli Lilly and Company
307 East McCarty Street
Indianapolis, Indiana 46285

24. Check One
- [x] Initial Report
- [ ] Follow-up Report

25. Reporter's Name and Address (In Confidence)

John Smith, M.D.
City Hospital
Any town, IN 46000

Mfr. Control No

Date Sent to FDA

NOTE: *Required of manufacturers by 21 CFR 310.300, 310.301 and 431.60. Manufacturers may attach additional clinical material and product analyses at their discretion.*

26. May the Source of this Report Be Released to the Armed Forces Institute of Pathology? [ ] Yes [x] No

FORM FDA 1639 (5/82) PREVIOUS EDITIONS ARE OBSOLETE

00 NS 9563 PRINTED IN USA

ORIGINAL

Figure 7

amendments to the package literature should be sought with the FDA and international regulatory agencies.

E. Drug Accountability

Accounting for clinical trial (CT) drugs, once sent by the sponsor to the investigator, is a shared responsibility between them. The investigator is accountable for the disposition of all CT drugs, and periodic auditing of records should be made by the sponsor. The task may be facilitated when the hospital pharmacy is responsible for dispensing the CT drugs; however, good record keeping still applies. A visit with pharmacy services at the start of the study will establish the procedures for accountability. The investigators should be supplied with a master drug inventory form, which is used to log in shipments of CT drug and is reviewed at each visit by the sponsor representative.

F. Clinical Trial Records Notebook

Organization of the documents relevant to the clinic trial will facilitate the clinical trial and aid with inspections by the sponsor or FDA. This can be done by the use of a loose-leaf clinical trial records notebook, which is designed to hold records, notes, and letters that pertain to the study. It should contain the following sections:

1. Study identification—CT drug IND No., Protocol No. and title, sponsor personnel, etc.
2. Record of visitors—A sign-in log to be used by visitors from the sponsor
3. Drug records—log of CT drugs received and used
4. Correspondence
5. Study documents—copy of protocol, FD 1572 or FD 1573, IRB approval, consent form, etc.

IX. COMPLETION OF THE STUDY

A. Close-out Visit

When a clinical trial is completed, the monitor or a member of the monitoring team should visit with the investigator to formally terminate the study. This is the time to discuss any pending matters and retrieve unused clinical trial drugs, forms, etc. The investigator should be advised that FDA regulations require the retention of all records of the study for a period of two years following approval of the NDA (the investigator is automatically notified by the sponsor when NDA approval is received) [4].

B. Monitor's Termination Report

Documentation to terminate a clinical trial can take any format, but it is necessary for the sponsor and investigator to determine that all FDA requirements have been satisfied. The final payment of the grant-in-aid should be discussed and agreed upon (see Table 3).

A final report by the sponsor consisting of data tabulated from the clinical trial results can serve as the framework for a manuscript if an investigator decides to publish the study findings.

The data and results from the "core" individual studies and collective findings from all pooled data will be used to form the corporate data base for support of claims in the NDA and international registration documents. The package insert, product monographs, and promo-

Table 3 List of Drug Informational Needs of the Sponsor

For sponsor management:

1. Status reports on the research plans (Table 1)
2. Budget and manpower needs reports
3. Publications in medical journal and books
4. Presentations at scientific meetings
5. Support advertising and promotion claims
6. Preparation of drug monograph for use by health professionals
7. Answer questions from health professionals
8. Respond to legal claims, actions, and suits
9. Innovation—leads for new claims, dosage schedules, modes of administration, and comparisons of safety and efficacy of competitive products

For the investigator:

1. Progress and final study reports
2. Reports from other centers regarding untoward events
3. Summarization of data for presentation at scientific meetings
4. Publication of study in medical journal or book
5. Grant payment information

For the FDA:

1. IND and annual reports
2. NDA (includes package insert) and periodic reports

tional and advertising materials are derived from and supported by this corporate data base.

X. EMERGENCY REQUESTS FOR CLINICAL TRIAL DRUGS

Occasionally a physician may request supplies of a CT drug for emergency use in a specific patient or patients. Such uses of the drug do not yield much useful information; however, they are usually honored for humanitarian reasons.

Obviously, the monitor must determine that the physician requesting the drug is qualified and has the laboratory facilities to properly monitor the patient receiving the drug. Further, a determination must be made that a bona fide emergency exists and that other available drugs have been unsuccessfully tried or are not appropriate for good medical reason.

The physician must (1) fill out the appropriate FDA forms (usually FD 1573) and provide a brief summary of the patient's condition; (2) agree to furnish a summary (CRF); and (3) comply with FDA requirements relating to the use and study of nonapproved drugs [see Form 1573 (Fig. 5)].

In these situations, approval by the IRB is not required prior to use of the drug; however, the proper forms should be completed and the IRB notified as soon as possible [6].

XI. COMMON PITFALLS

There are numerous pitfalls in the drug development process that the project-monitoring team must be aware of and deal with in order to effectively manage the monitoring process. Among the more common are the following:

1. Failure to define objectives of the clinical research plans that are explicit and achievable
2. Focusing attention on the expected results of the clinical trials rather than on their implementation
3. Faulty development of protocols and CRFs; unduly complicated protocols and CRFs (with unnecessary data) may appear scientific but often are detrimental to an orderly and economical research program
4. Failure to adequately communicate with investigators and their associates regarding their obligations
5. Inadequate review of the documents sent to the IRB by the investigator (particularly the completeness of the informed consent form)

6. Inadequate attention to the many regulatory issues that impact on the approvability of the NDA
7. Failure to provide for the management of the data so that needs other than those that are purely regulatory can be efficiently met
8. Insufficient or faulty data from the marketplace regarding the need for the drug under study and its desired characteristics
9. Inattention to the organization, staffing, and functioning of the monitoring team
10. Inappropriate or careless medical interpretation of data base results

XII. SUMMARY

Drug development is an important medical as well as societal activity since it puts powerful tools in the hands of physicians for the control of bodily processes in health and disease. Acquisition of knowledge, based on scientific methods, is the objective of the drug development process.

The monitoring process, a relatively new scientific management tool, is essential to the orderly and economical progression of research activities. Conclusions regarding the basic properties and clinical usefulness of a drug are dependent upon the generation, collection, retrieval, and interpretation of the data. Such a process is vital to meeting the informational needs for any drug. Therefore, it is essential that the data handling process be managed in a manner that will facilitate the sponsor's ability to meet his own informational needs and also those of the investigator and the regulator (see Table 2).

A complete drug development program has both regulatory and scientific objectives. Well-controlled clinical trials are, by FDA regulations, a required part of the substantial evidence of efficacy that must be shown before a "new drug" may be marketed in the United States. This requirement rests on proof that the clinical studies are "adequate and well controlled" [5].

Lest we forget, while the research efforts of drug development focus on specific plans, chance discoveries are almost certain to be encountered by those with curious, intelligent minds. Joseph Henry, an American physicist, touched on serendipity in research when he said, "The seeds of great discoveries are constantly floating around us, but they only take root in minds well prepared to receive them" [8].

In essence, the monitoring process is not measured so much by commercial considerations as it is by scientific ones. In the end, the monitoring process plays a vital role in the development of a drug by establishing a sound data base which, with use of proper scientific methods, will lead to credible conclusions regarding the basic properties and clinical usefulness of the drug.

REFERENCES

1. A. B. Hill, *N. Engl. J. Med. 248:*995 (1953).
2. F. Marti-Ibanez, (Ed.), *The Epic of Medicine*. Potter, New York, 1959, p. 256.
3. G. G. Udell, Federal Food, Drug and Cosmetic Act with Amendments; Drug Amendments of 1962, Public Law 87-781, 87th Congress, S-1552, Oct. 10, 1982. U.S. Government Printing Office, Washington, D.C., 1976.
4. Title 21, Code of Federal Regulations, Food and Drugs, subsection 312.1. U.S. Government Printing Office, Washington, D.C., 1981.
5. Title 21, Code of Federal Regulations, Food and Drug, subsection 314.111. U.S. Government Printing Office, Washington, D.C., 1981.
6. Protection of human subjects: Standards for Institutional Review Boards for Clinical Investigations. *Federal Register 46:*No. 8958 (Jan. 27, 1981).
7. Title 21, Code of Federal Regulations, Food and Drugs, subsection 310.301. U.S. Government Printing Office, Washington, D.C., 1981.
8. Samuel Rappaport and Helen Wright (Eds.), *Great Adventures in Medicine*. Dial Press, New York, 1952, p. 522.

13
DRUG SAFETY EVALUATION AND IMPLICATIONS FOR CLINICAL INVESTIGATION

Richard L. Steelman
McNeil Pharmaceutical
Spring House, Pennsylvania

I. INTRODUCTION

This chapter will be most meaningful if Chapter 1, "Pharmogenology: The Industrial New Drug Development Process," is briefly reviewed and if Chapter 14, "Special Considerations of Drug Disposition," is read in conjunction with it.

The purpose of this chapter is to present information on the number, quality, and timing of animal studies which precede or accompany clinical trials; also, on the use of animal toxicology data in conducting clinical trials; and, finally, on the areas where preclinical studies are of no value or can be misleading.

Drug safety evaluation or safety assessment or pathology and toxicology studies, as they are variously known, play an important role in every phase of drug development from the preclinical identification of early biological activity of possible therapeutic interest to postmarketing activities. Acute and subacute (subchronic) studies usually precede Phase I clinical trials. Chronic studies and special studies for carcinogenicity, mutagenicity, and reproductive studies usually precede New Drug Application (NDA) filing. Special studies on drug interaction, drug disposition, and overdose are often done before marketing. Postmarketing studies are done on the original dosage form following adverse reactions reported to the U.S. Food and Drug Administration (FDA), the pharmaceutical company, or international regulatory agencies. Experiments are sometimes conducted in response to questions from pharmacists, nurses, doctors, or patients. Examples of such experiments include: eye and skin studies relating to accidental contact of the drug to these tissues; compatibility studies of the drug with standard intravenous solutions; compatability studies with newly marketed agents to be given in conjunction with the drug in question; and, occasionally, studies in support of legal proceedings against the doctor, hospital, or drug firm relating to real or alleged adverse effects of the drug. Until a drug is withdrawn from all clinical use for economic, medical, or regulatory reasons, it cannot be forgotten by toxicologists and pathologists.

II. PRECLINICAL INFORMATION PACKAGE

When a clinician or clinical pharmacologist is first approached by a pharmaceutical company regarding his or her interest and willingness to conduct a clinical trial of a new drug, combination, or dosage form, an information package is usually available for discussion. This package is often called an investigational new drug brochure or preclinical information package. In addition to chemical structure, physical chemistry, and pharmacological or biochemical efficacy studies in vitro and in vivo, the package contains relevant toxicology data. An example of the contents of an investigational new drug brochure is shown in Table 1. It is important to understand that the efficacy and safety data presented are often not a complete representation of everything that has been done. Studies deemed not highly relevant are often omitted. Critical studies are often condensed by the original authors or by medical writers preparing the brochure. The final brochure is tailored to fit the perceived needs of the average clinician. If you want additional information about specific studies or want to know if additional work has taken place or has begun since the brochure was written, the medical monitor is the best person to put you in touch with the research or pathology and toxicology staff qualified to answer the question. These questions are welcomed because they demonstrate critical review of the brochure and help keep the preclinical research staff in touch with the real needs of clinical investigators.

Table 1 Investigational New Drug Brochure: Zomax (Zomepirac Sodium)

Table 1 (Continued)

A. Acute Studies

The toxicology and pathology portions of clinical brochures usually begin with a listing and discussion of acute toxicity tests. These include safety studies in which the drug is given once and animals are observed for 7-14 days for pharmacological signs and death. The most common test is the LD_{50}, which is the dose determined from a dose-response curve to kill 50% of a randomly selected population of animals. In addition to being the most common acute test, the LD_{50} is also the most misused and misleading test in the hands of some who conduct the test and some who interpret the results. A recent critical review by Zbinden and Flury-Roversi [1] identifies and discusses the problems, giving excellent examples. Humane groups, including humane scientists, are also concerned about the number of animals used (often 100 rodents or 10-15 dogs, cats, monkeys or other large species) to determine an LD_{50}. Despite this preamble, when conducted and reported properly, the LD_{50} can provide useful information. The minimal information accompanying the LD_{50} should be the species, age, strain, sex, number housed per cage, confidence limits, slope of the dose-response curve, relevant gross autopsy results, and duration of observation. Additional useful information includes onset and duration of physical signs, mode of death when known, average time of death, and comparative LD_{50} of a related drug or drugs. Remember that the whole purpose of the LD_{50} is to put the lethality of the new agent in perspective with similar agents from the same or related classes or with other agents for treating the same disease. Note also that variables affecting the LD_{50} include age [2], weight [3], temperature [4], sex [5], social factors (number per cage) [6], fasting, noise, and many more. Comparisons of LD_{50}'s between laboratories have been notoriously difficult [7].

Other acute studies which are done early in the development of a drug include skin and eye irritation tests and the Ames test [8-11] for mutagenicity. The skin and eye tests are often done on intermediates used in the manufacture of the final product, as well as on the final product. Such tests help to reassure those who handle the drug and its chemical intermediates that any hazard such as irritation has been studied and appropriate warnings affixed if needed. These studies do not predict skin sensitivities that may develop over longer exposure or photosensitization problems.

The Ames test is sometimes done early because it is a mutagen test, and a large series of known carcinogenic compounds have given positive mutagenic signs in this test. Those wishing to determine early in the development of a compound whether they may face extensive mutagenicity testing later on can conduct this test to exclude those chemicals that are mutagenic to certain special strains of *Salmonella typhimurium*. A negative response in this test does not necessarily mean that the compound will be free of mutagenic or carcinogenic effect in other in vitro or in vivo tests.

B. Subacute (Subchronic) Studies

The older toxicology literature described three general kinds of toxicity tests based on the length of dosing employed. They were acute (single dose to 1-week dosing), subacute (2 weeks to 3 months) and chronic (6 months to lifetime). Recently the term "subchronic" has been used to describe subacute testing because of the confusion sometimes caused by the word "subacute." Regardless of which term is used, prolonged studies of 2 weeks to 3 months are included.

In general, before a drug is tested in humans for repeated administration longer than 1 week, a subacute study in two or three other species should be completed. These studies are difficult to design and conduct and require all the skills possessed by statisticians, drug metabolism scientists, toxicologists, and pathologists. In most scientifically advanced countries these studies are regulated by health regulatory agencies and strict quality control measures are in effect.

Although rats, dogs and nonhuman primates are most often used for subacute studies, this is often a matter of convenience because of historical information and familiarity with these species. Emphasis recently has been placed on identifying the species which metabolizes the drug most like man. This is difficult because it requires early development of specific drug assays in biological fluids, often at nanogram or smaller quantities. Early synthesis of radioactive compound suitably designed for metabolism studies may also be required. Some of the drawbacks of this concept are well stated by Beyer [12]. He writes:

> There is a certain popularity to the notion of finding a species for toxicity studies that most closely approximates the metabolism of a specific drug by man. This gives an aura of authenticity or credibility that the results will more certainly resemble an effect in man. This illusion is apt to be more fiction than fact in my experience, for there is a great deal more to the interaction of subject and drug than the metabolism of the compounds. Indeed, something is to be said for selecting one of the species for the safety study that is dissimilar in the metabolism or overall elimination of the drug. Man is about as heterogeneous a species as we are likely to encounter with respect to both numerous genetic and environmental factors that affect drug disposition and action.

The next hurdle after species selection is route of administration. It should be as close as possible to the proposed human use. This is not always as easy as it sounds. Suppose the proposed route is an oral tablet for humans: Rats cannot swallow a tablet. The tablet can be crushed or the active ingredient used to prepare a solution or suspension which can be given by gavage to rats, or the material can be uniformly mixed in the food of rats. Which most closely mimics the human situation? According to Worden and Harper [13], it is generally accepted that the toxicity of a drug given by oral gavage is quite different from its toxicity when it is given by admixture in the diet. The examples in their paper included one in which the diet reduces the toxicity, one in which the diet increases the toxicity, and one in which the substance may render the diet "toxic" by rendering a vital component ineffective. As a result of such differences, it is sometimes necessary to try both gavage and drug in food to determine the best approach for subacute and chronic dosing.

Mimicking the human dose also involves pharmacokinetic considerations. Should intubation in the rat or tablet or capsule administration to dogs be done once, twice, three times, or four times per day? Factors such as time and degree of absorption, first-pass effect, half-life in plasma, area under the plasma curve, tissue distribution, and metabolism rate and pathways are all operative just as they are in the clinical setting, but often with quite different results. With the advent of long-acting preparations it may be proper to dose animals only once a week or every 2 weeks if plasma levels or adverse effects suggest this is appropriate and such a schedule is to be used clinically. Our personal experience has run the gamut from dosing every half hour for 24 hr with a topical eye preparation for herpetic keratitis to once a week with a long-acting oral neuroleptic.

Selection of dosage levels is based on all available pharmacological information, pharmacokinetic information, acute toxicity data, experience of the toxicologist, and some luck. It is nice if the range of dosage levels selected can include: (1) a dose which produces minimal toxicity that is above the proposed clinical dose; (2) a maximum tolerated dose in animals as measured by body weight loss, alterations in hematology or clinical chemistry, or histologic changes; and (3) a dose between these extremes. Because of species differences (see Baker [14], it is not unusual for one or more of the animal species to be more sensitive than humans. In these cases, the low dosage or even all dosage levels in prolonged experiments may be a *fraction* of the usual human dose. An example is the phenothiazine tranquilizers, which are not well tolerated by rodents because of effects on food intake related to catalepsy and are not well tolerated by dogs because of emesis and central nervous depression. A second example includes many nonsteroidal antiinflammatory drugs and the newer analgesics of this class because of the sensitivity to gastrointestinal ulcers which is more marked in rodents, dogs, and cats than in humans. So rules of thumb are out in selecting dosage levels and pilot studies and utilization of the available biological studies is preferred. It is quite permissible to alter the dosage level or schedule during the course of subacute experiments so long as this is done in a way which does not confound the results. It is sometimes necessary to lower the high dose in order to ensure survival of an adequate number of animals for statistical evaluation at the end of the experiment. Occasionally all three dose levels have to be reduced to obtain the objectives of the study. A poor study is one in which the dosage levels are adjusted in such a way as to obscure the time when toxicity occurs or to observe the extent of toxicity. Another hallmark of a poor study is one in which all of the doses are so low that little or no toxicity occurs at any level. This gives some information about safe levels in the species tested but none about overdose or sensitivity or possible target tissues when the data are extrapolated to man. It is also permissible to increase the high dosage or all of the dosage levels if this is necessary to reach toxic levels or to overcome tolerance to the drug.

Interpretation of the results of subacute studies has not received as much attention as it should in toxicology monographs and textbooks. Because the protocols are designed by statisticians and drug safety evaluation staff to minimize variation and obtain several sets of baseline values, the data are in many ways similar to those obtained in carefully conducted double-blind clinical trials. Clinical chemistry, hematology, and urinalysis values can be very closely controlled since the animals in a study have the same genetic background and are screened for abnormalities before dosing has begun. Hospital technicians coming into a pharmaceutical clinical laboratory must quickly learn that very small differences from baseline are important since they may confer statistical significance to a group when compared to controls. Quality assurance procedures prevalent everywhere in drug safety studies are particularly reinforced in clinical laboratories by utilization of several different survey programs and frequent control serum samples. The result of this attention of small changes mean that many comparisons of treated groups compared to control groups or other treated groups show statistically significant changes. Not all of these changes are clinically significant. The following are some pertinent questions to be asked: (1) Are the changes dose related? (2) Are they time related? (3) Are they within or beyond normal limits for the species, strain, sex, and age group studies? (4) Are there sex differences? (5) Has any attempt been made to determine reversibility of the change? If the reversibility of serious changes in clinicopathological tests has not been addressed in the main toxicology study, a separate study should be conducted to answer this question.

Subacute toxicity study design should include provisions that animals dying during the study have a necropsy performed while tissue is still suitable for critical evaluation. At the end of the study all animals should be necropsied. Histological examination of as many animals as practical should be done. At minimum, all control and high-dosage group tissues should be studied. When lesions are found in the high-dosage group in specific organs or tissues, the low- and middle-dosage groups should have those organs and tissue examined.

The value of organ weights obtained at necropsy is controversial. In order for the weights to have any usefulness, the organ must be carefully trimmed of fat and fascia and then weighed on a scale of high accuracy before drying occurs. Uusually, both absolute weight and organ to body weight or brain ratios are presented. Often the group values from such tables appear to show random changes from control group and to be not subject to explanation. Other times some useful information regarding fat deposition, hyperplasia, tumor incidence, or other general effects are present in organ weight data. Significant histological changes should be summarized by the pathologist in such a way as to highlight problem lesions or target organs. A separation should be made between spontaneous lesions occurring in the study and lesions which might be possibly, probably, or definitely related to treatment. Correlations should be made between during-life and postmortem events when possible. Physical signs, pharmacological signs, clinical laboratory values, gross necropsy findings, and histopathology findings should be correlated when possible. Speculation is legitimate only if clearly identified. Beware of phrases like "the [such and such] changes seen were believed to be unrelated to treatment" if the expert making the statement does not give the reasons why he or she believes they are unrelated. Remember that the scientists writing the final subacute toxicology report must deal with thousands of data bits, hundreds of correlative data points, and dozens of values statistically different from control. They will not all be resolved into a common explanation or postulate. You may have a different interpretation of the same data.

The during-life phase of these experiments contains much of value regarding onset, degree, duration of effects, cumulation, tolerance, and similar useful information. However, the events that most often cause problems in clinical trials are changes in clinical parameters that signal possible morphological changes. Thus clinical and histological changes are given great weight. If you are working or considering working for the first time in a clinical trial with a particular pharmaceutical firm, it might be valuable to request a complete copy of the text and tables of one or more of the subacute studies summarized in the investigational new drug brochure. Short of visiting the facility, there is no better way to assess factors such as completeness of necropsy procedures, detail of histological description, statistical treatment of the data, and credentials of the participants. Since it is true that no diagnosis can be made histologically on an organ lost at necropsy or a lesion trimmed into the waste can or a slide cut too thick or stained improperly, it is important to have confidence in the firm conducting the work.

C. Chronic Studies

The duration of chronic studies ranges from 6 months for registration in certain countries to 7-10 years in dogs or monkeys for special risk categories such as oral contraceptives and beta-blockers. Some laboratories prefer to conduct chronic studies and carcinogenicity studies separately. Others combine the chronic and carcinogenicity studies in rodents in the same experiment. Except that the doses may be slightly lower because of the possibility of chronic lesions, chronic toxicity studies are designed much in the same way as subacute studies. The intervals for periodic measurements are usually logarithmically spaced so that more frequent measures of clinical lab values, electrocardiograms, etc. occur early in the study and less frequent measures occur toward the end. In carcinogenicity experiment designs, the objectives are to minimize stress by reduced handling and to maximize survival so that an adequate number of survivors is present at the end of the study to assess relatively small differences in tumor incidence between treated and control groups. Of course, sufficient dosage must be administered to produce minimal toxicity in the high-dosage group. Often a 5-10% decrease in body weight is accepted as the measure of toxicity.

Interpretation of carcinogenic studies is the most important problem confronting regulatory and industry toxicologists and pathologists today. The root of the problem is the gradual extension of the length of carcinogenicity studies from 12 to 18 months or lifetime in mice and from 18 to 24 months or lifetime in rats by U.S. regulatory authorities. As the rodents age, the incidence and severity of many tumors and other lesions increase until the point when it is extremely difficult to measure real changes in incidence of drug-induced lesions. Salsburg [15] has stated the problem as follows:

> Long term chronic feeding studies in mice and rats pose serious statistical and toxicological problems in the analysis of their results. [We and Fears et al.] have pointed out that the use of statistical tests of hypothesis as widely applied are inappropriate if one seeks to detect the effects of treatment in terms of rare lesions and can lead to a high false positive rate if applied blindly to all observed lesions. Even when statistical tests are used with caution, the biological implications of those tests are difficult to interpret.

Salsburg goes on to develop the idea that the goal of subacute and carcinogenicity studies is to detect the shift in patterns of lesions that are dose related. Relatively simple patterns of lesions occur in subacute tests but as time is extended to 18 months, 24 months, or lifetime, the lesion patterns are much more complicated. The current statistical methods treat the results as if each lesion or tumor was a unique event, and so many individual comparisons are made that the probability of finding at least one false positive in a study may be over 50%.

It is interesting that in many carcinogenicity studies there are often groups with statistically significant reductions in lesions or tumors or increased life span and no attention is paid to these findings. But should one group have significantly greater incidence of tumors than the control group, the compound may be barred from human study or, at the least, carry serious warnings about carcinogenicity findings in the package insert and in all fair-balance advertising material. For true carcinogens, i.e., those that cause alterations in DNA leading to tumors, the evaluation is relatively easy: Chemical structure is similar to known alkylating agents or potent tumorigens. Mutagen studies in bacterial and mammalian cells are positive; tumors appear earlier than in controls or in higher incidence, or both. Often overall body tumor load is greater. Mortality is increased. Metastasis occurs in some animals. For the so-called epigenetic carcinogens where the mechanism is indirect, such as endocrine-mediated carcinogens or those promoted or enhanced by disease or other agents, the evaluation and determination of significance is more difficult. It seems likely that many chemicals causing an increase in hyperplasia, benign tumors, or even malignant tumors under the conditions of current animal carcinogenicity tests will not be risks to humans. Benefit-to-risk asessment must be done carefully to eliminate drugs which should not enter clinical trials, and postmarketing epidemiological studies can provide long-term answers to the correctness of the benefit-to-risk assessment.

D. Reproductive Tests

Tests for fertility, teratogenicity, and peri- and postnatal function are often done as separate experiments in several species. Summary reports of those studies that have been completed appear in the investigational brochure. Even with assurances that no adverse effects on reproductive function were seen in animal studies, most clinical protocols exclude pregnant women from the trials. The following statement is typical for rejection criteria:

> Pregnant women will be excluded. The responsible investigator will determine that female patients of childbearing capability are not pregnant prior to entry into the study, and [he/she] will take precautions to guard these patients against the possibility of becoming pregnant throughout the course of the study.

E. Mutagenicity Tests

The Ames test was described earlier because it is often done as part of the early screening of compounds intended for clinical use. Most firms doing mutagenicity testing conduct a battery of tests designed to examine various facets of genetic damage. There are presently about 80 in vitro and 36 in vivo tests to choose from. Most of these have not been validated, and for some which have been validated the meaning of a positive result is not understood. A typical minimal test battery might be:

Bacterial point mutation
Mammalian cell point mutation
In vivo cytogenetics

If positive results are found in two or more of these basic tests, further development of the compound is not likely to be continued unless the compound is considered a major advance in therapy or the firm is willing to conduct extensive further testing, including the so-called definitive tests such as the heritable translocation test or the specific locus test. Only about 20 chemicals have been evaluated to date in each of these tests because of their difficulty and expense.

III. SUMMARY AND CONCLUSIONS

Prior to and during the clinical trials of a new chemical entity or drug combination, about 10 years of toxicology, pathology, and metabolism work have been done. These studies represent several million dollars of drug company funds. The work has three major uses: first, to satisfy the company that the drug is safe for each of the phases of clinical trials; second, to satisfy regulatory authorities worldwide that the drug is safe for registration; and third and most important, to provide clinical pharmacologists and clinicians with data to evaluate possible clinical problems by extrapolation to human use.

Important issues disclosed by animal studies include onset, degree, and duration of adverse effects. Cumulation or tolerance may be revealed. Species differences or similarities are helpful in evaluating the likelihood that the data will extrapolate to man. Are adverse effects predictable by usual observation or clinical tests? Are adverse effects reversible before insidious damage can occur? Are overdosage problems to be expected and are treatment suggestions offered? Insight on dosage regimen may be gained from prolonged animal studies.

Finally, the clinician's difficult responsibility when administering an untried medicine is made less difficult as his or her total understanding of the preclinical information increases.

REFERENCES

1. G. Zbinden and M. Flury-Roversi, Significance of the LD_{50} test for the toxicological evaluation of chemical substances. *Arch. Toxicol. 47:*77-99 (1981).
2. S. H. Dieke and C. P. Richter, Age and species variation in the acute toxicity of alpha-naphthyl thiourea. *Proc. Soc. Exp. Biol. Med. 62:*22-25 (1946).
3. T. Balaz, Cardiotoxicity of isoproterenol in experimental animals: Influence of stress, obesity and repeated dosing. In *Recent Advances in Studies on Cardiac Structure and Metabolism* (E. Bajusz and G. Rona, Eds.), Vol 1. Urban & Schwarzenberg, Baltimore, 1972, p. 770.
4. M. L. Keplinger, G. E. Lanier, and W. B. Deichman, Effects of environmental temperature on the acute toxicity of a number of compounds in rats. *Toxicol. Appl. Pharmacol. 1*(2):156-161 (1959).
5. R. Kato, Factors influencing acute toxicity. *Biochem, Pharmacol. 11:*221-227 (1962).
6. M. R. A. Chance, Factors influencing the toxicity of sympathomimetic amines to solitary mice. *J. Pharmacol. Exp. Ther. 89-91:* 289-296 (1947).

7. J. F. Griffith, Interlaboratory variations in the determination of acute oral LD_{50}. *Toxicol. Appl. Pharmacol. 6:*726-730 (1964).
8. B. N. Ames, Identifying environmental chemicals causing mutations and cancer. *Science 204:*587-593 (1979).
9. B. N. Ames, W. E. Durston, E. Yamasaki, and F. D. Lee, Carcinogens are mutagens: A simple test system combining liver homogenates for activation and bacteria for detection. *Proc. Natl. Acad. Sci. (U.S.) 70:*2281 (1973).
10. B. N. Ames, H. D. Kammen, and E. Yamasaki, Hair dyes are mutagenic: Identification of variety of mutagenic ingredients. *Proc. Natl. Acad. Sci. (U.S.) 72:*2423-2427 (1975).
11. B. N. Ames, J. McCann, D. Yamasaki, Methods for detecting carcinogens and mutagens with the Salmonella microsome mutagenicity test. *Mutation Res. 31:*347-364 (1975).
12. K. Beyer, *Discovery, Development and Delivery of New Drugs,* Monographs in Pharmacology and Physiology, No. 13. Spectrum, New York, 1978, p. 116.
13. A. N. Worden and K. H. Harper, Oral toxicity as influenced by method of administration. In *Proceedings of the European Society for the Study of Drug Toxicity,* Vol IV: *Some Factors Affecting Drug Toxicity,* pp. 107-110 (1964).
14. S. D. De C. Baker, The problems of species difference and statistics in toxicology. In *Proceedings of the European Society for the Study of Drug Toxicity,* Vol. XI (1970).
15. D. Salsburg, The effects of lifetime feeding studies on patterns of senile lesions in mice and rats. *Drug Chem. Toxicol 3*(1): 1-33 (1980).

14

SPECIAL CONSIDERATIONS OF DRUG DISPOSITION

Eric C. Schreiber
The University of Tennessee
Center for the Health Sciences
Medical College
Memphis, Tennessee

I. PRECLINICAL STUDIES

Preclinical drug metabolism and pharmacokinetic studies should be initiated as soon as a drug candidate has demonstrated continued promise during second-level pharmacological testing and preliminary toxicity evaluation. Often, these essential operations can be quite time consuming and costly, and delays of the drug development program are to be avoided, if possible. Intelligently designed drug metabolism studies, when introduced early into the drug development program, can make major contributions which will help move the entire sequence of action ahead more rapidly.

A drug metabolism study begins with planning how and where to radiolabel the potential drug candidate and also with developing a nonisotopic blood level and/or urine assay for the parent molecule. These obviously are not mutually exclusive operations, and it may be desirable to conduct them simultaneously. If the radiochemical synthesis chosen is designed so that the radiolabel is introduced in the ultimate or penultimate steps of the synthetic sequence, a great deal of both time and money can be saved.

Once the labeled material is in hand, administration to a rodent species should be quickly undertaken. One-half of the animals would be given the pharmacologically effective dose by intravenous administration and the remainder by gavage. The animals should be maintained in metabolism cages for 1 week, with frequent collection of blood samples and excreta during the first 48 hr and at 24-hr intervals thereafter. The samples should be assayed for their undifferentiated radiochemical content at first and then later subjected to thin-layer chromatography in order to obtain some idea as to the extent that biotransformation occurs. This simple experiment will give an insight as to the half-life of the compound in the rodent, the extent to which oral absorption occurs, the importance of protein binding, how extensively it is metabolized, and the relative importance of the renal and hepatic clearance routes of elimination in this species. Such information is of great importance in the design and interpretation of toxicity experiments which are usually contemplated at this point or may already be underway. If the pharmacological testing was done in this species, the data obtained will also be of value therein.

After these preliminary data are obtained, a second experiment should be undertaken in order to learn about the tissue distribution of the compound and its metabolites. It is useless to do a tissue distribution at a single arbitrary time following drug administration because the rate of perfusion of a tissue, the tissue proteins, the lipid character of the tissue, as well as the physiocochemical characteristics of the xenobiotic molecules present all will interact and contribute to the half-life of the drug and its metabolites in any particular system or compartment. For this reason, tissue distributions should be determined at various time intervals after dosing; a reasonable schedule would encompass 5 half-lives of the drug, with tissue distribution to be determined in both sexes at 0.5, 1, 3, and 5 half-lives after dosing. Any well-executed drug metabolism study in rodents should be able to account for not less that 90-95% of the administered radioactivity, even when respiratory excretion of the label takes place. The obvious value of these data is that they indicate whether or not irreversible binding and potential buildup may occur in some essential organ system and what the half-life of the drug may be in various key tissues.

Once data in rodents have been obtained, the investigation would then progress to other species—most usually the dog and monkey, although some investigators prefer also to include guinea pigs, rabbits, or some other animal.

If significant plasma protein binding has been found, it is wise to embark upon a series of in vitro binding experiments using human and animal blood to establish the general character of the phenomenon. In the event that a drug interaction which displaces the drug candidate from its binding sites might present a potential clinical hazard, such as exists in the case of a drug candidate with a low safety index, the researcher is well advised to initiate a series of investigations designed to explore the likelihood of this occurrence. To this end, the investigator would incorporate into the protocol those drugs which a patient is likely to receive concomitantly with the drug candidate. In any event, salicylates should definitely be included in such an experiment.

Metabolite identification, often a difficult and laborious process—particularly when a large number of metabolites are formed—can be deferred until it appears likely that the compound has real promise of clinical success or scientific prudence dictates that such identification be undertaken.

It is reasonable to assume that the biological consequences which follow the administration of a drug to either experimental animals or humans are the sum total of the physiological actions of the drug itself and any of the metabolites, since no chemical is totally devoid of the ability—in one way or another—to influence its host organism. It may be that the effects are not obvious because the substance in question is not very potent or may not be present in a concentration which is sufficient (subthreshold level) to directly demonstrate its actions. Since biological response is due to the sum total of all the molecular species which are present, it is clear that the animal species whose metabolite profile, qualitatively and quantitatively, most closely resembles that of man should be considered to be most relevant to the human case, other things being equal. Unfortunately, the question of interspecies sensitivity to a particular agent is not often answered by metabolic data alone. It is rare to find another animal species which metabolizes a compound exactly as does man: nonagreement is by far the most common situation, and toxicological data must be analyzed on that basis. This being the case, it becomes important to find in which animal species does the biotransformation of the compound or its pharmacokinetic disposition most closely resemble these processes in man.

Human biotransformation data can be obtained *prior to the issuance of an effective Investigational New Drug (IND) application*. That is, by obtaining fresh human liver from an autopsy, one can perform a quick series of in vitro liver homogenate preparations using this material. One then simply repeats this in vitro procedure with livers of the various species of test animals which are being studied. Obviously, this approach cannot substitute for studies in an intact host, but it must be borne in mind that, although we are dealing with an incomplete biological system, whatever drug metabolizing enzymes are present in our in vitro human preparation are also present in the intact human. Thus, preliminary human information can be obtained many months before the compound is introduced into human clinical trials; it can provide early guidance in deciding how similar or dissimilar various animal species are with respect to man in the biotransformation of the drug candidate and, in this way, influence consideration of species to be used in designing toxicology studies.

II. PRELIMINARY HUMAN DRUG DISPOSITION STUDIES

The study of the metabolic disposition of xenobiotics in the human subject is an essential element in the early phases of investigation of substances which have potential clinical interest. An intelligently designed study, carefully conducted, will afford much valuable information from a few subjects.

In modern industrial practice, it is no longer uncommon for the first clinical introduction of a compound of interest to be designed as a drug metabolism study. Such a study, using oral and parenteral routes of administration, although probably conducted at a dose well below that which will ultimately be determined as the clinically effective dose, and although carried out in normal healthy volunteers instead of possibly debilitated patients, yields a great deal of preliminary clinical data which helps to guide the direction and design of further experiments.

The information, which can be obtained in a few days, includes: indications of the rate and extent of absorption; time of peak plasma level; approximate peak plasma level; protein binding; biotransformation; apparent volume of distribution; and the pharmacokinetic half-life of the substance and its metabolites (if any) in man—the ultimate species.

An important aspect of this first human metabolic study is the enormous comparative value of these findings as aids in evaluating and interpreting the data obtained during the course of animal toxicity studies. The findings of this study also serve as a source of valuable information which aids in the design of protocols for future clinical investigations. Both preclinical and clinical phases of the drug development path are benefited by such a study.

Clearly, early data obtained in volunteers, even with the caveats that the dose is probably not correct and the human population used in the test is not the same as those individuals who will ultimately receive the drug, does provide the planners of clinical investigations with an infinitely better knowledge base for rational design of subsequent trials than is provided by data from experimental animals alone. Even with the limitations outlined, such testing with human volunteers is a far more intelligent course of action than planning extensive Phase I or Phase II studies based only on animal data.

Included among these results are the first in vivo data of how man metabolizes the drug. At this point in the path of drug development, we still are not particularly concerned about the actual identity of the metabolites excreted by man per se, but we are interested in obtaining profiles of spots on thin-layer chromatograms and then comparing these spots (known or unknown) with similar chromatograms of the excretion products of the animal species used in the preclinical toxicity and screening studies. It is not necessary, or even particularly desirable, to known the structures of the metabolites when we make these comparisons, since we are primarily interested in learning how similar are the metabolite profiles obtained from the test animals to the profiles obtained from humans.

One of the more interesting and important applications of these early clinical findings is the ability to make an intelligent preliminary projection as to when plateau state blood levels will be achieved on multiple-dosing regimens and how these levels should vary between doses. None of these preliminary calculations can be taken as gospel, but they do usually afford reasonable first approximations of what we can expect to find in a given therapeutic situation.

Thus, this important first exposure of humans to a new drug offers an escape from the older Edisonian approach, which almost blindly embarked upon a brute force broadside of expensive and time-consuming clinical experiments whose protocols were based on possibly irrelevant animal data. Compared to other aspects of clinical investigation, the procedure is not expensive, and it does supply vital information with respect to human disposition of the compound very early in the clinical phases of the investigation. It is more fruitful to light one small candle than it is to curse the scientific darkness—or to blindly stumble on.

III. PHARMACOKINETICS

Pharmacokinetics is an important tool in modern drug disposition research. Unfortunately, the seductive promise of this technique has obscured a number of serious deficiencies which accompany its practice. Both potential and real sources of error seem to be consistently ignored in the estimation and use of average values, rate constants, and half-life values. Frequently, one encounters such statements as "the half-life of X in man is Y hours." It is a common practice to ignore the fact that a drug half-life, whether one is discussing a single patient or an entire population, is usually derived either by drawing the best-fitting line through a group of data points by visual approximation or by using a least squares fit regression analysis performed without proper weighting of data. What is ignored in both cases is the fact that the slope of the line is not estimated without error, that is, the slope of the line, like other statistics, has a standard deviation (S.D.). This S.D. is an inherent part of the rate constant of elimination (for example) and the half-life of elimination which is then derived from that value. It is the universal practice, however, to treat rate constants or half-lives as if they had been obtained from perfectly regular data, a procedure which implies that all points were exactly on the line. The net effect of this is that the intersubject variations of the parameters of interest have been entirely obscured; this is especially true in the development of "models." The value which should be used is $k_x \pm n$ (S.D.), where n should be selected so as to achieve a desired confidence level which will include at least 90% of the patients. The avoidance of the use of rate constants and their statistical limits in developing pharmacokinetic models can be due to the fact that many of our so-called models would not stand up to this type of mathematical treatment!

A brief example will make clear the magnitude of the error which can be introduced when the S.D. of a parameter is ignored; for the sake of simplicity, a normal distribution will be assumed. The half-life of desimipramine is reported to be 18 ± 5 (S.D.) hr [1]. This means that, in 68% of a normally distributed subject population, the half-life will vary from 13 to 23 hr. (In passing, it should be noted that a S.D. of ±35% is not at all uncommon.) A responsible clinician is not interested in what happens in only two-thirds of his patients; he wants to know what will happen in the vast majority of the cases he will encounter. Therefore, upon application of the criterion of two standard deviations, which would include about 95% of a population, we find that the half-life can be expected to vary from 8 to 28 hr! It may very well be that clinical failures will be more frequent in the 14% of the patient population where the half-life is only 8-13 hr and that toxic effects or increased side effects will be present in the 14% of the population where the half-life of desimipramine would be in the range of 23-28 hr. Either end of the spectrum would require altering the drug exposure of the patient.

The net effect of this is that the average concentration of drug, when the plateau state is attained during intermittent drug administration, may vary within a range of 350%, as in this instance.

Benet and Sheiner [2] have made a major contribution by compiling a great deal of pharmacokinetic data relevant to currently used drugs. These authors report eight key parameters (and obtain their S.D.) useful in the intelligent application of pharmacokinetics and discuss the limitations of these findings as well as the pitfalls inherent in such usage. The reader is urged to study their presentation.

It is not unusual for one of the two following practices to be followed during blood level studies: One procedure is to pool equal volume plasma samples, which of course affords an average blood level determination but unfortunately does not give the observer any means of estimating the range of values that were used to calculate the mean. This is obviously unsatisfactory. The second method is to determine the individual samples and then to average the individual values. In either case, the simple arithmetic average is obtained and is then used for pharmacokinetic analyses. The habit of reporting an average value and then listing the range of values that comprise the average is to be deplored.

The usual practice is to infer parameters for the population by estimating the parameters for each volunteer in the study and then arithmetically averaging across the entire subject group. This procedure is rarely ever based upon evaluation of the distribution characteristics of the individual pharmacokinetic parameters which comprise the population. Sheiner et al. [3] have published a method for estimating average population values and the intersubject variability of pharmacokinetic parameters without having to undertake intensive subject-by-subject pharmacokinetic characterization.

There does not appear to be any major textbook of pharmacokinetics which attempts to examine valid approaches in the analysis of arithmetic mean data [4-6]. Part of the problem may be that few scientists are aware of the techniques which are required for the treatment of statistical terms in various basic arithmetic operations (e.g., multiplying or dividing two numbers and their S.D.). Levy and Gibaldi [7] have shown some awareness of the problem and point out the hazard of basing pharmacokinetic inferences on ordinary arithmetic mean data. This paper suggests that it may be more appropriate to use the arithmetic means of the logarithmic values of the plasma drug concentrations which have been obtained from the individuals for application in the analysis of pharmacokinetic data and in the calculation of population parameters. In an earlier paper, Levy and Hollister [8] felt that pharmacokinetic models which had been obtained by the use of average data may be *entirely without value*, because the model may only fit those average data and none of the individual data. After all, each individual has his own unique volume of distribution and clearance characteristics; body water compartments vary markedly with age; and so on.

The half-life of a drug is an essential component in the calculation or projection of the body burden under the conditions of a multiple-dosing regimen, but the questions of *inter*subject and *intra*subject variation [9] and dose-dependent response make the blind use of pharmacokinetic equations and models a procedure which has many pitfalls.

The question of intrasubject variation in the disposition of xenobiotics requires some further comment. Previously, we had advanced the thesis that intersubject variation must be considered in the calculation and management of biological half-life data and in the estimation of apparent volume of distribution (AVD) and pharmacokinetic models. Surprisingly, a number of studies [10-17] have shown that *intra*subject variation of half-lives can be greater than *inter*subject variation in the use of a number of clinically important substances. A number of the studies cited also demonstrate that the route of administration also can greatly affect the half-lives of some drugs.

Novak et al [10] reported serum concentrations of clindamycin in 10 male volunteers who had received from 1.07 to 3.72 mg/kg of the antibiotic by intramuscular administration in 1.0, 1.5, or 2.0 ml of vehicle (100-mg base equivalent per milliliter). Statistical analysis demonstrated that the results were not dose dependent but the coefficient of intrasubject variation was approximately 20%. In one subject, the $t_{1/2}$ ranged from 2.5 to 4.7 hr; another showed a range of 4.9-7.7 hr. The overall mean $t_{1/2}$ was 4.8 hr in this study.

Wagner et al [12] administered 150-mg doses of clindamycin base equivalent to 12 subjects in a dosage form crossover study (capsules vs. tablets). The half-life determinations of the two dosage forms showed no significant differences and had an overall average of 2.4 hr but were in sharp contrast with the 4.8 hr $t_{1/2}$ reported by Novak et al [10] following intramuscular administration. When Wagner and colleagues calculated the ratio of the half-lives obtained for each subject during the course of the crossover study, the ratios varied from 0.68 to 1.33, with an average of 1.0.

Furthermore, the average area under the serum concentration curve, after the 150-mg dose, was 10.9 μg-hr/ml per mg of administered antibiotic base. The doubling of the half-life of clindamycin after intramuscular administration, when contrasted to the orally ingested materials, results in nearly doubling the area under the serum concentration curve. With this drug, at least, an altered route of administration can lead to markedly different consequences.

The foregoing discussion has been confined to clindamycin, but similar findings have been reported for warfarin [13], ephedrine [15], lincomycin [11], and ethosuximide [16]. The phenomena reported are not limited to substances whose elimination follows a first-order kinetic pattern but are also seen in the zero-order elimination of ethanol [14, 17].

Wagner and Patel [14] measured capillary blood ethanol concentrations in a normal male volunteer during five controlled trials at three levels of ethanol administration. The volume of distribution had a coefficient of variation of 26.7%. These authors applied the integrated form of the Michaelis-Menten equation to calculate V_m and K_m. The estimated V_m values had coefficients of variation of 98.3% and 97.3%, depending upon which estimates were used as the basis of calculation. The authors pointed out that the intrasubject coefficient of variation

(26.7%) for the apparent zero-order rate of ethanol metabolism was greater than the coefficients of variation of 18.4% and 14.8% calculated from Vessell et al [17] from six subjects and an S.D. calculated from the difference (before minus after) of 0.055 with an accompanying grand mean of 0.198 zero-order reaction rate (mg-hr/ml); this gives a coefficient of variation of 28%, which closely agrees with the value obtained by Wagner and Patel [14].

In this instance of ethanol metabolism, clearly, the *intra*subject variation of the apparent zero-order reaction rate is greater than is the *inter*subject variation.

IV. ALTERED DRUG DISPOSITION

As already stated, it is not unusual to conduct absorption, distribution, biotransformation, and elimination studies in healthy, young, ambulatory individuals as part of Phase I clinical studies. Data obtained in the course of such studies are frequently treated, without consideration of the possible or probable effects of a disease state as though these findings would also apply to a patient. Often little thought is given to the possibility that an entirely different situation may exist in the real world when one treats older, bedridden—often debilitated—patients suffering from a disease state and who, in addition, may be suffering from renal impairment or cirrhosis as well.

The study of the disposition of drugs after their introduction to the human has resulted in thousands of scientific papers. Once exposure to the agent has been effected by either topical or parenteral administration, ingestion, or inhalation, only a limited number of possible fates can befall the drug substances: it may undergo absorption, followed by distribution; it may or may not undergo any of a wide variety of enzymatic or hydrolytic changes which enhance or reduce its biological activity; ultimately it and/or its biotransformation products will be excreted. Also, it may spend part of the time it is present in the body bound to one or a number of endogenous proteins. Although the sequence outlined is a very simple one (cognoscenti will consider it highly simplistic), the actual disposition of a drug is, in fact, extremely complex. Many factors influence the fate of a drug and its pharmacokinetic behavior. It is unwise to ignore the fact that the disease state itself, or complications which are present commonly but are not related to the disease state, may result in drug disposition findings widely at variance with earlier data.

A. Effect of Age

Comparatively little work has been directed toward the study of drug biotransformation in the newborn human infant. It is well-known that certain conjugation reactions cannot take place because the prerequisite enzyme systems do not develop in utero. This observation has unfortunately been frequently extended, at least by inference, to include many—or all—drug-metabolizing enzyme systems. Such generalizations demonstrate ignorance of the fact that, at the end of embryogenesis, a 6- to 7-week-old human fetus, weighing only a few grams, already contains enzymes capable of hydroxylating aromatic compounds, N-dealkylation reactions, sulfoxidation, N-oxidation, epoxide formation, etc. [18].

A variety of human fetal tissues are capable of metabolizing xenobiotics; these include tissues of the liver, lung, skin, and adrenals. The concentrations of cytochrome P450 in the liver of the human fetus is surprisingly high when compared to that of test animals, and the values in the adrenal gland seem to be even higher yet [19]. Thorgeirsson [20] reported that the level of NADPH-cytochrome P450 reduc tase in human fetal liver is approximately 40% of that of adult human activity. In many ways, the fetal hepatic enzyme system is closely analogous to that of the human adult.

Although animal studies can often be of great value in predicting the human disposition of a compound, this may not be the case in studies intended to elucidate the biotransformation of a drug by an animal in utero. Apparently, the lower animals do not seem to appreciably develop some important drug-metabolizing enzymes until after parturition. For example, Nicholas and Schreiber [21] have shown that the newborn mouse cannot metabolize phencyclidine before it is more than 1 week old. It is obvious that one cannot safely conjecture about the possible consequences of human placental transfer of a drug based upon animal data.

This is a matter for concern, because the relatively nonpolar parent drug molecule may be able to cross the placenta and enter into the fetus while more polar metabolites cannot cross this barrier into the fetus. The human fetus, however, contains low levels of drug-metabolizing enzymes which can create these same polar metabolites. If these metabolites are formed and cannot enter the maternal circulation, they will continue to concentrate in the fetus with possibly harmful effects. This is a question which has not often been considered in the safety assessment of drugs.

In the very young human (less than 4 months old), body water is 70% of the body weight, whereas it ranges from 42 to 53% in adult females and from 54 to 58% in adult males. Therefore, in a newborn infant, the higher level of water content will increase the AVD and this, in turn, will result in decreased blood concentration of drug. In such infants, renal blood flow—as related to the total body water content—is also less than that of adults. These factors combine to reduce the glomerular filtration process and, in so doing, slow any renal clearance of drugs whose primary route of elimination is through the kidney. Insulin clearance in the newborn takes about three times longer than it does in the adult. Lastly, the drug-metabolizing enzyme systems do not fully develop until the infant is several months of age. The oxidative pathways and the conjugative mechanisms are particularly affected. Any multiple-dosing regimen involving the newborn should consider the possibility that accumulation of drug can occur and hence wider spacing of doses than is used with adults may be advisable. Since biotransformation may be occurring only slowly and elimination of the drug is not as rapid, it is likely that no loss of therapeutic effectiveness will be seen.

The incidence of adverse drug reactions is, among other things, an age-dependent phenomena. Among patients of advanced age (70-79 years), there is a seven-fold greater occurrence of adverse drug reactions than in a patient population which is 50 years younger. Although the aging process per se is still poorly understood, it is reasonable to suggest that aging brings about a general deterioration of many physiological functions. In the same instance of renal function, mean glomerular filtration rates decrease by one-half, renal blood flow is reduced by a similar amount, and tubular excretory capacity falls also—perhaps by 40% between the second and eighth decennium. Other changes are also taking place: hepatic microsomal enzyme activity is gradually diminished; cardiac output is lessened; intestinal absorption is impaired; and the relative fat burden is elevated while lean muscle is reduced.

It is a matter of no little interest that the incidence of achlorhydria is age dependent and seems to increase from about 5% in younger populations to 35% in the aged. What is important is that an element of drug absorption variability, which may influence blood levels and is common in "normal" populations, is neglected in the interpretation of clinical findings or in the consideration of data obtained from bioavailability studies. The question of how naturally occurring achlorhydria may be influencing the findings is rarely, if ever, considered in the analysis or comparison of data.

The level of acidity of the stomach can strongly influence the rate and extent of absorption of drugs, since it is well known that absorption usually occurs best in a nonpolar (nonionic) form. In the normal stomach, the absorption of mildly acidic substances is favored because ionization is repressed. With basic substances, the opposite is true because of salt formation and the consequent presence of the base in an ionized form.

Achlorhydria in a patient can lead to decreased absorption of weakly acidic drugs. Conversely, the decreased level of acid in the stomach favors the faster absorption of all basic drugs, especially those which are relatively weak bases. On the basis of this information, it should not be surprising, therefore, that achlorhydria, which is certainly not a rare phenomenon in the senior population, might distort absorption and blood level studies in older age groups and affect clinical response.

B. Protein Binding

The binding of drugs to proteins may vary from zero to 99%. It has long been accepted that drugs bound to plasma proteins are, in the bound state, therapeutically and physiologically inactive. The major plasma and tissue protein which is generally involved in nonspecific binding of xenobiotics is albumin.

Although nonspecific protein-binding studies are almost routinely carried out in the course of clinical pharmacology investigations and AVD studies, seemingly little attention is being paid to the effect of disease states or changed normal physiological function on the amount of free (active) drug substance which is available to cause an enhancement of therapeutic effect or to exacerbate existing side effects (or lead to new ones).

Anything which can alter the amount of drug which is nonspecifically bound to albumin can, theoretically at least, cause a changed clinical picture to appear. It is not likely that significant clinical differences will be brought about in those instances where very little binding occurs, since only a relatively small increase in free drug occurs. For example, if a drug which is 25% bound were, for some reason or another, to become unbound, the relative increase in free drug would only be about 35%—an increase which in most cases probably would not be of clinical importance. However, in the case of highly bound substances, even a small decrease in binding can lead to an enormous increase in blood level of free drug. If the amount of a drug which is bound is reduced from 99 to 95%, a *400%* increase in free drug comes about.

Neither the gastrointestinal tract nor the kidney exerts a dominant role in the disposition of albumin in healthy subjects. Renal disease or disease of any portion of the gastrointestinal tract may cause rapid elimination of albumin. Nephrosis, especially, is directly associated with marked hypoalbuminemia; in 30 nephrotoxic patients, Jensen et al. [22] found that 80% of the patients had serum albumin levels which were less than one-half of the clinical norm for that laboratory. Nephrosis results in an increase in the catabolism of albumin and also a reduction in the intravascular and total exchangeable mass of albumin.

Casey et al [23] have reported an extensive study which shows that albumin may be lowered very markedly in an overwhelming number of disease states including among others, pulmonary collapse, bowel obstruction, autonomous growth, primary invasive and metastatic cancer, as well as many infections by viruses and bacteria. Eighteen percent of the patient admissions to their hospital demonstrated lowered albumin. In some conditions, elevation of serum albumin was noted, but the increase in serum albumin was not nearly as dramatic.

These findings of Casey et al. make it abundantly clear that nonspecific protein-binding measurements made in normal volunteers can be entirely misleading if these data are applied to patients or during the late stages of pregnancy; normal delivery markedly lowered serum albumin.

A higher incidence of adverse drug reactions have been noted in patients having impaired renal function. It was thought that these adverse drug reactions were primarily the result of increased drug accumulation in the body, arising from the decreased excretion of drug by the kidneys. This, however, is not the case, since the binding of drugs is less in uremic patients than in normal subjects.

Andreasen [24] demonstrated decreased binding of a number of acidic drugs (salicylic acid, acetylsalicylic acid, phenytoin, phenylbutazone, thiopentyl, and sulfadiazine) in renal impairment arising from acute trauma. The decreased binding ranged from 22 to 60%, as compared with normal controls; it is of interest to note that the observed decreased could be only partially accounted for by the lowered concentration of albumin. The effect seems to be more prevalent among acidic substances, since Reidenberg and Affirme [25] found no alteration of the binding of some basic drugs (quinidine, dapone, desmethylimipramine) in uremic patients.

Other investigators have also concluded that, in the uremic patient, decreased binding of drugs cannot be accounted for on the basis of albumin levels alone. Using equilibrium dialysis to determine protein binding, Craig et al. [26] investigated sulfamethoxazole, dicloxacillin, penicillin G, phenytoin, salicylate, and digitoxin binding in treated and untreated sera from uremic patients and normal subjects. They undertook measurements in serum before and after charcoal treatment at pH 3.0. Treatment of the sera with charcoal did not affect the binding in any way when the serum was from normal subjects. However, charcoal treatment of the serum from uremic subjects led to significantly increased binding in almost every case. Obviously, charcoal treatment had removed one or more substances which either directly or indirectly interfered with protein binding.

Addition of creatinine, urea, phosphate, or uric acid to normal serum did not change the binding of these drugs, nor did excessive quantities of uremic toxins such as guanidinosuccinic acid, guanidinoacetic acid, and methylguanidine.

C. Renal Impariment

The findings obtained with a normal subject who received quinidine orally [27] or by parenteral administration [28] were in sharp contrast with the results obtained when the drug was used to treat patients in congestive heart failure. Analysis of these studies [29] demonstrated that patients had significantly higher blood levels than did the normal subjects, although absorption was considerably slower in the patient population. Peak plasma times were increased from 2 to 4 hr. The amount of orally absorbed quinidine excreted in patients' urine was only about one-half that of the normals. The elimination rate in the population was virtually identical, and there was good agreement between the results of these two groups of investigators. A possible explanation for these observations is that the AVD of quinidine was greatly altered by the advent of congestive heart failure. Following oral administration to humans, the AVD in normals was 33.3 liters as opposed to 9.6 liters in patients. Similar results were noted upon parenteral administration. The smaller volume of distribution, which results in higher blood levels of quinidine, is possibly due to decreased perfusion of the body tissues. Thomson et al [30] observed a similar situation with lidocaine. The altered AVD in patients can present a toxicological hazard to the patient, since a greater portion of the dose remains in the vascular system. The heart, brain, and other organs which are still well perfused may now be exposed to excessively high total amounts of drug, leading to potential toxicity in the central nervous system or cardiac toxicity. These investigators suggested this as a possible cause of lidocaine toxicity in congestive heart patients.

The examples just cited suggest that, when elimination of a drug or its metabolites occurs preponderantly as a result of renal elimination, monitoring of drug excretion may be very misleading. In disease states where the absorption is reduced (with, coincidentally, reduced drug elimination) urinary excretion determinations would indicate that increasing the dosage should be considered. However, this is precisely the *wrong* procedure to follow, since the blood levels are already very high. Another complicating factor is that, since the rate of absorption has also been depressed, blood levels measured at a projected peak time determined in normal subjects are misleading, because in the patient the peak blood level does not occur until several hours later. In the instance of quinidine, blood levels measured at 2 hr (normal peak time) would be the same for both normal subjects and patients in congestive heart failure, but after 4 hr that of the patients would have increased another 50% while the normal population values would have declined.

It should be noted that in these studies the differences seen between the populations were found where complicating factors such as hepatotoxicity or renal damage were *not* present.

When renal function is compromised either by chronic or acute renal disease, those drugs whose removal from the body is preponderantly via the renal pathway can be seriously affected. The half-lives of the parent substance and its metabolites may be considerably prolonged, and repeated administration of the drug can lead to the accumulation of toxic levels, as evidenced by the appearance of new side effects or increased severity of existing ones. The pharmacodynamic response may be exaggerated. Also, increased accumulation of metabolites may result in displacement of the parent molecule from protein binding sites. Further drug metabolism may become inhibited [31]. In an excellent review [32], Pagliaro and Benet report on a number of drugs that have been extensively studied.

Almost invariably, the half-life is increased by either renal or hepatic dysfunction. Renal impairment is considered to be mild (creatinine clearance 50-80 ml/min), moderate (10-15 ml/min), or severe (less than 10 ml/min). When a subject is on a multiple-dose regimen, the interval between doses usually remains unchanged, although some investigators appear to increase the interval by 1.5-2.0 times. Doubling of the time interval between doses is more commonly applied in instances of moderate impairment. Since impairment of creatinine clearance can force withdrawal of the drug entirely, followed by substitution of another agent of that therapeutic class which undergoes elimination via the alimentary canal rather than the kidney, it is difficult a priori to generalize as to how severely a particular degree of renal impairment will effect any drug. Warfarin, diphenylhydantoin, tricyclic antidepressants, penicillins, some barbiturates, and the great majority of the narcotic analgesics do not appear to be eliminated more slowly because of renal failure. Other compounds, including tetracycline, kanamycin, and streptomycin, may require greatly increased dose intervals even in cases of quite mild renal impairment.

D. Hepatic Dysfunction

Liver diseases are among the most prevalent and complex chronic disorders that affect man. The normal liver synthesizes a great deal of the plasma protein, and a compromised hepatic function can—and often does—result in diminished protein levels in the body. The diminution of hepatic function can result in significant alteration of the host's ability to metabolize drugs, leading to higher concentrations of circulating free drug than would otherwise be anticipated, altering AVD, and prolonging drug half-life.

Unfortunately, there does not appear to be any means of precisely assessing hepatic dysfunction on a quantitative basis. Antipyrine, d-propanolol, and indocyanine green clearances were studied in patients with chronic liver disease [33]. Antipyrine has a low intrinsic hepatic clearance and is slowly, but completely, metabolized prior to excretion [34]. Indocyanine green is not metabolized but is eliminated exclusively by active uptake into hepatic parenchymal cells. Indocyanine green is excreted via bile and does not undergo reabsorption after excretion into the small intestine. These properties make it suitable for the estimation of liver blood flow using the Fick principle with the direct measurement of the hepatic extraction ratios.

Since antipyrine has a hepatic clearance which is less than 2% of hepatic blood flow, its hepatic clearance is essentially the same as its intrinsic hepatic clearance and is independent of hepatic blood flow; this, then, reflects only the activities of the drug-metabolizing enzymes which are present.

The kinetics of these drugs were compared with d-propanolol, which is devoid of beta-adrenergic blockade potential and is eliminated by hepatic metabolism with a high intrinsic clearance.

The clearance of all three drugs was lessened by the presence of liver disease. The greatest reduction in clearance was seen in those patients who exhibited raised bilirubin levels, prolonged prothrombin times, and low serum albumin concentrations. There was no significant degree of correlation with other tests of liver function, such as serum glutamic-oxaloacetic transaminase (SGOT), serum glutamic-pyruvic transaminase (SGPT), or alkaline phosphatase.

The data obtained suggest that, since antipyrine elimination is dependent upon drug metabolizing enzyme activity only, the observations made could have arisen from reduced enzyme activity. Reductions in both hepatic blood flow and intrinsic hepatic clearance could lead to reduced clearance of d-propanolol and indocyanine green. Despite the diverse modes of elimination of the three compounds, it is of considerable clinical significance to note that their clearance correlated positively.

As chronic liver disease progresses, protein synthesis in the liver is diminished, especially the albumin production. Since the protein pool is smaller, less total binding of drugs can take place. Small changes in protein binding can have a considerable effect on drug distribution, since protein binding acts to keep the drug in the vascular compartment and, in this manner, causes less drug to be available for tissue distribution.

V. ADVANCED HUMAN DRUG DISPOSITION STUDIES

The original drug metabolism study in man was probably undertaken as a single low dose administered to healthy young volunteers. The data thus obtained may be extremely different from those gained from a multiple-dose situation where patients of varying age are receiving what are often substantially higher doses. It is not at all unusual to obtain subsequent pharmacokinetic and drug biotransformation data which are in considerable variance with the findings of the earlier study.

Although the single-dose drug metabolism study performed in the early days of a Phase I investigation can supply much badly needed information, such a study does not address itself to several important factors. These factors include the effect of route, dose, and regimen on drug disposition, the effect of saturation of protein drug binding sites, and the question of tolerance or sensitivity developing as a result of drug enzyme induction or inhibition.

It is, therefore, highly desirable to undertake one or more additional drug disposition studies during Phase II of the clinical research program. Here, intensive studies are conducted in a population that is representative of the ultimate clinical pool which will receive this drug after its eventual approval. A range of doses that are clinically effective is employed. Importantly, these studies should be done under "equilibrium conditions," in a multiple-dosing situation, that is, after a series of initial doses have led to a stabilization of the blood level at the plateau state.

Whether or not one wishes to use a radiolabeled dose or is content to measure blood levels by nonradioactive assays is a selection which must be made on an individual case basis. Certainly if the goal is to examine the biotransformation of the compound, usually the administration of labeled drug is the method of choice, except in the simplest of cases.

If the development of drug tolerance has been experienced during the clinical trials, it is well to consider whether or not concomitant medication may be exerting effects that induce drug-metabolizing enzyme or whether the drug candidate is itself capable of acting as an enzyme inducer. In this case, one would administer the drug to a patient population which has not been exposed to the drug before, does not have a recent history of widespread drug ingestion and is not currently taking other medication. Such patients are then started on their drug treatment regimen, but a drug disposition study (absorption, blood levels, elimination, biotransformation, etc.) is carried out immediately. After tolerance has been noted, a second such study is done in the same individuals and the findings are compared with the first. If there are no significant differences in the findings of the two studies, then an explanation other than altered drug disposition must be entertained.

The development of a promising new drug is made difficult by the large number of clinical variables which can so greatly affect the disposition of the administered material. Investigations of a clinical pharmacological nature are greatly complicated by the fact that many things which can strongly influence the findings are often difficult to quantify, for example, liver disease. No one would seriously propose detailed human investigation of every permutation and combination of age, sex, dose level, drug interactions, diet, disease state, and so forth. Brute force approaches are very costly and are not particularly fruitful. A good deal of common sense must be used in deciding what special studies of drug disposition are required. It is unreasonable to undertake an investigation of the effects of decreased protein synthesis or protein binding of the drug, or displacement from protein as a result of drug interactions, in a case where the drug candidate is only slightly bound, and especially so when a relatively large safety margin exists.

A not dissimilar situation exists with respect to drug interactions which may have clinical significance, and often the consequences of drug interactions can be predicted a priori by a competent pharmacologist. After all, one should not be overwhelmed with surprise to learn that two drugs which have opposite therapeutic uses may exhibit antagonism when administered together, or that two drugs which have the same activity may each reinforce the other by synergism or potentiation. An intelligent investigator can make a reasonable prediction as to what the effect would be on a patient who is on anticoagulant therapy and begins receiving repeated doses of a drug that stimulates the production of drug-metabolizing enzymes in laboratory animals.

There is no golden road map, nor guideposts engraved in stone pointing the way, for the conduct of drug disposition studies. Each compound is a unique case, and programs must be designed to fit the physiocochemical, pharmacological, and toxicological properties of the potential drug candidate, as well as the idiosyncracies of the patient population who will be exposed to it. There will never be a substitute for common sense, which is not common.

REFERENCES

1. B. Alexander, *Eur. J. Clin, Pharmacol. 5:*1 (1972).
2. L. Z. Benet and L. B. Sheiner, in *The Pharmacologic Basis of Therapeutics* (A. B. Gilman, L. S. Goodman, and A. Gilman, Eds.). Macmillan, New York, 1980, Appendix II.
3. L. B. Sheiner, B. Rosenberg, and V. V. Marathe, *J. Pharmacokinet. Biopharm. 5:*445 (1977).
4. M. Gibaldi and D. Perrier, in *Pharmacokinetics: Drugs and the Pharmaceutical Sciences* (J. Swarbrick, Ed.), Vol. 1. Dekker, New York, 1975.
5. J. G. Wagner, *Biopharmaceutics and Relevant Pharmacokinetics.* Drug Intelligence Publns., Hamilton, Ill., 1971.
6. J. G. Wagner, *Fundamentals of Clinical Pharmacokinetics.* Drug Intelligence Publns., Hamilton, Ill., 1975.
7. G. Levy and M. Gibaldi, *Concepts in Biochemical Pharmacology.* Vol. 28 of *Handbook of Experimental Pharmacology* (J. R. Gillette and J. R. Mitchell, Eds.). Springer-Verlag, New York, 1975, Chap. 59.
8. G. Levy and L. E. Hollister, *J. Pharm. Sci. 53:*1446 (1973).
9. J. G. Wagner, *J. Pharmacokinet. Biopharm. 1*(2):165 (1973).
10. E. Novak, J. G. Wagner, and D. J. Lamb, *J. Clin. Pharmacol. 3:*201 (1970).

11. J. G. Wagner, J. I. Northam, and W. T. Sokolski, *Nature 207:* 201 (1965).
12. J. G. Wagner, E. Novak, N. C. Patel, C. G. Chidester, and W. L. Lummis, *Am. J. Med. Sci. 256:* 25 (1968).
13. J. G. Wagner, P. G. Welling, K. P. Lee, and J. E. Walker, *J. Pharm. Sci. 60:* 666 (1971).
14. J. G. Wagner and J. A. Patel, *Res. Commun. Chem. Pathol. Pharmacol. 4:* 61 (1972).
15. P. G. Welling, K. P. Lee, J. A. Patel, J. E. Walker, and J. G. Wagner, *J. Pharm. Sci. 60:* 1629 (1971).
16. R. A. Buchanan, L. Fernandez, and A. W. Kinkel, *J. Clin. Pharmacol. 9:* 393 (1969).
17. E. S. Vessell, J. G. Page, and G. T. Passonanti, *Clin. Pharmacol. Ther. 12:* 192 (1971).
18. O. Pelkonen and N. T. Kärki, *Biochem. Pharmacol. 22:* 1538 (1973).
19. O. Pelkonen, E. H. Kaltiala, T. K. I. Larmi, and N. T. Kärki, *Clin. Pharmacol. Ther. 14:* 840 (1973).
20. S. S. Thorgeirsson, Ph.D. thesis. University of London, 1972.
21. J. M. Nicholas and E. C. Schreiber, *Am. J. Obstet. Gynecol. 143:* 143-146 (1982).
22. H. Jensen, N. Rossing, S. B. Andersen, and S. Jarnum, *Clin. Sci. 33:* 445 (1967).
23. A. E. Casey, F. E. Gilbert, H. Copeland, E. L. Downey, and J. G. Casey, *Southern Med. J. 66:* 179 (1973).
24. F. Andreasen, *Acta Pharmacol. Toxicol. 32:* 417 (1973).
25. M. M. Reidenberg and M. Affirme, *Ann. N.Y. Acad. Sci. 226:* 115 (1973).
26. W. A. Craig, M. A. Evenson, K. P. Sarver, and J. P. Wagnild, *J. Lab. Clin. Med. 87:* 637 (1976).
27. S. Bellett, L. R. Roman, and A. Boza, *Am. J. Cardiol. 27:* 368 (1971).
28. E. L. Ditlefsen, *Acta Med. Scand. 159:* 105 (1957).
29. W. G. Crouthamel, *Am. Heart J. 90:* 335 (1975).
30. P. D. Thomson, K. L. Melmon, J. A. Richardson, K. Cohn, W. Steinbrunn, R. Cudihee, and M. Rowland, *Ann. Intern. Med. 78:* 499 (1973).
31. D. Perrier, J. J. Ashley, and G. Levy, *J. Pharmacokinet. Biopharm. 1:* 231 (1973).
32. L. A. Pagliaro and L. Z. Benet, *J. Pharmacokinet. Biopharm. 3:* 333 (1975).
33. R. A. Branch, J. A. James, and A. E. Read, *Clin. Pharmacol. Ther. 20:* 81 (1976).
34. B. B. Brodie and J. Axelrod, *J. Pharmacol. Ther. 98:* 97 (1950).

15
PRODUCT DEVELOPMENT

Lester Chafetz and Theodore I. Fand
Warner-Lambert/Parke-Davis Pharmaceutical Research Division
Warner-Lambert Company
Morris Plains, New Jersey

I. INTRODUCTION

Preclinical studies of drugs are performed using extemporaneous admixtures or solutions of chemicals shown to have potential value as drugs, for administration to animals or testing in vitro in pharmacodynamic models. Formulation and manufacture of drug products becomes necessary if the results of preclinical research warrant beginning clinical trials in humans. These functions are the primary responsibility of the Product Development (or Pharmacy) Division. The organization, relationships, and functioning of product development in the clinical research process are discussed in detail in this chapter.

Pharmacy may be defined in a broad sense as those arts and sciences directed toward or concerned with the discovery, development, formulation, testing, utilization, manufacture, and distribution of drugs. Pharmacy cannot be defined as what pharmacists do, for people trained in pharmacy are found in virtually every discipline and organizational division in a pharmaceutical company. Conversely, the people engaged in product development may not necessarily have degrees in pharmacy; they may have degrees in chemistry, engineering, or biological sciences. Because it is a more restrictive term, we prefer to use "Product Development" to designate the division that is charged with research and development (R&D) on drug delivery forms.

Product development provides the interface between preclinical research and the manufacture and distribution of a pharmaceutical product in commerce, the culmination of clinical research and the endpoint of pharmaceutical R&D. If it fulfills its role effectively, product development can enhance the value of the drug in therapy; if ineptly, it may negate good preclinical research and confound clinical trials. The potential problems that arise with incorporation of a chemical with drug acitivity, i.e., a drug substance, into a pharmaceutical dosage form (drug product or drug delivery system) are numerous. Avoiding or minimizing these problems requires both science and art in pharmaceutics and continual interaction with experts in other areas of pharmaceutical R&D.

In itself, the production of an active ingredient is not a negligible achievement. It requires proof of structure, identification and estimation of synthesis precursors and process contaminants, development of appropriate specifications and tests, and development of a process for repeatedly making kilogram quantities of the drug substance for use in preclinical and clinical studies that minimizes cost and optimizes such physical properties as crystal habit, particle size, bulk density, and color. Product Development is required to use this drug substance, which is usually at least 98.0% pure by best estimate, in more or less complex dosage forms. At the simplest, the dosage form may be a solution of the drug in water. Most often, solid oral dosage forms are needed for clinical study and projected for eventual marketing, and this requires addition of excipients to add bulk, aid flow, provide lubrication, impart color, and affect drug dissolution rate. A drug substance may be affected adversely by heat, light, air, moisture or reaction with a component of its immediate container. Formulation of the dosage form from it involves adding excipients, each of which is a potential reactant that may affect its chemical stability, physical properties, and biological activity. Formulation of drug dosage forms requires that attributes of drug identity, strength, dose uniformity, stability, and bioavailability be taken into account, along with those properties that permit their manufacture on high-speed machinery, shipment in commercial channels, and storage. In the course of clinical studies leading to New Drug Application (NDA) approval in the

United States (or the equivalent license to market in another country), a large number of formulations of the same drug may be provided both for clinical and pharmaceutical study. (In one instance, a marketed estrogen tablet, Estrovis, more than 100 formulations were prepared and documented. The marketed 100-μg tablet derived from the eighty-ninth experimental formulation.) A minimum number of distinct formulations is needed at the outset of clinical studies to accommodate dose finding in the clinic, and so several strengths may be required. Different types of dosage forms may be prescribed by the clinicians, e.g., oral liquids, oral solids, or injections. After the types and strengths of products are defined in the course of clinical studies, formulation changes may be necessitated by the results of stability studies, by the type of equipment selected for manufacture, by the availability or nonavailability and cost of certain excipient materials and tooling (punches and dies), by quality assurance considerations in providing physical distinction among products made at the same site, and by marketing considerations on product image. Since preparation of formulations involves not only the labor of the manufacturing operation but entails extensive testing and ties up storage facilities, it requires skillful management to keep the number of formulations under active study down to a minimum and to focus on the potential market formulation(s) in the market package(s) as quickly as possible.

In addition to working on new formulations developed from new chemical entities from its own company's preclinical research, Product Development works on products based on drugs licensed from other firms, foreign or domestic, line extensions of already-marketed drugs, and new drug delivery systems. This work may use some of the resources of preclinical research but not require all of them. In some of these instances, preformulation data and prototype formulations may be available for study.

The end product of R&D is a pharmaceutical dosage form along with the process for its manufacture, the documentation covering its manufacture and the specifications and test methods relevant to identity, strength, quality, stability, bioavailability, and other attributes which may affect its usefulness in therapy. At that stage in clinical research where enough information is available to justify a marketing decision and the submission of an NDA, the responsibility of Product Development enlarges to include effecting a smooth transfer of its technology to a manufacturing division. This includes providing its experts to observe or participate in plant scale-ups, furnishing the documentation needed for preparation of manufacturing and analytical technical directions used within the company, and supplying all of the information needed for successful application for NDA approval from government licensing authorities.

II. ORGANIZATION

Where Product Development fits in the table of organization of a pharmaceutical company and how its internal structure is set up differs among companies and, at different times, within the same company. The size of a company, its history and traditions, the number of people employed, their level of expertise, and even the personalities of key staff members are determining factors. Growth in technology and increased depth of governmental regulation are additional factors affecting organization. Ultimately, the governing criterion in organization is pragmatism. An organization that accomplishes its mission well tends toward stability; one that does not is restructured.

Because the range of product development activity is often split evenly between working up new chemical entities from its own preclinical research and working on line extensions, generic products and new drug delivery systems, it is equivocal whether Product Development should report to the head of R&D or to the head of manufacturing and marketing. Swarbrick and Stoll [1] surveyed 12 United States and 5 British companies and found that 16 of the 17 have clinical supplies prepared in Product Development and have the latter incorporated in the R&D organization. Regardless of the reporting relationship, Product Development must maintain an effective working relationship with all of the departments in the company which provide information and services to it and which receive them from it. There are several instances where R&D activities are divided between locations varying from a few to several hundred miles apart and where these may be separated from manufacturing and marketing functions. It is essential to establish personal contact among key people in the various departments and divisions, regardless of location. Advances in telecommunications technology have minimized the effect of separattion.

At minimum, Product Development comprises formulation and analytical functions. Because the technology of making tablets and capsules, for example, differs appreciably from that used in making sterile products for injection, a division of responsibilities within the formulations department by product type is a natural one. Table 1 shows one approach to organization of formulation groups on this basis. Another approach to organization can be based on function rather than product type. This division of responsibility can exist in parallel with the first system; however, it is superimposed on it. These functions are listed in Table 2.

In most companies, the analytical research function in Product Development is a department equal in status, but less often equivalent in size, to the formulations department, with the heads of both reporting to the same division head. Since the analytical research department is responsible for the quality control of products used in clini-

Table 1 Organization of Formulations Activities by Product Type

- I. Liquid and semisolid products
  - A. Oral liquids
    - 1. Elixirs
    - 2. Suspensions
    - 3. Syrups
  - B. Dermatologicals
    - 1. Creams
    - 2. Lotions
    - 3. Ointments
  - C. Sterile products
    - 1. Injections
    - 2. Nasal solutions
    - 3. Ophthalmics
    - 4. Inhalations
  - D. Miscellaneous
    - 1. Enemas
    - 2. Aerosols
    - 3. Transdermal devices
    - 4. Suppositories
- II. Solid products
  - A. Tablets
  - B. Capsules
  - C. Sustained-action oral products
  - D. Powders

Table 2 Organization of Product Development by Function

- I. Formulations
  - A. Physical pharmacy
    - 1. Preformulation studies
    - 2. Kinetics
  - B. Stability
  - C. Clinical supplies preparation
  - D. Pilot plant—unit operations
  - E. Research
    - 1. Materials evaluation
    - 2. Innovative drug delivery
- II. Analytical research
  - A. Methods development
  - B. Stability

cal research, the organizational independence of those who test product from those who manufacture it is consonant with good manufacturing practices. Swarbrick and Stoll [1] note, however, that 6 of the 17 companies in their survey place the responsibility for clinical supplies release with a quality assurance or production quality control unit external to Product Development. They comment that "this appears to be a worthwhile arrangement that should provide for an independent and useful check on the product's specifications prior to its release into a clinical trial situation." While, indeed, the analytical research operation may be within and report to a different company organization, these arrangements suffer the crucial disadvantage that they inhibit a close working relationship between formulators and analysts. Effective development of clinical drug products demands that these groups be interactive and interdependent. The enormous growth in analytical technology and the monstrous expansion of government regulation which has accompanied it during the past two decades has led to an increase in the extent and depth of testing protocols for a tablet formulation from 1962 and 1982. It can be seen that content uniformity assays, requiring 10 to 30 units, are used in place of weight control for release of product; dissolution testing,requiring analysis of 6 to 24 units, is used both for product release and during stability studies; and the assay and limit testing technology used in 1982 is much more discriminating and informative than that used 20 years earlier. Owing to the prevalence of sophisticated analytical instruments, like high-pressure liquid chromatographs fitted with automated sample introduction devices and computerized data reduction, the time spent in methods development is likely to be less now than in the past. On the other hand, more time is required for validating and documenting methodology.

It was not uncommon in the past to have formulators performing testing during development of a new formulation and using the analytical research group as a service function. As a mere scorekeeper for formulation development, the analyst in this working arrangement tended to develop an adversary relationship to the formulators. It is much more efficient to establish relationships where both groups have equal responsibility for successful formulation and are given equal recognition for successful products—or for product failures. Cooperation in defining critical factors affecting the chemical and physical properties of the drug in different formulation approaches can serve to cut down the number prepared to pharmaceutical evaluation and clinical supplies to manageable size. Although a great input of art is required for successful formulation, interjection of as much science as available can lead to production of better products faster.

Preformulation studies have become an essential activity for Product Development. These may be accomplished within a physical pharmacy group, a separate preformulations group, or by an ad hoc project team approach using formulations and analytical people. A physical pharmacy group can be given primary responsibility for physical measurements, such as thermal analysis, determination of particle size, microscopy, kinetic studies, and phase solubility. The activities are essential; how they are divided among groups is less important.

Stability studies are vital for predicting product performance and choosing among alternative formulations, for use in establishing expiration dating and for obtaining data to satisfy regulatory authorities. A separate group may be established to plan and perform stability studies for Product Development, or this activity may be integrated in the divisional activities using one or more people to handle the scheduling and storage of samples as well as the submission for testing and recording of results. Computer systems have proved worthwhile for these functions. It has been found useful to produce a computer printout monthly for products to be tested in the stability program within the following 2 months. This "stability alert," however it is generated, is a management tool which allows frequent review of clinical dosage forms in the stability program in the light of constantly changing interest, so that some may be discontinued and the status of others may be changed (Fig. 1). It also allows the analytical department the opportunity to group like products for testing by moving forward or delaying testing dates. Stability data are monitored usually by the department heads, but it is useful to appoint the formulator of each product as the "stability monitor" responsible for recommending the testing schedule and protocol and performing the physical observations at each test period.

Product development is a scientific activity! Whether or not there is a Research Group established formally within the organization, R&D toward new pharmaceutical products through work on drug delivery systems, manufacturing processes, and testing methodology is performed in the course of generating new products. Research activities that are not undertaken in conjunction with a specific project but may enhance the technology of the group should be encouraged. Some pharmaceutical R&D organizations have had formal policies which allow senior staff members a stated percentage of their time for research that is not project oriented, and others have designated certain senior scientists on their staff as full-time researchers who may serve as in-house consultants. It would be difficult for management to countenance staff members who abandon product-directed activities for their own projects during critical times when problems arise and deadlines must be met. It might also be awkward for Product Development management to defend its staff "Herr Professor" in times of company retrenchment. Perhaps a better policy is one which recognizes that senior staff members have time and facilities—if they choose to take advantage of them—that can be used in nondirected activities that may lead to patents, publications, and peer-group recognition. Such work should be rewarded. Staff members who do everything they are asked

ALERTING SYSTEM PAGE 1
PHARM. R & D STABILITY ASSAYS 11/23/81
DECEMBER

XXXXXXXXXXXXXXX 0.8 MG/ML 10 ML AMPULES	W	1273-49	AIK05358	06	12/23/81	A
XXXXXXXXXXXXXX 0.9 MG/ML 10 ML AMPULES	W	1273-49	AOL-5350	06	12/23/81	A
XXXXXXXXXXX 7 MG SUPPOSITORIES	W	1435-113	SOB05201	12	12/12/81	A
XXXXXXXXXXX 7 MG SUPPOSITORIES	W	1435-113	SOB05203	12	12/12/81	A
XXXXXXXXXXX 7 MG SUPPOSITORIES	W	1435-115	SOB05335	06	12/01/81	A
XXXXXXXXX 400 MG SA TABLETS	W	1654-62A2	JRD05479	03	12/16/81	A
XXXXXXXXX 400 MG SA TABLETS	W	1654-62A2	JRD05480	03	12/16/81	A
XXXXXXXXX 400 MG SA TABLETS	W	1654-62A2	JRD05481	03	12/16/81	A
XXXXXXXX 10 GR ENTERIC COATED TABLETS	W	1989-52	MAK05220	12	12/19/81	A
XXXXXXXX 10 GR ENTERIC COATED TABLETS	W	1989-52	MAK05221	12	12/18/81	A
XXXXXXXX 5 GR ENTERIC COATED TABLETS	W	1989-54A2	MAK05488	03	12/29/81	A
XXXXXXXX 10 GR ENTERIC COATED TABLETS	W	1989-55A2	MAK05489	03	12/29/81	A
XXXXXXXX 15 GR ENTERIC COATED TABLETS	W	1989-56	MAK05366	06	12/26/81	A
XXXXXXXXXXXXXXXXXXX 50 MG TABLETS	W	2234-10A1	FSW05497	02	12/23/81	A
XXXXXXXXXXXXXXXXXXX 50 MG TABLETS	W	2234-10A1	FSW05498	02	12/23/81	A
XXXXXXXXXXXXXX HYDROCHLORIDE 25 MG CAPSULES	W	3839A-35	JPM05363	06	12/24/81	A
XXXXXXXXXXXXX 25 MG CAPSULES	W	4269-3	NLM05346	06	12/10/81	A
XXXXXXXXXXXXX 25 MG CAPSULES	W	4269-3	NLM05352	06	12/20/81	A
XXXXXXXXXXXXX 25 MG CAPSULES	W	4269-3	NLM05353	06	12/20/81	A
XXXXXXXXXXXXX 50 MG CAPSULES	W	4269-4	NLM05347	06	12/10/81	A
XXXXXXXXXXXXX 50 MG CAPSULES	W	4269-4	NLM05354	06	12/20/81	A
XXXXXXXXXXXXX 50 MG CAPSULES	W	4269-4	NLM05355	06	12/20/81	A
XXXXXXXXXXXXX HYDROCHLORIDE 500 MG SR FC TABLETS	W	8213A-10A1	RCP05464	03	12/01/81	A
XXXXXXXXXXXXX HYDROCHLORIDE 500 MG SR FC TABLETS	W	8213A-10A1	RCP05465	03	12/01/81	A
XXXXXXXXXXXXX HYDROCHLORIDE 500 MG SR FC TABLETS	W	8213A-10A1	RCP05466	03	12/01/81	A
XXXXXXXXXXXXX HYDROCHLORIDE 750 MG SR TABLETS	W	8213A-5	ATG04896	24	12/07/81	A
XXXXXXXXXXXXX HYDROCHLORIDE 750 MG SR TABLETS	W	8213A-5	ATG04897	24	12/07/81	A
XXXXXXXXXXXXX HYDROCHLORIDE 750 MG SR TABLETS	W	8213A-5	ATG04898	24	12/07/81	A
XXXXXXXXXXXXX HYDROCHLORIDE 750 MG SR FC TABLETS	W	8213A-5	RCP05458	03	12/01/81	A
XXXXXXXXXXXXX HYDROCHLORIDE 750 MG SR FC TABLETS	W	8213A-5	RCP05459	03	12/01/81	A
XXXXXXXXXXXXX HYDROCHLORIDE 750 MG SR FC TABLETS	W	8213A-5	RCP05460	03	12/01/81	A
XXXXXXXXXXXXX HYDROCHLORIDE 250 MG SR FC TABLETS	W	8213A-9A1	RCP05461	03	12/01/81	A
XXXXXXXXXXXXX HYDROCHLORIDE 250 MG SR FC TABLETS	W	8213A-9A1	RCP05462	03	12/01/81	A
XXXXXXXXXXXXX HYDROCHLORIDE 250 MG SR FC TABLETS	W	8213A-9A1	RCP05463	03	12/01/81	A
XXXXXXXXX 100 MG CAPSULES	W	8495-22	MJC05482	03	12/22/81	A
XXXXXXXXX 100 MG CAPSULES	W	8495-22	MJC05483	03	12/22/81	A
XXXXXXXXX 100 MG CAPSULES	W	8495-22	MJC05483	03	12/22/81	A

Figure 1 Specimen page of computer-generated stability alert providing (from left to right) product title and strength, formulation code (W or WL number), initials of stability monitor with unique stability number, period of test (in months), due date, and status.

to do skillfully and efficiently are valuable; those who go beyond this to do worthwhile things that are not asked of them are priceless.

Swarbrick and Stoll [1], in their survey of organization in 17 pharmaceutical companies, divided these into small, medium, and large firms on the basis of sales. Interestingly, they reported that all of them had total staffs of about 100 people in the production of clinical supplies, although there were qualitative differences in distribution of scientists and technicians and of B.S., M.S., and Ph.D. degrees among them. Some proportion of any staff of that size has to be devoted to administration, as well as secretarial and documentation services. Office automation has trailed laboratory automation; however, a new era of office word processors with electronic filing, with simplified revision of reports and manuscripts and rapid telecommunication of data, bids to make changes as revolutionary in the life of the Product Development manager as have occurred in the laboratories.

It is impractical to attempt to describe an ideal Product Development organization. An organizational scheme that flourishes in one company may fail in another. People may be fitted into an organizational mold, but good ones mold an organization. Product Development organizations, like all other industrial divisions, are dynamic, changing with the times and the technology available.

III. INTERRELATIONSHIPS AND FUNCTIONS

Many of the functions and interrelationships of Product Development in the clinical research process have been mentioned already in this chapter and are alluded to elsewhere in this book. These can be discussed in detail using the stages of clinical research as a framework.

The clinical research process begins with the planning of the use of the first dosage forms in humans. Before this can happen, an active compound must be identified in preclinical studies. A "lead compound committee" may be formed to select among potential candidate drugs. It is useful to have a Product Development representative at this stage to evaluate whatever information is available on the chemistry and potential routes for administration and give advice on extemporaneous preparations to be used in further pharmacological testing and in toxicology. A brief prepharmacology study of a candidate compound may be performed in Product Development laboratories, where solubility and short-term stability are assessed and potential problems such as solubility, pH, moisture absorption, and chemical instability may be identified. Pharmacy considerations may be useful in choosing among alternative lead compounds for further study and in fostering the preparation of additional salts or other derivatives of an active compound at a stage before a heavy investment in toxicology studies has made changes in chemical form excessively expensive. Information on the physicochemical properties of a lead compound is generally sparse at this stage of development, and it is well to open avenues of communications on properties and analytical methods between Product Development scientists and their colleagues in preclinical research departments, especially those in the analytical/physical chemistry area and in drug metabolism.

Formation of a planning team has proved useful in coordinating the development of a clinical candidate compound. This team is composed of senior staff members of the various disciplines needed for guiding the new drug candidate from the mere pharmacologically active chemical substance to a drug product. The team is usually chaired by a pharmacologist or medicinal chemist at the beginning stages of clinical research, then by the physician monitoring clinical studies as they progress, and ultimately by a Product Development scientist or a marketing product manager at its final stages. The team reports to the management group of R&D and should not include anyone who is a member of that executive directorate. This group appoints the planning team members from among its senior staff. The team includes representatives from chemistry, pharmacology, toxicology, drug metabolism and bioavailability, the medical research group, regulatory affairs, product development, biometrics, marketing, the research planning and coordination function, and usually medical and marketing people from the firm's international division. It may include representatives from a licensor firm where a cooperative agreement exists between companies. The planning team chairman is assigned responsibility for coordinating the efforts of its members and, through them, their departments, in beginning clinical studies and recommending changes in plan as they progress.

Interaction with marketing personnel was mentioned above. Product managers have important input into the types of dosage forms selected for study, their strengths, taste, color, shape, packaging, and other attributes as well as into the types of clinical studies and laboratory data that might yield claims useful in product introduction and promotion. They are prolific sources of new product ideas involving line extensions or novel drug delivery forms. Interaction between product management and product development is continual. New product ideas should flow in the other direction as well. It is usual to have some kind of committee established to evaluate new product ideas, from whatever the source, to determine if a planning team should be established and recommend whether or not company resources should be committed. This committee may comprise people from a new products acquisition group, medical research, product development, marketing, market research, and regulatory affairs. Preformulation studies are essential whether a new clinical candidate compound derives from company R&D or is obtained from outside the company. The Product Development member of a planning team should insist that time be allotted for this activity in the plan of a clinical study. At its best, a preformulation study is not a mere mechanical adherence to a recipe for tests to be performed and data to be gathered. An intelligent study takes account of the type of dosage forms needed for study, the information on physicochemical properties already available from prepharmacology study and from medicinal chemists, chemical development, and drug metabolism, and it also focuses on developing information on the pharmaceutics of the compound. Bulk density of a compound, for example, may be of little consequence to scientists in preclinical research, but it is an important consideration in product development work, especially if capsule products are needed for clinical study. Information on the stability or reactivity of a new chemical compound usually must be developed, because the opportunities for reaction during manufacture and storage of dosage forms are much greater than with the drug substance per se. Preformulation studies require interaction with the formulation groups and close cooperation with the Product Development analytical research department. Indeed, this last should be included in the preformulation team in planning these studies, performing the tests and assays required, and using the methods and information obtained therefrom as the basis for product specifications and stability tests prior to release of clinical supplies.

Clinical study of a new drug requires filing an Investigational New Drug (IND) application in the United States or an equivalent preclinical submission document in another country (or, in some other countries, at least having the necessary information available in case it is requested). The required information is assembled and submitted by a regulatory affairs department. Product Development provides information and methodology for the manufacture and quality control of clinical dosage forms during clinical study and, ultimately, for the product to be marketed. Because manufacturing and control information on the drug substance and the drug product are reviewed by the same government regulatory official, information on both are included in the same document and must be compatible. This requires cooperation between analytical groups in Product Development and in Chemistry to eliminate seeming inconsistencies in standards and tests. If the bioavailability group or department is not included in Product Development, similar cooperation between groups is necessary from a regulatory affairs standpoint as well as mutual interest in developing good products. Requirements for initiating clinical study and for licensing products for sale differ in other countries. Product Development is often the resource for information on the manufacturing processes, controls, and stability data used by its international corporate colleagues. Since development timetables differ in an international company, information can flow in both directions to mutual benefit. Not infrequently, questions and comments are generated by regulatory officials on the documents submitted by a firm. Responses to those on the manufacturing and control section are provided to Regulatory Affairs for transmittal.

Formulation, manufacture, and distribution of clinical supplies usually is a function of Product Development. To draw an analogy, we can use the game of seven-card stud poker as a metaphor for clinical research in the pharmaceutical industry. It requires an investment to get to play the game and there are competitors at the same table. Both have several decision points along the way where each player must either increase his investment or withdraw from the game, using knowledge of his holdings at that time and incomplete knowledge of the holdings of competitors in the game as the bases for decisions.

A winner does not play out all his hands. The development of a product during clinical study parallels this increase in investment. Relatively small lots of dosage forms are needed in the initial dose-finding and safety clinical studies. Occasionally, small lots of dosage forms of known drugs in unique dosage forms such as sustained action tablets may be made for bioavailability testing to attempt to establish correlation between dissolution and bioavailability. The later clinical studies may require lot sizes that approximate production lots. Thus the scale increases from, say, 50,000 tablets in a lot at the beginning of a clinical study to more than 1 million tablets with some drugs in the last stages. (A production lot of a generic or a drug with limited indications, however, may be more on the order of 50,000 or 100,000 than 1 million.) After a product is formulated, with prototypes available and documentation on the formula and its direction for manufacture provided, clinical supplies might be made on a larger scale by a group set up for that purpose. At the same time, development activities needed for increasing scale must be carried on. Sufficient drug substance must be ordered, excipients must be obtained, tools and dies must be designed, and packaging planned for the machinery selected. The equipment used for clinical supplies manufacture is the same type as will be used for manufacture of product for sale if clinical studies are successful. Although, in general, the number of units or volume produced during manufacture of clinical supplies are smaller than those that will be produced for sale, there should be no qualitative difference in the product itself. Most often, the equipment used is identical, the lower volume needed for clinical studies being obtained by using fewer machines or operating them for shorter periods. Where special equipment will be needed or modifications to existing equipment made, the manufacturing division must be apprised. Manufacturing divisions almost invariably have a process improvement group which studies the application of new equipment and technology to its operations with a view toward improvements in productivity. This group interacts continually with Product Development on selection of equipment. Improved technology flows in both directions.

Near the end of the clinical research process, Product Development must assist in the transition which takes place when manufacture of product is moved from its pilot plant to one or more production facilities. A "Request for Plant Scale-up" might be sent to the manufacturing plant (Fig. 2). For those products which do not require an NDA, a "Technical Profile" which provides all of the information and directions needed for manufacturing a product is prepared. Even with this or the manufacturing directions which are provided in the NDA, it has been found useful to bring plant operators to the Product Development laboratories for training in the processes of manufacture for a new product before they undertake this in the first plant scale-ups. Another useful practice in attaining a smooth transition has been to

Product ______________________ W No. __________

Required Batch Size ____________ Required Batch Units ______

Material Supplied By: Finished Product Disposition By:

Production ____________ Production ____________

Development ____________ Development ____________

Test Objectives ______________________________

____

____

Equipment Required and Location ______________________________

____

____

Authorized Observers: Production ______________________________

Development ______________________________

Packaging ______________________________

Charges: Account No. ______________________________

Date: ____________ Requested By: ______________________________

Comments:

cc:

*Data Sheet Enclosed
Rev. 6/81

Figure 2 Request for plant scale-up.

have personnel from both Product Development and manufacturing plant observing the first scale-ups, recording their impressions and signing off that the process can be transferred successfully in an "Evaluation Summary" (Fig. 3).

Just as manufacturing directions for clinical supplies become technical directions for manufacture in the plant, the tests and standards used for release of clinical supplies evolve into quality control procedures for the plant. This requires a similar degree of interaction between research and production.

IV. CLINICAL PHARMACEUTICAL OPERATIONS

Most companies have a clinical coordinator or a group responsible for arranging for the timely supply, packaging, coding, record keeping, and distribution of drug products under clinical investigation. This function, which may be called Clinical Pharmaceutical Operations,

DATE OF TEST ____________

TEST FORMULATION: ____

DATA SHEET#: ____________ LOT NUMBER ____________ BATCH SIZE ____________

TEST OBJECTIVES: ____

____

____

AUTHORIZED OBSERVERS: ____________________ ____________________

(PRODUCTION) (DEVELOPMENT)

- -

(SUMMARY OF RESULTS) Prodn. Devel.

1. WAS TRIAL SATISFACTORY: (SEE OBJECTIVES) (YES) (NO)

2. IS FORMULA SUITABLE FOR ROUTINE PRODUCTION? (YES) (NO)

PRODUCTION COMMENTS AND RECOMMENDATIONS:

OBSERVATION AND COMMENTS BY:

PRODUCTION: ____________

DEVELOPMENT: ____________

DATE: ____________

Figure 3 Scale-up evaluation summary.

operates at the interface of Product Development and Clinical Research and may report to either.

Clinical Pharmaceutical Operations participates in the review of draft protocols for clinical study or investigational new drug products. It anticipates the need for bulk drug substances for preparation of these supplies and assists in arranging for their timely receipt by Product Development. Requisitions for investigational new drug products, including reference drug products and placebos, required for a protocol may be issued by staff members of Clinical Research or by the clinical coordinator; however, the latter necessarily participates in the process.

Drug product supplies are assembled for each protocol, and packaging and labeling are provided along with documentation systems that assure accountability of each package dispensed. Medication is assembled for each patient, taking in account the specific visit and the specific dose, in unit-for-use prescription format. We may compare this operation to a professional pharmacy service which, in a large research-based pharmaceutical company, may dispense upwards of 50,000 "prescriptions" per year.

Solid oral dosage forms may be received by Clinical Pharmaceutical Operations in plastic-lined fiber drums and dispensed in standard production type bottles and/or high-quality blister packages; liquids and parenterals may be requisitioned in the container types and sizes required for the study. A transfer memo, which is attached to all of the relevant documents on the manufacture and control of clinical products, is shown in Fig. 4. Because most clinical studies involve a double-blind design, the packaging is arranged to provide dosage forms of the same appearance and size for the investigational drug product, the reference drug product, and the placebo. (Where color, odor, and taste may be factors, Product Development attempts to match these as closely as possible.) Comparison of an investigational drug product with a projected twice-a-day (b.i.d.) administration and a reference product with a three times a day (t.i.d.) schedule, for example, might require directions that both be administered t.i.d., with the center dosage unit in the investigational drug product unit-for-use a placebo. Comparison of a 50-mg dose of an investigational new drug product tablet with a 100-mg dose of a reference drug product might be handled by packaging one tablet of the investigational new drug tablet and one placebo tablet in one compartment of a blister package and two 50-mg tablets of the reference product in another, the two packages differing only in coding.

A label for clinical supplies is shown in Fig. 5. This label comprises a main portion and a smaller, tear-off section, which is detached at the time the product is given the patient and affixed to the patient case report form by the clinician. Both portions of the label bear the date, protocol number, the patient number, the patient visit number, the code designation for the drug product, the strength of the product and pharmaceutical type (e.g., tablet or capsule), the lot number, the control number, and the name of the sponsor company. In addition, the main portion of the label, on the container given the patient, bears the clinician's name and telephone number, space for the patient's name to be entered by the physician, directions for use, the statement, "Caution—New Drug Limited by Federal Law to Investigagational Use," and the mailing address of the sponsor. The control number uniquely identifies each unit-for-use dispensed.

DATE: ____________________

From: Product Development
Morris Plains, NJ

To: Clinical Research Coordinator
Ann Arbor, MI

Preparation # ____________________

Name: ____________________

Lot #: ____________________

Date of Manufacture: ____________________

Quantity Shipped: ____________________ Log #: ____________________

Quantity Requested: ____________________

Requisition #: ____________________ Date: ____________________

The yield from this investigational drug lot is being transferred to your attention and is approved for clinical use.

This lot has been prepared according to CGMP regulations and meets all requirements for human use.

Attached are copies of all manufacturing records.

____
(for Clinical Supplies Quality Control)

____
(for Clinical Supplies Production)

Figure 4 Investigational drug transfer memo.

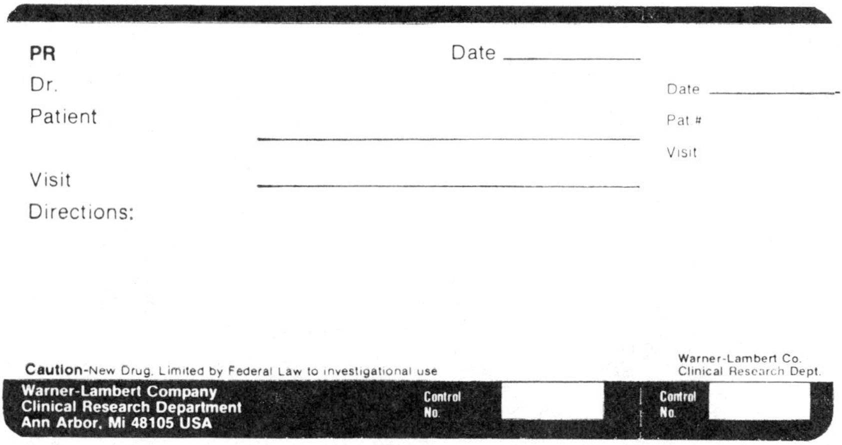

PR Date ____________

Dr.

Patient

____

Visit ____________

Directions:

Date ____________

Pat #

Visit

Caution-New Drug. Limited by Federal Law to investigational use

Warner-Lambert Company
Clinical Research Department
Ann Arbor, Mi 48105 USA

Control No.

Warner-Lambert Co.
Clinical Research Dept.

Control No.

Figure 5 Label for clinical supplies. (Courtesy of Warner-Lambert Company, Clinical Research Department, Ann Arbor, Michigan.)

The accountability and integrity of clinical supplies is assured by documentation systems and testing protocols, including random confirmation of identity by laboratory testing and occasional reassays of clinical stocks. Product Development maintains stability testing surveillance of all lots undergoing clinical investigation for periods as long as any supplies remain extant.

V. CULMINATION: A MARKETED PRODUCT

After a clinical research product receives approval for marketing, it is no longer the responsibility of the R&D divisions of a company. Yet, if problems arise during production or if questions on a marketed product are directed to the medical affairs group responsible for marketed products, R&D usually is the resource needed and called on in troubleshooting production problems and responding to medical and pharmaceutical inquiries.

The function of the Product Development Division in a pharmaceutical company is to transmute a pharmacologically active substance into a manufactured drug product for use in medical therapy. There are many ways in which this function can be organized. In the terminology, examples, organization, and interrelationships used in this chapter, the intention has been descriptive—not prescriptive.

REFERENCE

1. J. Swarbrick and R. G. Stoll, *Organization within the pharmaceutical industry: A survey. Drug Dev. Pharm. 7:*633-644 (1981).

BIBLIOGRAPHY

Adamec, R. J., *How to improve your new product success rate. Manage. Rev. 70:*38-42 (Jan. 1981).

Bromer, William W., and Charles E. Redman, Research. In *Remington's Pharmaceutical Sciences* (Arthur Osol, Ed.), 16th ed.) Mack, Easton, Pa., 1980, pp. 59-66.

Dailey, Robert C., The role of team and task characteristics in R&D team collaborative problem solving and productivity. *Manage. Sci. 24:*1579-1588 (Nov. 1978).

Howton, F. William, Work assignment and interpersonal relationa in a research organization. In *Administering Research and Development* Irwin (Dorsey Press), 1964, pp. 389-404.

Katz, Ralph and Michael Tushman, Communication patterns, project performance, and task characteristic: An empirical evaluation and Integration in an R&D setting. *Organ. Behav. Hum. Performance 23:*139-62 (Apr. 1979).

Lasagna, Louis, Problems of drug development. *Science 145:*362-367 (July 24, 1964).

Ley, H. L., Jr., and F. H. Desmond, The role of regulatory agencies and industry in assessment of the safety of drugs for use in man: United States Food and Drug Administration's viewpoint, *Can. Med. Assoc. J. 98:*318-321 (Feb. 10, 1968).

Meyer, Fred H., and T. R. Mika, R&D management in the pharmaceutical firm. *Drug Cosmet. Ind. 128:*46, 48, 50, 52, 54, 56, 60, 62, 114-115 (Apr. 1981).

Oberholzer, R. J. H., The clinical trial of new drugs. *Pharm. Acta Helv. 39:*465-479 (Aug. 1964).

Smith, Edward A., Organization and method in research and development. *Mfg. Chemist Aerosol News 36:*62,64 (Sept. 1965).

Thomas, K., Drugs and their dosage forms. *Drugs Made in Germany 11:*(3)124-133 (1968).

Tushman, Michael L., Technical communication in R&D laboratories: The impact of project work characteristics. *Acad. Manage. J. 21:* 624-645 (Dec. 1978).

Tushman, Michael L., Managing communication networks in R&D laboratories. *Sloan Manage. Rev. 20:*37-49 (Winter 1979).

16
OVER-THE-COUNTER DRUG RESEARCH

William Feinstein
Whitehall International, Inc.
New York, New York

I. INTRODUCTION

Over-the-counter (OTC) drug research deals primarily with active ingredients and combinations of active ingredients that have been shown to be safe and effective for their intended use and can be labeled with appropriate directions and warnings such that the consumer can utilize the product for the purposes of self-medication. Generally, OTC products do not contain new chemical entities but are, in fact, old ingredients or new combinations of old ingredients and/or new brand names [1]. The ingredients currently being utilized in OTC products have been under review since the early 1970s by a number of expert advisory panels commissioned by the U.S. Food and Drug Administration (FDA) [2,3]. A brief discussion of this review appears in Sec. IV, below.

II. OTC COMPANY DISCIPLINES

The development and marketing of products for OTC use typically ranges from 1 to 3 years in duration (i.e., from idea generation to ultimate consumer sale). A team effort is involved in identifying a concept or claim which has a unique selling point (USP) and in formulating and making available for sale a product which delivers the USP. Typically, people having expertise in the following disciplines are involved:

Pharmacy
Chemistry
Engineering
Medicine
Law
Dentistry
Microbiology
Marketing
Finance
Art
Statistics
Medical writers

As can be seen, this is quite a broad range of disciplines. Many of these personnel may be employed in any of the following departments, which are typical of OTC drug research companies:

Product Development
Analytical Research and Development
Pharmacology/Toxicology
Medical
Regulatory Affairs

Pilot Plant
Packaging
Statistics

In addition, they work closely with other departments in the organization such as:

Market Research
Marketing
Legal
Manufacturing
Finance

It should be emphasized that this description of an organization is general in nature and, depending upon the size and needs of a particular company, it may have all of the aforementioned departments or only some of them. Many smaller companies depend on consultants and outside resources for certain types of expertise if they do not have such expertise within their organization.

III. OTC COMPANY ORGANIZATION

The unique characteristics of the OTC market in the pharmaceutical industry can be demonstrated by evaluating the organizational makeup or structure of competing companies. In addition, OTC drug research organizations have either created separate divisions for their OTC products or have created separate companies [4].

A typical OTC drug research organization would be part of an overall organization which may be represented as shown in Fig. 1.

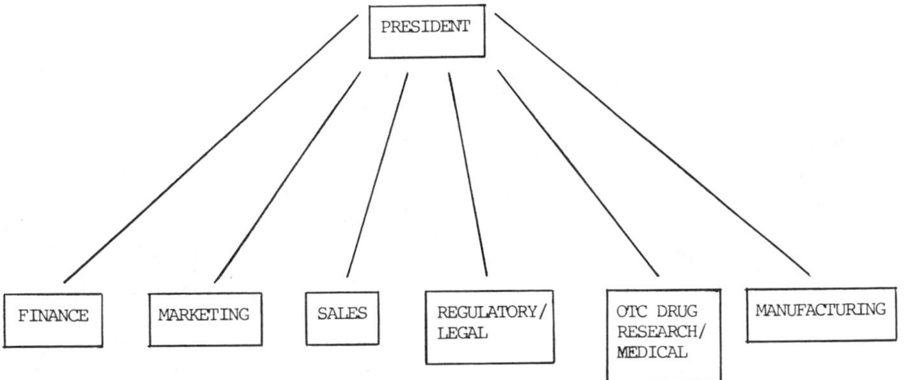

Figure 1 A typical OTC drug research organization as part of an overall pharmaceutical company organization.

IV. U.S. FOOD AND DRUG ADMINISTRATION OTC HISTORY

In 1972, the FDA established a number of expert advisory panels to review published and unpublished information on various active ingredients within a number of pharmacological classes. Each panel was charged with an objective evaluation of each drug and with the eventual preparation of reports for the FDA. From these reports the FDA developed monographs for various categories of drugs. Active ingredients were to be categorized as Category I (safe and effective), Category II (unsafe and/or ineffective), and Category III (more data needed to finalize classification).

The FDA divided the OTC drugs into 26 pharmacological categories [4,5], and these were reviewed by the following 17 advisory review panels:

Antacids
Antimicrobial I
Internal Analgesics
Cough and Cold Products
Sedatives and Stimulants
Topical Analgesics
Dentifrices
Laxatives, Antidiarrheals, etc.
Ophthalmic Products
Contraceptives and Vaginal Products
Hemorrhoidal Products
Oral Cavity Products
Antiperspirants
Vitamins and Minerals
Antimicrobial II and Topical Antibiotics
Miscellaneous External Products
Miscellaneous Internal Products

Each panel's final recommendations were submitted to the FDA Commissioner as the panel's final report. This report became the preamble to the Proposed Monograph now referred to as the Advanced Notice of Proposed Rulemaking (ANPR). The Proposed Monograph contains only the Category I conditions, as "proposed regulations" to eventually appear in Title 21, Code of Federal Regulations (21 CFR).

At this stage, comments are invited; and subsequently FDA responds to these comments in the preamble to the Tentative Final Monograph (or Notice of Proposed Rulemaking which, for all practical purposes, represents the actual proposed regulations). After sufficient time for objections, FDA responds and publishes the Final Monograph (FM). The FM is "the law." To date, only a few FMs have been published. An FM can be changed only by submission of a petition to that FM. A marketer wishing to sell an OTC drug with some conditions, i.e., ingredients, label claims, doses, etc., at variance with the FM can do so only by means of this petition to the FM or by submission of a New Drug Application (NDA) for approval. Petitions are published in the *Federal Register* for comment. The FDA must eventually either grant the petition or disallow it.

A large number of Proposed Monographs and several Tentative Final Monographs have been published. The OTC drug industry has, in general, taken the view that, although the Proposed and Tentative Final Monographs are not yet the law, they *will* be eventually and therefore most new products are developed within the scope of the monographs. Existing products which do not conform to the monographs are, in general, being modified.

V. OTC COMPANY DEPARTMENTS

The following outlined descriptions indicate some of the responsibilities of the departments generally involved in OTC drug product research:

A. Product Development Department

Product Development, or Pharmaceutical Research and Development, is a laboratory environment dedicated to the formulation of safe, effective, and stable drug dosage forms. Pharmaceutical chemists utilize the latest state-of-the-art equipment and techniques to provide a wide range of products, from tablet and capsules, to liquids, creams, and ointments. Some of the objectives of this group follow:

1. Develops prototype formulations for marketing approval
2. Develops suitable manufacturing processes and procedures
3. Monitors initial production batches and modifies formulations and procedures accordingly
4. Establishes stability evaluation programs (by establishing product and technical requirements)
5. Provides samples for stability testing program in order to determine shelf-life characteristics of (a) the formulation and (b) package compatibility
6. Prepares samples suitable for consumer/clinical testing programs

B. Analytical Research and Development Department

This department is distinguished from a quality control group in that its efforts are much more intensive with respect to chemical and physical testing. New methodology is very often generated by this department, which may be utilized later by quality control. New analytical equipment is often first evaluated by Analytical R&D. This group normally is under the Pharmaceutical R&D umbrella and, with respect to OTC drug product development, it:

1. Establishes assay development programs for raw materials
2. Assays competitive products
3. Develops stability-indicating assays for active ingredients in the final product
4. Performs assays on samples stored under stability storage conditions

C. Pharmacology Department

This department serves many functions, but, with respect to OTC drug product development, its responsibilities are somewhat limited in that the active ingredient is a known entity. That is, its usefulness has already been established. Much of the effort of the Pharmacology Department, especially in larger firms, is directed toward the screening of new chemical entities, for safety and efficacy.

Using established procedures involving animal testing, proposed OTC drug formulations intended for topical use are evaluated for:

1. Dermal irritation
2. Inhalation
3. Dermal LD_{50}
4. Eye irritation
5. Repeated insult or human sensitivity testing

Products intended for oral administration are tested for:

1. LD_{50}
2. Gingival irritation (for example, orally applied dental products)

Closely associated with the aforementioned testing is certain specialized testing that is warranted for particular situations. The following are done with multiple generations of several species of laboratory animals and/or by the use of several types of recognized in vitro methods:

1. Carcinogenicity
2. Teratogenicity
3. Mutagenicity

D. Medical Department

The Medical Department is composed of physicians specialized in clinical medicine and those specialized in clinical and experimental therapeutics. This group:

1. Provides medical evaluation of new marketing and product development concepts
2. In conjunction with the Legal Department and Regulatory Affairs, reviews promotional materials and advertising
3. Designs and monitors human clinical studies in order to confirm the safety and efficacy of the tested products
4. Conducts comparative efficacy testing against competitive products to provide data to support advertising claims
5. Designs and monitors bioavailability studies on solid dosage forms

E. Drug Regulatory Affairs Department

This department is responsible for monitoring all government regulations pertaining to drug products, for liaison with the appropriate government agencies, and for making submissions to the agencies. In the case of OTC drug products, this group:

1. Reviews the status of each monograph for each pharmacological category
2. Reviews with marketing the proposed/modified OTC drug product for compliance with monographs
3. If not in compliance, advises on appropriate regulatory route to marketing, i.e., (a) Petition to Monograph or (b) Investigational New Drug/New Drug Application (IND/NDA) routes
4. Reviews labeling for compliance with monographs *and* with general OTC drug product regulations in 21 CFR
5. If an IND is involved, assists in developing, compiling, and reviewing: (a) protocols, (b) methods, facilities, controls (M-F-C), and (c) sponsor's and investigators' commitments (Forms FDA 1571, 1572, 1573) and biomedical research monitoring; submits to FDA and monitors progress through the FDA review of maintaining communications between FDA and the sponsor
6. Develops NDA based upon IND; submits to FDA, follows progress through reviews by maintaining two-way communication, and "predicts" approval date (for marketing launch purposes)
7. If approved NDA product, maintains required submissions of stability data (as regards the expiration date), any M-F-C changes, label changes, adverse reactions, and periodic reports, seeking approval where required
8. Ensures compliance with Drug Listing Act
9. "Drug vs. cosmetic" issues explored, where appropriate [6]
10. Maintains interdepartmental communications (New Products-Marketing-R&D-Manufacturing & Quality Control-Legal)
11. Prepares appropriate field force communications where regulated products are involved

F. Pilot Plant Group

The primary function of the pilot plant group is to concern itself with the evaluation of the manufacturing techniques performed by the Product Development Laboratory and to scale-up to the standard manufacturing procedures and equipment. The major responsibilities are as follows:

1. To serve as a liason between the Product Development Laboratory and all manufacturing functions
2. To evaluate the procedures developed for the new products, ensuring that they are applicable to the manufacturing of the new product in the production environment
3. To monitor the initial production batches of the new product

G. Packaging Department

This group is concerned with designing the appropriate package for the new product. Such considerations as child-resistant closures, tamper-resistant packaging, folding cartons, and shipping cartons all require design for marketing consideration. In addition, this group may also arrange for compatibility studies between the new formulation and various alternative packaging materials being considered.

H. Statistics Department

This group assists in designing and evaluating various consumer or market research studies, clinical studies, and other experimental activities where the results require mathematical analysis.

I. Market Research Department

The following are some functions of this group:

1. To evaluate the market place in terms of *what*, *when*, *where*, and *why* the consumer is buying in any given category of products
2. Once a particular product profile has been prepared, to design and then assist in evaluating the appropriate consumer study
3. To assist the Marketing Department in determining the most appropriate new product, including the dosage form itself and product claims, for launching
4. To evaluate various new product categories which may present themselves as corporate opportunities for increasing sales
5. To analyze sales trends through sales data

J. Marketing Department

The following are some functions of this department:

1. To decide the product category to pursue for the new product based on an evaluation of competition in each category [7-9], current sales information, and various information concerning product-category potential for growth
2. To review and evaluate product concept and product objectives
3. To develop product and packaging, and coordinate new product activity with all the groups, but especially with R&D (in some companies steps 1-3 are designated as functions of the New Planning Department
4. To develop marketing strategy and marketing programs including advertising campaigns
5. To determine distribution and introduction plans
6. To plan various test markets and evaluate the results
7. To develop media plans, advertising strategy, and new product label copy
8. To forecast new product sales and potential profitability associated with this project

K. Legal Department

This department's responsibilities are as follows:

1. To coordinate all contractural arrangements, including new product ideas, acquisitions, and contract manufacturers' arrangements
2. To review the new product entity in terms of compliance with the various regulations concerned
3. To determine whether adequate substantiation has been provided to support labeling claims for the new product

L. Manufacturing Department

This group's major responsibilities as regards new products are as follows:

1. To prepare a production plan to meet forecasts
2. To manufacture and package within the standard costs developed
3. To maintain minimal inventories
4. To manufacture in compliance with current Good Manufacturing Practices

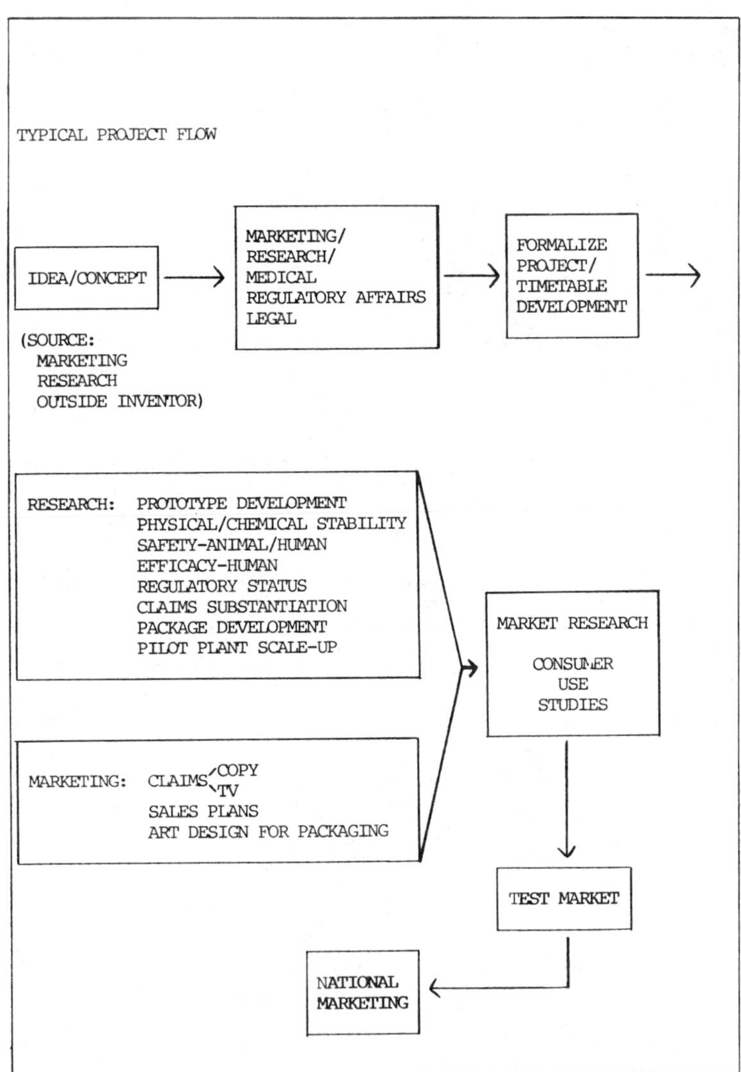

Figure 2 Typical OTC project flow.

M. Finance Department

This department's functions are as follows:

1. To coordinate the various forecasted expenditures into a new product budget
2. To monitor and determine whether the new product entity has achieved its forecasted profitability

VI. OTC PROJECT FLOW

The OTC project flow is schematized in Fig. 2. The idea, concept, or product is presented from a source (e.g., marketing, research, or outside inventor). The Medical Department evaluates the project's feasibility. The Marketing and Finance Departments evaluate the financial outlook were the corporation to proceed with such a project. The Analytical R&D Department initially evaluates the technical feasibility of such a product or the chances of developing a product to meet the idea or concept. The Regulatory Affairs and Legal Departments present their points of view with regard to legal status and regulatory compliance (i.e., whether the proposed product is in compliance with the appropriate OTC drug product monograph, whether it is a new drug (IND/NDA) situation, or whether there is a cosmetic-drug issue). Following a review of all of the various inputs, a decision is made to proceed or not proceed with the project. If a decision is made to proceed, all of the necessary disciplines then meet and establish a timetable for the project.

REFERENCES

1. M. C. Smith, *Principles of Pharmaceutical Marketing,* 2nd ed. Lea & Febiger, Philadelphia, 1975, Chap. 10.
2. *Federal Register 37:* 9464-9465 (May 11, 1972).
3. David C. Oppenheimer, The over-the-counter drug product reviews. *U.S. Pharmacist* (Oct. 1979).
4. S. St.P. Slatter, *Competition and Marketing Strategies in the Pharmaceutical Industry.* Croom Helm, London, 1978, Chap. 7.
5. Code of Federal Regulations, Title 21, Food and Drugs, Office of the Federal Register, General Services Administration, Washington, D.C., 1981, Sec. 330.5.
6. David C. Oppenheimer, The OTC drug product reviews and the cosmetic vs. drug issue. *Cosmetic Technol.* (Jan. 1980).
7. *Handbook of Nonprescription Drugs,* 6th ed. American Pharmaceutical Association, Washington, D.C., 1979.
8. *Physicians' Desk Reference for Nonprescription Drugs.* Medical Economics, Oradell, N.J., 1981.
9. A. C. Nielsen Co., *Drug Index,* Northbrook, Ill.

17
POSTMARKETING SURVEILLANCE OF DRUGS*

E. Keith Borden
The Upjohn Company
Kalamazoo, Michigan

---

*This chapter is based to a considerable extent on E. Keith Borden, Post-marketing surveillance: Drug epidemiology. *J. Intern. Med. Res.* 9:6 (1981).

I. INTRODUCTION

At the time of this writing it is fair to say that there is no specification of what constitutes postmarketing surveillance (PMS). There is currently considerable interest in monitoring medications after they are marketed, but there is certainly no consensus about what should be monitored or how it should be done. The primary concern is to be certain that no unexpected catastrophes occur as a result of the introduction of a medication. However, beyond that the substantive issues of PMS have not been established; nor is there a unanimity of opinion about the need for systematic drug monitoring. Nevertheless, PMS is an idea whose time has come. At this time it would seem that the major issues are not whether PMS is appropriate, but rather who should do it, how it should be done, what should be the substance, and how it should fit into the drug development process.

II. THE NEED FOR POSTMARKETING SURVEILLANCE

The idea of monitoring drugs is not new. Finney described a proposal for a drug monitoring system in 1965 [1]. Cluff et al. [2] and Borda et al. [3] initiated and described methods of intensive hospital monitoring during the 1960s. Voluntary reporting of adverse drug reactions has been widely accepted for years.

The current issues are the need for formalized systematic evaluations of medications after they are marketed and the appropriate place of such evaluations in the regulatory process for approval of new medications.

Some of the major factors which motivate for such systematic evaluation are presented below [4].

A. The Amount of Therapy

According to the Joint Commission on Prescription Drug Use it is estimated that there are approximately 54,000 drug products on the market. These consist of 1,900 active ingredients. In the past 40 years, 1,024 chemicals have been introduced [5]. In 1980, there were an estimated 1.4 billion prescriptions dispensed from retail pharmacies in the United States [6]. This amount of medication and the very large number of therapeutic exposures quite appropriately constitute a matter of public health interest.

B. Attitudes Toward Therapy

Over the past several decades major advances have been made in therapeutic interventions, resulting in significantly diminished risks from many diseases and a change in the perception of risk from disease. At the same time increased awareness of the amount of potential adversity from therapeutic intervention has increased the sensitivity to risk from such treatment and raised questions about the appropriateness of therapeutic exposure.

C. Distinction Between Premarketing Studies and Postmarketing Use

Before a drug is marketed, it is investigated intensively in experimental animals and human subjects to determine its safety and efficacy. During the Phase III development period it is tested in carefully designed prospective randomized controlled clinical trials in as many as 3,000 subjects. These clinical trials have a rigorous design in which the participants must satisfy specific inclusion and exclusion criteria. For instance, children and pregnant women (or women who might become pregnant) are likely to be excluded. Patients with a broad range of comorbid conditions of varying severity are specifically excluded. Dosing and length of prescription are carefully predetermined. The result is that the observations and conclusions made under rigorous scientific conditions may not specifically apply to the broad subsequently treated population. Nor are events that may occur when much larger populations are treated likely to be observed among the limited-sized population which is studied.

After a drug is marketed it is used in much larger numbers of patients; dose size and dosing schedules may be modified by experience. The patients need not satisfy the rigid inclusion and exclusion criteria of a clinical trial. Visit schedules are not likely to be as rigid as in the study period. With experience, physicians may better define the optimum usage, and identify new indications.

D. Lack of Systematic Collections of Information About the Broad Ongoing Experience

Perhaps one of the most important and least recognized reasons for systematic evaluation of the customary use of drugs is the current failure to exploit the cumulative experience of the many practitioners of medicine. There are no collections of data for feedback to physicians about information on what medications they are using and how they are using them.

Undoubtedly, the acceptance of the need for monitoring medications after they are marketed was precipitated by the occurrence of serious unexpected adverse effects (e.g., Thalidomide and phocomelia; MER-29 and cataracts; practalol and oculomucocutaneous syndrome; chloramphenicol and aplastic anemia; sudden death associated with certain aerosol inhalers). The magnitude of the adverse drug reaction (ADR) problem in the U.S. is unclear. The Medicine in the Public Interest (MIPI) monograph *Adverse Drug Reactions in the United States* concluded: "current estimates of the magnitude and cost of the adverse drug reaction problems are completely unreliable because they are derived from a database which is incomplete, unrepresentative, uncontrolled and operationally not defined" [7]. Given the importance of the debate and its impact on attitudes which affect drug development, regulatory decisions, and appropriate therapy, it seems imperative that an effort should be made to supply sound information.

Because of the great concern about the instances of catastrophic but unexpected adverse effects noted above, the major preoccupation has been with monitoring new drugs. Currently, most efforts in PMS are directed at the design of methods to collect information on newly marketed drugs. However, in the review by MIPI the majority of fatal adverse reactions seemed to be due to older, standard drugs [7].

Irey reported on a review of 827 fatalities from adverse drug reactions of which he classified 220 as ADRs "in the usual sense." The leading 10 drugs involved were introduced between 1899 (aspirin) and 1962 [8]. The need for PMS extends beyond "new drugs." As an information source for guidance in therapeutic decision-making it should supply data about all currently used medications.

III. THE ART AND SCIENCE OF DRUG EPIDEMIOLOGY

Over the years there has been considerable conflict over the appropriate methods for conducting PMS. This seems to have been the result of several factors: (1) the commitment of the pharmacological sciences to the prospective randomized controlled clinical trial as the sole scientific method; (2) differing assumptions about the goals and purposes of PMS; and (3) disagreement about the merits and feasibility of various methodological approaches, and the perception of various methods as being mutually exclusive rather than complementary. These points will be discussed in detail in the following paragraphs:

1. The standard methodological tool among scientists in pharmacology is the prospective randomized controlled clinical trial. This is certainly the most powerful study method, but for the most part

it is not readily applicable in PMS. A controlled trial is designed specifically to test a hypothesis which has been postulated from other information sources. In PMS the purpose is to make observations about therapeutic agents in their customary use situation in such a way as to generate hypotheses and draw conclusions about the therapy. Obviously, PMS requires the application of observational study methods rather than experimental methods. However, observational methods have been viewed by many scientists in the pharmacological sciences as nonscientific. It must be recognized that there are scientifically sound observational study approaches, and these in fact are the methods of drug epidemiology.

2. Regarding differing assumptions about PMS goals and purposes, it is clear that being alerted to uncommon and unexpected adverse drug reactions is different than trying to estimate an incidence of more common adversity, or measure how a drug is used, or look for unexpected benefit. Obviously, the methods employed should be appropriate to the questions asked. It would seem that at least some of the disagreements about methods have been due to assumptions about the questions to be answered without expressly indicating what the questions are.

3. Finally, there has been considerable discussion about the appropriate method. This kind of debate was frequently based on comparing the value of operations which were established and in place. There is no single method and no single resource. The correct method is situation dependent, and different questions or objectives are likely to require different approaches. The various epidemiological approaches should be seen as complementary, answering different aspects of compound questions and satisfying multiple objectives.

Postmarketing surveillance would be better named "systematic postmarketing drug evaluation." Surveillance suggests a specific study method, whereas multiple complementary methods are required. The development and application of these methods is the art and science of drug epidemiology. A more specific discussion of methods will be presented later in terms of the goals and functions of postmarketing drug evaluation.

IV. THE SUBJECT MATTER OF POSTMARKETING DRUG EVALUATION

There is no generally accepted specification of the subject matter of postmarketing drug evaluation. (The Joint Commission on Prescription Drug use has addressed the question in some detail [9]. A proposed outline has been presented by the present author and will be developed here [4].)

A decade or more ago, Sir Richard Doll said that, given the great advances made in the pharmacological sciences, what we must now do is learn how to "maximize the benefit and minimize the risk" from the many therapeutic agents introduced [10]. If one perceives postmarketing drug evaluation as one more information resource for accomplishing this, then the subject matter is the therapeutic intervention and its outcome.

The appropriate use of a medication is based on an evaluation of the risk, discomfort, or dysfunction of disease or some health issue, compared to the expected benefit and potential adversity from the intervention. This requires a judgmental assessment for which we have coined the term the "therapeutic value" of a medication. This concept has two basic components—a value judgment and the information upon which a sound judgment can be based. Systematic postmarketing drug evaluation should supply information about the impact of treatment drawn from the broad experience of physicians treating patients in a customary use situation. (Antagonists who argue that PMS is not needed are taking the position that the additional information will not add enough information to warrant the cost.)

The appropriate content of PMS can be specified in terms of the answers to a set of four questions [4]:

1. *Why is the drug used?* This refers to information about the indications in terms of signs, symptoms, and diseases for which patients are treated. It includes specification of the disease characteristics which motivate for or against the selection of this drug, e.g., the stage of the disease or the severity of the disease.
2. *How is the drug used?* This includes questions about the dose, the dosing schedule, the length of treatment, whether the drug is used alone or in combination, whether treatment is continuous or intermittent, etc.
3. *In whom is the drug used?* The answer to this question should supply information about the treated populations in terms of demographic characteristics, comorbid conditions, hospitalization status, etc., with specification of those characteristics that motivate for or against the selection of the drug.
4. *What happens?* The answer to this question should supply descriptive and quantifiable information about the subsequent health course of treated subjects in order to determine the effects of treatment, either beneficial or adverse.

Of these four questions, the first three deal with descriptive information about drug use and the population treated. The fourth addresses the effects of treatment and requires some elaboration, because of the implications in selection of methodology and the likelihood of collecting information about certain kinds of effects. Below is a series of dichotomous classes into which treatment effects can be grouped. They are not mutually exclusive. Actually, the classes are not all dichotomous; however, this format is useful for purposes of discussion:

Beneficial effects vs. adverse effects
Observable events vs. nonevents
Early effects vs. late effects
Immediate vs. delayed effects
Expected vs. unexpected effects
Common vs. rare effects

These classes will be discussed in the following subsections.

A. Beneficial vs. Adverse Effects

It is useful to recognize that there is nothing intrinsic about benefit or adversity which affects study design. From a purely study point of view, effects can be observed, measured, quantified, etc., independently of whether they are adverse effects or beneficial effects.

The important implications for study design are derived from the sociomedical acceptance of different levels of evidence for conclusions about benefit vs. risk. A conclusion of benefit requires significant, rigorous scientific evidence. A decision of hazard is based on much less rigorous evidence. In fact, a credible question of hazard without qualifying information must lead to a conclusion of hazard.

One aspect of this distinction which does have specific design and methodological implication is the assumption that benefit from therapy is common whereas adversity is uncommon.

B. Observable Events vs. Nonevents

This classification distinguishes between effects which result in observable health events as opposed to those which are preventive. These effects can be either beneficial or adverse and either expected or unexpected. An example of an unexpected beneficial preventive effect would be the protective effect of aspirin in stroke.

The decision about which kind of effects one wishes to study has important implications about the choice of method and the study design.

C. Early Effects vs. Late effects

Early effects are those which occur shortly after drug is started. Late effects are those which occur as a result of chronic therapy. Obviously, this has both design and implementation implications. To observe late effects one must make observations in a chronically exposed population.

D. Immediate vs. Delayed Effects

This distinguishes effects which occur in a time frame which is proximal to drug taking as opposed to those which take place in a time frame remote from drug administration. Perhaps one of the most complex examples of a delayed effect is the occurrence of vaginal adenocarcinoma in the children of women exposed to diethylstilbestrol.

E. Expected vs. Unexpected Effects

This classification is self-evident but is very important. A useful monitoring system must identify unexpected adverse effects. This requires a system which can collect, tabulate, and monitor health events in an open-ended way.

There has been considerable discussion about the possibility of collecting information about expected benefits. In fact, expected benefits have been established in clinical trials before a drug is marketed, and it is not appropriate to try to establish efficacy in postmarketing drug evaluation. One can, however, observe the overall response rate or changes in response rate from therapy when a drug is dispensed in a customary use situation.

One would hope that appropriate systematic monitoring of drugs will result in a timely identification of unexpected benefits. Currently, there is no experience, but systems should be designed to maximize the opportunity of identifying unexpected benefits. Exactly how to accomplish this is unclear and requires further development and experience.

F. Common vs. Rare Effects

This distinction has significant implications for the choice of method. However, one must recognize that, given the health impact of an event, the rarer the event the less important it is in the terms of the therapeutic value of the drug. As well, it is important to have a database which allows one to estimate the upper limit of incidence. If some major adverse reaction is identified by the voluntary adverse reaction reporting system, it is useful to have made systematic observations in a sizeable population even if the adverse reaction has not been in that population.

V. FUNCTIONS OF POSTMARKETING DRUG EVALUATION

To establish a system of postmarketing drug evaluation it is essential to consider the purposes of the system and how the collected information is to be used.

In order to be useful the system must accomplish three functions: First, the system must draw attention to unexpected drug effects in a timely fashion, i.e., an alerting function. Secondly, the system must allow an estimation of the incidence of the observed effect. Finally, the system must provide a data resource which can be used to investigate the postulated drug effect and allow for hypothesis testing.

It is important that all of these functions be readily available. Especially in the case of possible major adversity, one must be able quickly to determine whether the effect is happening frequently enough to affect the appropriate use of the drug, i.e., the therapeutic value. If it does, then there is a need to be able to determine the probability that the effect is truly a drug effect and to describe the population at risk.

This kind of approach, i.e., (1) alerting, (2) quantification, and (3) investigation and hypothesis testing, is distinctly different from the usual scientific approach of establishing a hypothesis and then testing it. In the latter approach the primary interest is to determine causality. The purpose of postmarketing drug evaluation is to maximize the benefit and minimize the risk of therapeutic intervention. Hence, the goal is pragmatic rather than basic. If an event is occurring so infrequently that it is not important in deciding on the appropriate use, then the determination of a causal association, although important in terms of basic knowledge and understanding, is secondary in postmarketing drug evaluation.

Based on the collections of descriptive information about drug use and the treated population, the system should exploit the expertise of the many practicing physicians to identify potentially better ways of using the drug and better specification of the population most likely to be benefited.

Finally, the system should supply information about treatment in situations where drugs are ordinarily not tested, e.g., during pregnancy, in children, in patients with various comorbid conditions, and in patients being treated with other medications.

VI. CHARACTERISTICS OF SYSTEMATIC POSTMARKETING DRUG EVALUATION

In prior sections we have attempted to outline the functions and substance of systematic postmarketing drug evaluation. Following are some of the important characteristics of any useful system:*

1. The methods must be appropriate for application to large numbers of patients.

---

*Adapted from an article by E. Keith Borden and J. G. Lee, *J. Chron. Dis.* 35:803-816 (1982).

2. In order for the alerting function to be useful, the collection of data must be concurrent with drug use. The accumulated data should draw attention to unexpected effects, especially major adversity at the earliest possible time in the use of the drug.
3. The data collection should quantitatively parallel the amount of drug exposure. For drugs used to treat very large numbers of people, it is important to identify uncommon effects. Alternatively, because an unexpected effect has not been observed with a less commonly used drug, it cannot be concluded that none will occur if its use is expanded. This is very important because more adversity is likely to be seen with the more widely used or studied drugs than with drugs less widely used or studied. A switch to the less commonly prescribed drug may result in either more or less adversity than was occurring previously.
4. The collected data must be referable to broad, definable populations of patients treated in customary use situations.
5. For quantification and hypothesis testing, one must have a denominator which is representative of some definable population and a numerator which includes all the events of some set that occur in the denominator population. There must be a high level of confidence that all health events of some definable and meaningful nature are included in the numerator.
6. Health event description, coding, and classification must allow the formation of logical sets for meaningful evaluation and interpretation. This is perhaps the most difficult aspect of postmarketing drug evaluation since one is dealing with clinical information which is not recorded with a view to classification needs.
7. In clinical trials the study is designed to have a fixed sample size with analysis at the end of the study. In postmarketing drug evaluation the sample size may not be rigidly fixed and the system must be structured to monitor the data as it accumulates, with thought given to a priori decision-making rules.
8. For purposes of operational feasibility, some priority must be established to distinguish between important and unimportant events or observations. Some limitations must be imposed on what kinds of information can be collected and what can and should be investigated.
9. For purposes of interpretation, all of these studies require the application of two distinct forms of logic—statistical analysis of aggregated data sets and medical evaluation of case reports. This is discussed in further detail in the following paragraphs.

As just noted, two distinct kinds of logic are required in the evaluation of medical events occurring during systematic observation of a drug. This important point is reiterated because it requires the simultaneous acceptance of two forms of evaluation which are ordinarily perceived as antagonistic.

The first form of evaluation is the standard scientific study approach of aggregating like events into sets and applying statistical methods for analysis. For example, one might consider the proportion of myocardial infarctions occurring on drug A compared with those occurring on drug B. Assuming the treated populations are at comparable risk, one can appropriately test for statistically significant differences between the groups or estimate the true incidence of myocardial infarction on either drug. However, this approach has two basic requirements:

(1) The event set must be constructed by combining what are perceived as like events. This requires a judgment about likeness. In the example given above, myocardial infarction forms a neat and acceptable diagnostic set. The greater the forcing to form a set, the more likely is the analysis to be misleading. Alternatively, some events do not fit neatly into sets. For example, a patient who presents with nausea, vomiting, hematemesis, and a generalized maculopapular eruption and whose condition continues to deteriorate until death without any further definitive diagnosis will not fit into a diagnostic set.

(2) The numerator events must be observed frequently enough to supply adequate numbers to be meaningful for analysis. The isolated or uncommon event will not supply adequate numbers.

In neither of these cases will ongoing observation allow for the formation of logical sets to which statistical methods can be meaningfully applied. (See Sec. VII.E. below, for the use of the case comparison method for rare events.)

In both cases an alternative approach must be used. This is the medical case evaluation of a single case. A medical investigative logic is applied to the individual case to assign a level of concern that a given drug may be causing a given event. This judgment is based on such factors as the prior knowledge of the association between the drug and the event; the time/sequence relationship; the response to dechallenge and rechallenge; and alternative etiologies. Proposals have been made to standardize this approach, although it remains subjective and fraught with potential flaws. Nevertheless, if during a monitoring study one observes a unique or bizarre serious event, it cannot be ignored because statistical methods cannot be applied to yield a meaningful conclusion. This problem is seen in clinical trials where the efficacy data are subjected to standard statistical analysis but the less commonly occurring untoward health events are evaluated singly.

Exactly how these methods should be applied in an ongoing way during the collection of observations on a marketed drug is an important, difficult, and as yet unresolved question.

VII. CONSIDERATION OF SOME COMMONLY USED OR PROPOSED METHODS

All currently used approaches are directed at accumulating information about adversity.

A. Voluntary Reporting of Adverse Drug Reactions

This consists of collections of reports of cases of drug-induced or suspected drug-induced health events by individual health professionals or patients. It is subject to significant biases. It is dependent on a subjective determination of causal association by an observer followed by a decision to report. (Algorithms have been proposed for evaluating reports of adversity; however, they do not resolve the problems of interpretation. They are applied only in cases where the same judgmental material has already been used in a less specified way to conclude that there is a potential causal association. All events are not subjected to the algorithm. Algorithms should increase agreement in categorizing events that have already been proposed as potentially drug related. However, they do not increase the likelihood of identifying adverse drug reactions or the likelihood of a correct conclusion about causal associations.)

The observer bias in the voluntary reporting method is very important. The unusual and bizarre is likely to be blamed on drugs, while common clinical events (e.g., myocardial infarction) will almost never be reported. The reporting itself is subject to significant bias. It is affected by publicity or inclusion in the *Physicians' Desk Reference* (PDR), by the novelty of the reaction, by comfort in being able to manage the reaction, etc. The approach supplies neither a complete numerator nor any denominator data.

Alternatively, the voluntary reporting method is based on collecting the observations of a very broad and large group of sophisticated observers—the treating physicians as well as other health professionals. For this reason it is a very useful alerting system.

In short, the voluntary reporting system is a useful alerting tool especially for the unusual or bizarre drug-induced health event and can draw attention to rare events. It has no value in quantifying, hypothesis testing, or supplying descriptive information about the population at risk.

B. Collections of Clinical Experience

One approach which has gained some popularity is to invite a group of physicians to complete a standard form on a number of patients whom they are treating with a drug of interest. Data on several thousand patients can be accumulated in this way. This approach supplies both numerator and denominator data.

It is a very weak approach, however. The prospective collection of data in a manner which allows the formation of both the numerator and the denominator implies a scientific validity that is not warranted. One has no knowledge of which patients were included from a physician's practice or why they were included or excluded. Hence, the denominator is undefinable. The events reported in the numerator are based on the reporter's observation and decision to report. The content of the numerator is logically similar to the reports in a voluntary reporting system, but it is confounded by an ill-defined reporting requirement which makes interpretation difficult. At the same time the reporting is from a much smaller number of physicians.

This approach is undesirable. As an alerting mechanism it has limited value. For quantification it has both an ill-defined numerator and denominator.

C. Comprehensive Reporting

This approach is particularly applicable to newly marketed drugs. It consists of a reporting requirement by physicians on all patients treated during some defined period of time after the drug is released. Several such proposals have been made in Great Britain [11-13]; however, none have been implemented. The approach supplies both numerator and denominator data. The numerator consists of subjective observations as in the voluntary reporting system, but to the extent it requires reporting of all events it increases the likelihood of completeness and of reporting usual clinical events as well as the unusual or bizarre.

The alerting function is diminished as compared to the voluntary reporting system because of the limitation on the number of observers and the length of observation, but it is significantly superior to the clinical experience approach. To the extent that the denominator is a census of treated patients among reporting physicians, the data are representative of some definable experience and rates are extrapolatable to a definable population. The method offers a useful approach to quantification. There is no opportunity for hypothesis testing or investigating causal associations.

D. Cohort Studies

This approach is comparable to the clinical trial except there is no randomization. A definable group of patients, representative of some population treated in a customary manner, is followed forward over time and observed for health effects. This approach will allow answering all the previously presented substantive questions of systematic postmarketing drug evaluation and is not limited to questions of adversity.

The usefulness as an alerting function is limited by the size of the cohort. However, one can conceive of developing large registries of patients which could be used as cohorts for follow-up. With appropriate design there is clear definition of numerator events, not limited to subjective judgments of ADRs and including both usual as well as peculiar events. With appropriate comparison groups the investigator can compare differences in rates between groups as well as estimating incidence within groups. The method is appropriate for hypothesis testing and investigation for causal associations.

This is not a method useful for being alerted to or investigating rare events because of the size of the cohorts which would be required.

E. Case Comparison Method

This is a hypothesis testing method. The method has little value as an alerting mechanism. It cannot draw attention to potential associations de novo. In fact, the entire procedure is based on historical data. The method is not concurrent with drug use. For purposes of quantification the method supplies an odds ratio rather than an estimate of incidence. The design and implementation of the case comparison study is complex. On the other hand, this is certainly the method of choice for hypothesis testing when dealing with rare events.

In short, this method supplies no alerting function and a limited quantifying function. Most experts would agree that for hypothesis testing it is not as powerful as a prospective cohort study; however, it is the method of choice for rare events or when a new, urgent question arises about events resulting from historical drug exposure. For instance, if one estimates that some cancer may be associated with some remote drug exposure with a latency of perhaps 10 years, and there is a 10 year or greater experience, the method of choice to test the hypothesis would be a case comparison study. One could not ethically set up a cohort study and follow patients for another 10 years.

Slone et al. [14] have described unique applications of this method for screening. In essence this consists of cross-classifying a number of disease entities with a number of drug exposures and seeking potential associations. Any that are found are subjected to further hypothesis testing using the case comparison method. This approach is not nearly as simple as outlined and involves a sophisticated operation. The reader is referred to the aforementioned paper of Slone and colleagues. Even in their approach there is the requirement for postulating a hypothesis.

Table 1 Functional Evaluation of Some Common Methods

Method	Functions		
	Alerting	Quantifying	Testing for Associations
Voluntary ADR reporting	+++	0	0
Collections of clinical experience	+	+	0
Comprehensive reporting	++	+++	0
Cohort studies	++ → ++++	++++	++++
Case comparison studies	0	++	+++

Table 1 graphically presents the functional utility of the methods described and how they complement each other. The weightings are arbitrary.

VIII. WHO SHOULD DO POSTMARKETING DRUG EVALUATION?

It is not in the purview of this chapter to address the specifics of who should conduct postmarketing drug evaluation. Rather, it should be understood that this is a new area of investigation which will undergo significant evolution and that we are attempting here to lay down some principles.

Drug epidemiology is a new science. Its development requires the collaboration of physicians, clinical pharmacologists, epidemiologists, and biostatisticians. It requires the innovative application of multiple methods and multiple resources. For these reasons, the development of drug epidemiology should be promoted as a scientific discipline to be developed wherever there is interest, expertise, and resources, whether this be in academia, the pharmaceutical industry, large group practices, or at the community, state, or federal governmental level.

It is important to avoid the temptation to look for the single center that will conduct the conclusive study. A number of studies have already resulted in conflict, e.g., the reported association between coffee and pancreatic cancer, or between reserpine and breast cancer. Drug epidemiology should develop so that there are multiple independent investigators who can collect evidence and resolve conflicts in a scientific forum. One could hardly conceive of a center for pharmacology or chemistry which conducted all studies and issued all reports.

On the other hand, since systematic postmarketing drug evaluation impacts on an important public health issue, there is need for some centralization where issues can be resolved in a scientific environment.

It would seem that there is need for both centralized and decentralized components. The role of the centralized component would be to promote the development and teaching of the art and science of drug epidemiology, to identify issues of current concern and possibly fund studies to resolve them, to provide a scientific forum for the presentation of study results, to interpret the scientific debates,and to recommend appropriate public actions based on accumulated information. One other function that might appropriately be centralized is the development of national registries which could be used as a resource for investigation by a variety of approved scientists.

The development of the best science, the opportunity for innovative approaches and unique observations, and the acquisition of the best information depends upon decentralization.

A rational evaluation of information for policy decisions, maintenance of the public health interest, and the establishment of national investigative resources require centralization. Exactly where to draw the line is unclear.

It is important that the centralized component be free of any appearance of a vested interest or drawn into an adversarial position by affiliation. It would seem best that a center be funded jointly by government, industry, and private foundations.

Organizationally, the center should be independent of industry, government, and academia. This does not mean that the personnel could not be drawn from all of these, but as an organization it should not be affiliated with any of them. The interests and adversarial roles of government and industry are clear. Much less obvious, but equally important, are the vested interests and potential adversarial postures of academic institutions.

The question of whether such studies should be done in the pharmaceutical industry has been raised. There would seem to be no reason why they should not, assuming the data be made available for review. In fact, the industry, because of its extensive experience and expertise in conducting clinical trials with drugs, may be a most fertile place for the development of drug epidemiology.

The conduct of studies by a governmental regulatory body seems more problematic. It is essential that the regulator should have access to the information necessary to do his job. However, it is not appropriate that the regulator should have a vested interest in the information which he uses for decision-making. Even worse, the regulator should not be the sole generator of the data. The decisions about a drug's therapeutic value should be based on scientific debate; the regulator should implement the decisions but must remain objective in the evaluation of the scientific merits.

REFERENCES

1. D. J. Finney, The design and logic of a monitor of drug use. *J. Chron. Dis. 18:*77-98 (1965).
2. L. E. Cluff, G. F. Thorton, and L. G. Seidl, Studies of the epidemiology of adverse drug reactions. I. Methods of surveillance. *JAMA 188:*976-983 (1964).
3. I. T. Borda, D. Slone, and H. Jick, Assessment of adverse reactions within a drug surveillance program. *JAMA 205:*99-101 (1968).
4. E. K. Borden, Post-marketing surveillance: Drug epidemiology. *J. Int. Med. Res. 9:*401-407 (1981).
5. Press Release—Joint Commission on Prescription Drug Use. Stanford University Medical Center, News Bureau, Stanford, Calif., Jan. 1980.
6. Richard Hampton, I.M.S. America, personal communication.
7. *Adverse Drug Reactions in the United States: An Analysis of the Scope of the Problem and Recommendations for Future Approaches.* Monograph published by Medicine in the Public Interest, 1974.
8. N. F. Irey, Adverse drug reactions and death. *JAMA 236:*575-578 (1976).
9. Summary of the Final Report of the Joint Commission on Prescription Drug Use. Submitted on Jan. 23, 1980.
10. R. Doll, Report on international meeting of medical advisors in the pharmaceutical industry, London, 1972. In *International Aspects of Drug Evaluation and Usage.* Churchill-Livingston, Edinburgh and London, 1973.
11. W. H. W. Inman, Recorded release: A proposal for post-marketing surveillance of new drugs. In *EED Symposium on Drug Surveillance Techniques Held at the Maris Negri Institute in Milan, May 4-6, 1977.* PSG Publ., Littleton, Mass., 1977.
12. A. B. Wilson, Post-marketing surveillance of adverse reactions to new medicines. *Br. Med. J. ii:*1001-1003 (Oct. 1977).
13. D. H. Lawson and D. A. Henry, Monitoring adverse reactions to new drugs: "Restricted release" or "monitored release"? *Br. Med. J. i:*691-692 (1977).
14. D. Slone, S. Shapiro, and O. S. Miettinen, Case-control surveillance of serious illnesses attributable to ambulatory drug use. In *Epidemiological Evaluation of Drugs* (F. Colombo, S. Shapiro, D. Slone, and Tognoni, Eds.). PSG Publ., Littleton, Mass., pp. 59-70.

18
RESEARCH QUALITY ASSURANCE

John J. Donahue
Hoffmann-La Roche Inc.
Nutley, New Jersey

I. HISTORICAL PERSPECTIVE

Prior to 1975, the term "research quality assurance" was unheard of. During Senate subcommittee hearings in 1975 and 1976, however, the Commissioner of the U.S. Food and Drug Administration expressed concern for the integrity of animal safety data being submitted to the agency in support of regulated products and the need for better control of preclinical animal experiments. The commissioner's testimony indicated that FDA experiences had revealed numerous apparent deficiencies in animal testing procedures [1]. The following are some of the deficiencies cited:

1. Experiments were apparently poorly conceived, carelessly executed, or inaccurately analyzed or reported.
2. Technical personnel were apparently unaware of the importance of adherence to protocols and to accurate observations, as well as the need for accurate administration of the test substance and accurate record keeping and record transcription.
3. The Commissioner alleged that management did not ensure a critical review of the data or proper supervision of personnel.
4. Studies were apparently impaired by protocol designs that did not allow the evaluation of all available data.
5. Assurance could not be given for the scientific qualifications and adequate training of personnel involved in research studies.
6. There was apparently a disregard for the need to observe proper laboratory, animal care, and data management procedures.
7. Sponsors failed to monitor adequately studies that were being performed in whole or in part by outside testing laboratories working under contract.
8. Firms failed to verify the accuracy and completeness of scientific data in reports of nonclinical laboratory studies before submission to the FDA.

In support of these allegations, the FDA cited specific examples of studies for which no complete records for any animal existed; numerous study reports submitted to the agency did not agree with the raw data; and animal tissues, which had been examined histopathologically, could not even be found by contractor laboratories that had submitted reports to the agency.

As a consequence of this testimony, Congress approved a substantial budget increase for the FDA of approximately $16.6 million for fiscal year 1977. This increase was approved with the understanding that the FDA would utilize these new resources to ensure the quality and integrity of data submitted to the agency in support of products. In response to this increased support from Congress, the Commissioner established the Bioresearch Monitoring Program. Not only did this program include preclinical studies being submitted to the agency, but it also embraced clinical research relating to FDA-regulated products. A Toxicology Laboratory Monitoring Task Force was also established by the FDA. This task force was assigned the responsibility for developing an agency strategy to ensure the quality and integrity of nonclinical laboratory studies. To meet these goals, the task force proposed that the following steps should be taken:

1. They would develop Good Laboratory Practices (GLP) regulations analogous to the existing Good Manufacturing Practices (GMP) regulations.
2. They would develop an agencywide compliance program that would include regular inspections of nonclinical laboratories.
3. They would establish a specific course for training agency investigators in evaluating testing facilities and their compliance with GLP regulations.
4. They would identify facilities involved in nonclinical testing of regulated products.

As a result of the efforts of the Toxicology Laboratory Monitoring Task Force and the FDA experience in reviewing preclinical laboratory studies, the FDA issued proposed GLP regulations on Friday, November 19, 1976 [2]. In the preamble to these regulations the agency indicated that the regulations were meant to be process oriented and would not deal with the scientific aspects of preclinical safety studies. While the major requirement of the proposed GLP regulations was good record keeping, there were numerous other requisites of the proposed regulations that would help ensure the proper maintenance of records. These included the establishment of a quality assurance unit, the establishment of an archive system and archivist responsible for retaining all original raw data generated in studies, and a requirement for test facility management to name an individual responsible for the conduct of the study (i.e., the study director). Based on its concern for standardizing testing procedures, the FDA also proposed that standard operating procedures be established for all operations in a preclinical testing facility.

Following the publication of these proposed regulations, the FDA in February 1977 held open public hearings on the proposed regulations [3]. These hearings lasted for 2 days. While specific concerns with the regulations were addressed by many participants, the overall feeling by those presenting their opinions was that the proposed rules should not be published as formal regulations but should only appear as guidelines. However, on December 22, 1978, the FDA published final regulations effective June 22, 1979 [4].

At about the same time that the GLP regulations were being formulated, the FDA also issued several reports dealing with the conduct of clinical investigations as well. Based on these reports, the FDA, on September 27, 1977, issued proposed regulations detailing the obligations of sponsors and monitors who conduct clinical investigations. In proposing these regulations, FDA noted that while inspections of several clinical trials revealed no gross violations of generally accepted clinical practices, there were numerous minor deficiencies [5]. These deficiencies centered around areas of patient consent, protocol adherence, study control, and records availability and accuracy. In the preamble to the regulations, the Commissioner also cited the General Accounting Office (GAO) report entitled, "Federal Control of New Drug Testing Is Not Adequately Protecting Human Test Subjects and the Public" [6]. In this report, dated July 15, 1976, the GAO recommended that the FDA improve regulations governing clinical investigations to protect the rights of the patients involved.

As with the GLP regulations, the proposed clinical regulations did not address the scientific aspects of clinical studies but dealt more with the procedural aspects of monitoring clinical trials. Within the year, the agency followed with two additional sets of regulations. On August 8, 1978, the FDA published proposed regulations concerning the obligations of clinical investigators of regulated articles [7] and also published proposed regulations dealing with standards for institutional review board approval of a clinical investigation as well as standards for obtaining patient consent [8].

To date, the proposed regulations on the obligations of sponsors, monitors, and clinical investigators have not been published in their final form. However, on January 27, 1981, the FDA did publish final regulations concerning the protection of human subjects, informed consent, and standards for institutional review boards for clinical investigations. These regulations were effective July 27, 1981.

II. GOOD LABORATORY PRACTICES REGULATIONS

The scope of GLP regulations is not limited entirely to pharmaceutical products. Table 1 lists the products which are covered by GLP. Safety studies done in support of applications to FDA for research or marketing permits for these particular products, must be conducted under GLP regulations.

The GLP regulations are concerned with seven areas: laboratory organization and personnel; testing facilities; equipment used in the

Table 1 Products Covered by Good Laboratory Practices Regulations

Food and color additives	Medical devices for human use
Animal food additives	Biological products
Human and animal drugs	Electronic products

testing procedures; the operation of test facilities; test and control articles; protocol for and conduct of a nonclinical laboratory study; and records and reports.

A. Organization and Personnel

Under organization and personnel, the responsibilities of the testing facility management, the study director and the quality assurance unit are delineated insofar as the conduct of nonclinical laboratory studies is concerned. Specifically, testing facility management is charged with the following tasks: designating a study director for each study before it is initiated; replacing the study director promptly should this become necessary; making sure that the quality assurance unit is established; and, finally, making sure that the resources necessary to conduct the study properly are available to the study director and the scientific personnel involved.

The overall responsibility for the scientific conduct of a study is placed with the study director. This individual is responsible for interpretation, analysis, and documentation of data as well as reporting of the results. While the study director may not directly carry out each of the individual functions required in a particular study, he does serve as the focal point for the study. He must see to it that all operations are carried out in a timely manner and that the data generated in a study are reported to the FDA in a timely manner.

The quality assurance unit as established by the GLP regulations serves as a means by which management ensures that the facilities, equipment, personnel, methods, practices, records, and controls used in a study are in conformance with the regulations. Specifically, the quality assurance unit is charged with maintaining a master schedule of all nonclinical studies conducted at the particular testing facility. The master schedule must be indexed by test article, and for each study it must contain the information listed in Table 2.

Table 2 Content of the Master Schedule

1. Nature of the study
2. Test system used (animal, plant, or microorganism treated with the test or control articles)
3. Name of the study director
4. Date the study was initiated
5. Current status of the study
6. Status of the final study report

The quality assurance unit must also maintain copies of all protocols related to preclinical studies carried out at the facility it is responsible for.

Periodically, the quality assurance unit must inspect each phase of a nonclinical laboratory study and maintain written records of these inspections. Specifically, the regulations require that for "studies lasting more than 6 months, inspections shall be conducted at intervals adequate to assure the integrity of the study." Significant problems uncovered in these inspections must be reported to the study director and management as soon as possible.

The quality assurance unit is also responsible for reviewing the final study report to verify that the report adequately and accurately reflects the raw data that has been generated in the study as well as the methods and standard operating procedures used to generate this data. In performing this task, the quality assurance unit is also required to prepare and sign a quality assurance statement which specifies the dates on which the unit performed its inspections as well as the dates that the findings of such inspections were reported to facility management and the responsible study director. This statement must be included in the final study report.

While the FDA does not insist on routine access to the written inspection reports of the quality assurance unit, the agency does have access to the written procedures which the unit uses for its inspections and reserves the right to ask management to certify that inspections are being conducted properly as required by regulation.

B. Facilities, Equipment, and Test Facility Operation

The sections of the regulations dealing with facilities, equipment, testing facility operation, and control articles are very similar to what we find in the GMP. Basically, these sections are concerned with the housekeeping and record-keeping procedures which must be followed to assure the adequate conduct of a preclinical animal safety study. These sections of the regulations, fortunately, are written in general terms thus allowing the management of a test facility to adapt the regulations to their own unique systems and operations.

The GLP regulations dealing with protocol design are very specific in terms of the requirements for a study protocol. The requirements that one must follow in designing a protocol which meets GLP are listed in Table 3.

Once a protocol has been defined and approved by the study director, all changes and revisions in that protocol must be made with the approval of the study director and must be maintained with the original protocol. In addition, the reasons for all changes in the original protocol must also be documented in writing.

Table 3 GLP Protocol Requirements

1. Descriptive title and objective of the study
2. Identification of the test material
3. Name of the sponsor and name and address of the testing facility
4. Proposed starting and completion dates for study
5. Justification for test system used
6. Descriptive information for the test system
7. Procedure used for identification of the test system
8. Description of the experimental design
9. Description and identification of the diet used in the study to administer the test or control article
10. Route of administration of the test and control article and the reason for that route
11. Dose level of test article to be administered as well as the method and frequency of administration
12. Method by which the degree of absorption of the test and control articles by the test system will be determined
13. Type and frequency of tests, analyses, and measurements to be made in the study
14. Records to be maintained
15. Date of approval of the protocol by the study director and the sponsor
16. Statement of the proposed statistical methods to be utilized

C. Records and Reports

The section of the GLP regulations dealing with records and reports is fairly specific about the content of a final study report. Among the items which must be included in the final study report are the following: name and address of the facility performing the study; dates on which the study was initiated and completed; the objectives and procedures of the study as stated in the approved protocol; the statistical methods employed in analyzing the data; the name of the test and control articles; the stability of the test and control articles; the description of the methods used in a particular study; a description of the test system used in terms of the experimental animals' sex, species, body weight range, source of supply, strain and substrain, age, and procedure used for identification. A description of the dosage regimen used, the route of administration, and the name of the study director must also be included in the final report as well as a description of all transformations, calculations, and operations performed on the data. The signed and dated reports of any consultant scientists or other professionals involved in the study should be included. One is required to list the locations where all specimens, raw data, and the final report are to be stored and, as mentioned earlier, the quality assurance unit's statement must also be attached to the final report.

In addressing the storage and retrieval of records, the GLP regulations require that an archive be established and that a single individual be named who is responsible for that archive. The archive must be under positive control so that any materials removed or deposited are recorded and documented. In addition, the regulations establish time periods for which these records must be maintained in the archives. Records and specimens must be kept for whichever of the following periods is the shortest:

1. A period of at least 2 years following the date on which an application for a research or marketing permit is approved by the agency
2. A period of at least 5 years following the date on which the results of the study are submitted to the agency in support of a research or marketing permit
3. In situations where the study does not result in the submission of an application for a marketing permit, a period of at least 2 years following the date of completion of the study

III. GOOD CLINICAL PRACTICES REGULATIONS

As part of its Bioresearch Monitoring Program, FDA has proposed four different sets of regulations which are related to the conduct of clinical trials in humans. The regulations delineate responsibilities of the sponsors and/or monitors of clinical trials, the responsibilities of investigators conducting clinical trials, the regulations for Institutional Review Board (IRB) approval of clinical studies, and standards for obtaining patient consent from individuals involved in studies. To date, only regulations dealing with IRBs and patient consent standards have been finalized and implemented. Proposed regulations dealing with investigator and sponsor/monitor obligations are still being evaluated by the agency.

A. Sponsor/Monitor Regulations

The proposed regulations dealing with the obligations of sponsors and monitors who conduct clinical studies serve to reinforce existing FDA regulations in more specific detail. To the point, the proposed regulations spell out the specific obligations of a clinical monitor in monitoring the progress of clinical studies. The agency requires that the monitor make preinvestigation visits and periodic visits throughout the course of the study. During these visits the monitor is charged with determining that the following four obligations are being met by the investigator: (1) that the facilities of the investigator continue to be acceptable in terms of the studies being conducted; (2) that the investigator is following the protocol or investigational plan and that he is also following applicable FDA regulations; (3) that adequate records of the study are being kept, particularly in terms of subject identification, clinical observations, laboratory tests, and drug receipt and disposition; and (4) that the investigator is submitting timely reports to the sponsor related to the safety and/or effectiveness of the drug being tested.

The proposed regulations also require that the monitor review source documents which are relevant and necessary to assure the accuracy and completeness of the data being submitted to the sponsor on case report forms (CRFs). This particular requirement has caused a great deal of concern, particularly in terms of the confidentiality of patient records. It remains to be seen whether the FDA will continue to insist on the review of source documents in the final regulations. The review of source documents is a particular problem for FDA when one considers FDA's position on foreign studies. Should FDA accept foreign data, there are several foreign countries which may not allow access to patient records.

As part of the requirements for the monitor of a clinical trial, the FDA mandates that written records be kept for each visit to a clinical investigator. These records must contain the information listed in Table 4.

The sponsor/monitor regulations are also specific about the requirements for test article receipt and disposition records. The proposed regulations indicate that the records for test articles should include the dates, quantity, and batch identification number of the test article units received by the investigator; the name of the investigator who received the test article; and, if the test article was used for other than clinical purposes, the circumstances and reasons for such use should be documented.

Prompt review of all new data received from an investigator and periodic evaluation of all data received from all investigators in a clinical investigation are required. This review is particularly important in terms of the safety and effectiveness of the test article. Current

Table 4 Monitoring Reports Requirements

1. Date of the visit
2. Name of the monitor
3. Name and address of the investigator visited
4. Summary of the findings made by the monitor
5. Statement as to any corrective actions taken by the monitor or investigator

regulation requires that information on all unexpected side effects or adverse reactions, whether related to the administration of a drug or not, must be submitted to the agency within 15 working days of receipt of such information by the sponsor.

The records retention requirements of the proposed sponsor monitor regulations are identical to those of the GLP regulations. Current regulations require that the investigator maintain the records of drug disposition and case reports for a period of at least 2 years following the approval of the New Drug Application (NDA); or, if the application is unapproved, until 2 years after the investigation itself is discontinued.

B. Clinical Investigator Regulations

The proposed regulations covering the obligations of investigators of regulated articles which were issued on August 8, 1978, primarily deal with the responsibilities of a clinical investigator in terms of the IRB review, test article accountability, protocol compliance, patient consent records, and retention of records and reports.

The investigator is required to submit the proposed clinical investigation protocol and materials to be used in obtaining patient consent to a duly constituted IRB. He is also required to report to the IRB any proposed changes in study protocol or study deviations which may occur, if the change or protocol deviation increases the risk to or affects the rights of the human subjects in the study or if it adversely affects the validity of the investigation.

Finally, the investigator is required under the regulations to maintain records of all submissions to the IRB as well as the actions taken by the IRB concerning his particular clinical investigation.

The proposed regulation on test article responsibilities requires the clinical investigator to allow the test article to be administered only by authorized persons who are directly responsible to him. In administering the test article, records of the receipt and disposition of all test articles are required. These records include such details as date of receipt, lot numbers, serial numbers, quantities of drug received, and quantity of drug dispensed per patient. Finally, under this regulation, the investigator is required either to return the unused or reusable test article to the sponsor or to get authorization in writing from the sponsor to dispose of it.

The investigator is required to obtain informed written consent from each patient using materials which have been reviewed and approved by the IRB. Under the proposed regulation on records and reports, the investigator is required to make accurate and adequate reports to his IRB and to the sponsor at least annually. The investigator must also make a final report to the sponsor and to the IRB within 3 months after the completion, termination, or discontinuation of the clinical investigation. Any death, life-threatening problems, or other serious adverse effects that may be reasonably associated with or caused by the test article must be reported by the investigator to the sponsor and to the IRB. These reports should be made as soon as possible but no later than 10 working days after the investigator discovers the problem. The investigator is also required to retain records associated with the study for the same periods of time outlined in the GLP and proposed sponsor/monitor regulations. Furthermore, if the investigator retires, relocates, or for any other reason can no longer be responsible for maintaining records, he must transfer the custody of the records to another person who is willing to accept the responsibility, such as the sponsor, IRB, or another investigator. This transfer must be in writing, and a copy of the transfer memo must be given to the sponsor.

C. Institutional Review Board

While the agency has required IRB approval of studies in the past, the new IRB regulations further delineate the responsibilities of the board in terms of its organization, functions, and operations and its record-keeping and reporting requirements. These regulations, which were implemented on July 27, 1981 [9], require that a duly constituted institutional review board have a minimum of five members. The membership of an IRB may not consist entirely of men or women, or entirely of members of one profession. In addition, it should have at least one member whose primary concerns are in nonscientific areas, such as a lawyer, clergyman, or ethicist. The IRB must also have at least one member who is not affiliated with the particular institution in which the IRB serves. This individual may not be part of the immediate family of a person who is affiliated with the institution. The IRB is allowed to invite consultants to serve as advisers, but these consultants may not vote at board meetings.

The regulations require that there be written procedures which define the functions and operations of the IRB. Basically, an IRB must consider seven different criteria in approving research proposals. They must be certain: (1) that the risks to subjects are minimized; (2) that the benefit/risk ratio for the patient is reasonable; (3) that the study design provides for the equitable selection of patients; (4) that the investigator is aware of the need to obtain informed consent from each prospective subject; (5) that informed consent will be appropriately documented; (6) that the research plan provides for adequate monitoring of the data collected to ensure the safety of subjects; and (7) that there are adequate provisions to protect the privacy of the subjects and to maintain the confidentiality of the data.

The IRB is also required to maintain seven different types of records: (1) copies of all research proposals which are reviewed; (2)

minutes of the board's own meetings which delineate the attendance at the meeting, the actions taken by the board, and the vote on these actions, recording both the number for and against; (3) records for continuing review of ongoing research activities; (4) copies of all correspondence between the IRB and the investigators; (5) a list of board members, identified by name, earned degree, representative capacity, etc.; (6) written operating procedures; and (7) statements of significant new findings provided to subjects. Unlike the previous retention requirements for proposed sponsor/monitor and GLP regulations, the IRB regulations require that the board retain its records for at least 3 years following completion of the research.

D. Patient Consent Requirements

The patient consent regulations implemented in July 1981 [10], deal with two major issues. The elements of and documentation of informed patient consent. These regulations require that informed consent be documented using a written consent form which has been approved by a duly constituted IRB and which is signed by the patient or subject involved in the investigation or the subject's legally appointed representative. A copy of the signed consent form must also be given to the subject who is involved in the study. The consent form should contain the following items: a statement that the study involves research; a full explanation of the purpose of the research; the expected duration of the subject's participation in the research; a description of the procedures to be followed and an identification of any procedures which are experimental; information concerning any foreseeable risks or discomforts; benefits to the patient which might be expected from the research; alternative procedures or courses of therapy which would be advantageous to them as patients; a statement addressing to what extent the records of the patient will be held confidential; and a statement that there is a possibility that the FDA may inspect the records. For research, where more than a minimal risk to the patient exists, there must also be some explanation as to whether any compensation is available to the patient for medical treatments should injury occur. The patient must also be given the name of an individual who can be contacted to answer questions pertinent to the study. Finally, the consent form should indicate that participation in the trial is voluntary and that refusal to participate or withdraw will not result in a loss of benefits to which the patient is otherwise entitled.

IV. QUALITY ASSURANCE AND CLINICAL RESEARCH

While the role of research quality assurance in preclinical safety studies has been mandated by regulation, the role of research quality assurance in clinical trials has not. Indeed, there is no proposed regulatory requirement mandating the existence of a research quality assurance unit. Many pharmaceutical companies, however, have seen fit to establish a clinical quality assurance unit as well as a preclinical quality assurance unit. The reasons for establishing these units in the clinical research area are twofold: (1) they are a natural outgrowth of the quality assurance requirements established by GLP; and (2) pharmaceutical research management has recognized the role of research quality assurance as an effective management tool, particularly in the area of maintaining adequate research records. At a meeting held in 1979 by a major pharmaceutical corporation to discuss the impact of proposed clinical regulations, it was determined that at least 70% of the companies present (15 out of 22 companies) had some type of quality assurance operation in their clinical area [11]. These quality assurance operations at that time took different organizational forms and were charged with various responsibilities. Organizationally, some of them were one-person operations reporting to the medical department or they were much larger groups with four to six people reporting to the research quality assurance department along with the GLP quality assurance unit. The activities and responsibilities of these various quality assurance groups offered some rather interesting contrasts as well. At one end of the spectrum, certain groups were not only involved in auditing their own internal records but were also making routine on-site visits to their clinical investigators and were conducting FDA-type audits using the FDA compliance manuals as guidelines. Some groups were only concerned with on-site visits when it was apparent that for-cause inspections or audits were necessary. Others were not auditing investigators' data but were concentrating on the housekeeping and record keeping at their own facility.

While the role and responsibilities of quality assurance in clinical research may not be uniquely defined, it is becoming more and more apparent through experience that the research quality assurance unit serves a vital function as a mangement tool for both clinical management and general management. While the major function of clinical quality assurance units has been to audit internal records and clinical investigator records, the quality assurance unit has provided other important services. When monitors encounter difficulties with particular studies, the quality assurance unit can provide the monitor and clinical research management with an unbiased evaluation of the conduct of the study, often leading to decisions related to continuation or termination of a study. The quality assurance unit can also provide an awareness to clinical research personnel of regulatory and corporate record-keeping policies. Based on the broad scope of quality assurance responsibilities, both in preclinical and clinical research, research quality assurance can also provide senior research management with an unbiased evaluation of the efficiency of systems and procedures used throughout research operations. The research quality assurance unit can assure clinical research management that policies and procedures are consistently carried out among all clinical research units, whether they are involved in Phase I, II, III, or IV studies. Finally, the research quality assurance unit may serve as an educational resource for medical monitors and clinical research associates insofar as it can point out commonly encountered problems in monitoring and work with the medical monitors to try to solve these problems.

While research quality assurance is not mandated by regulation, it can be a very powerful management tool for both clinical research personnel and senior management.

V. CLINICAL AUDITOR/MONITOR QUALIFICATIONS AND TRAINING

A. Qualifications

The major qualifications one looks for in clinical auditors [12] and monitors are formal degree in a scientific discipline and the ability to interact and deal effectively with people. Most of the personnel utilized in clinical auditing within the pharmaceutical industry have at least a B.S. degree in a relevant scientific discipline and several years' experience in a laboratory or hospital environment. The need to deal effectively with people is obvious. The clinical auditor is in a position where tact and strong negotiating skills are necessary to get medical monitors and/or clinical investigators to correct deficiencies observed during the course of an audit. While entry-level auditors may have strong backgrounds in science and human relations, there is still a need to train these individuals in the techniques of auditing as well as the overall pharmaceutical research process itself. While the auditor or monitor may only be directly involved in the final rate-determining step of the drug development process, it is important that he or she understand what goes into the development of the drug prior to studying it in the human model.

B. Training Programs

An effective training program for clinical auditors and monitors should address three basic areas: the program should provide an overview of the entire research and development (R&D) process; it should address federal regulations dealing with the conduct of clinical studies; and, finally, the program should concentrate on the specific corporate policies regarding the conduct of clinical trials.

An overview of the pharmaceutical research process should afford the auditor an awareness of his or her place in the overall process relative to pharmacology, toxicology, clinical pharmacology, drug regulatory affairs, and postmarketing studies. The auditor should also be exposed to the federal regulations dealing with the conduct of clinical studies. Here it is imperative that the auditor be made aware of the responsibilities of sponsors and monitors as well as clinical investigators, IRBs, and patient consent. The major portion of the clinical program should be devoted to corporate policies regarding the conduct of clinical trials. Specifically, the questions of protocol and case report form design should be addressed in relation to the pitfalls that one encounters in addressing the issues of patient population, exclusions, disease, and concomitant therapy. The logistics of clinical trials should also be addressed. Such topics should be discussed as choosing investigators, shipping and blinding drugs, setting up blinding codes, monitoring frequency and reports, records to be reviewed, common problems associated with records, final report design, NDA formatting, and finally drug reconciliation and returns.

VI. CLINICAL AUDITING PROCESS

A. Initiating the Audit

The first step in any clinical auditing process is to establish which studies should be audited [13]. The decision as to whether or not to aduit a particular study may depend on answers to any of the following questions: What priority does the particular drug being studied have in the overall corporate R&D plan? How close is the product under study to being filed as an NDA? Is the study in question so pivotal as to support the efficacy and safety of that product? Is the monitor having difficulty in getting data in a timely manner from the investigator, or is the investigator not providing the quality of data which the monitor would like to see?

Having chosen a specific study or studies for audit, the auditor continues the process by notifying the responsible medical monitor that a particular investigator and clinical study will be evaluated. His next step is to begin a review of the intramural records which are available in the corporate medical records center. These records usually include the following: monitoring records; protocols and protocol amendments; investigator correspondence; drug shipment records; all CRFs received from the investigator; and any correspondence with the FDA concerning these specific studies.

Concurrent with this inspection of internal records, the investigator should be notified by letter that an audit will take place. Basically, this letter explains the functions of the quality assurance department, what the auditor would like to accomplish during the audit, which specific study is to be audited, and whom to contact should the investigator have any questions concerning the study or audit procedures. The investigator is also informed that he will be contacted by phone in 2 weeks to set up a definite appointment for the audit.

Once this 2-week period has elapsed, the auditor will normally telephone him to establish a convenient date and time for the visit. Usually, this date is set in conjunction with several other audit dates within the same geographic region. The auditor normally sends a follow-up letter to the investigator that summarizes the telephone conversation and confirms the date and time for the audit. The investigator is also told exactly what items will be reviewed. These include protocols, CRFs, evidence of IRB approval, the investigational drug brochure, informed consent forms, reports of serious or adverse reactions, drug accountability records, drug storage and security information, patient exclusion records, records of monitoring visits, record retention requirements, adequacy of facilities such as the pharmacy and clinical laboratory, and patient records. The letter finally requests that work space be made available for reviewing records and that either the investigator or any associate be available should clarification of the records be necessary.

B. On-Site Audit

In conducting the audit, the auditor makes sure that the currently approved protocol and all amendments are in the medical monitor's file, in the medical research record retention room, and in the investigator's study file. The protocol and subsequent addenda must be common to these three sources. The auditor will verify whether the protocol and all amendments have been submitted to the FDA.

In keeping with the Institutional Review Board regulations, the auditor makes sure that IRB approval has been obtained for the study and that the investigator is keeping the IRB up to date with annual reports. In addition, the auditor will try to determine whether the IRB is meeting current federal regulations regarding the composition of the board.

He will also determine that an informed consent form exists and that the investigator is obtaining informed written consent from each subject in the study.

Drug accountability records are usually examined by determining whether the records contain the following information: the shipping dates and quantities of drugs shipped; the identities of patients; and the date on which the drug was dispensed to these patients. The auditor should determine that the drug dispensing log is up to date and, if the study has been blinded, whether patients have received the drug according to the properly assigned sequence. If the study has been completed, the auditor will verify that the investigator has returned all unused drugs or disposed of unused drug as requested by the medical monitor. If the study is still ongoing, the auditor will ascertain whether the drug being used is properly stored, particularly when drug stability may be a problem or the drug may be subject to

Drug Enforcement Agency regulations. With regard to the administration of the test drug, the auditor will also verify that the drug was administered by the principal investigator or by coinvestigators authorized to do so in accordance with the signed FD1572 or FD 1573 form, which outlines the investigator's responsibilities. The auditor will also determine whether the investigator has an up-to-date patient exclusion record and whether a current investigational drug brochure describing preclinical and clinical findings is available. If such a drug brochure has not been written, then the auditor will try to ascertain whether the investigator has the required drug safety information. If the drug is a marketed product, the auditor will determine if the investigator has a current package insert. The certification of the clinical laboratory being used by the investigator is also of concern to the auditor. The auditor must verify that all laboratory tests are being performed at the intervals required by protocol and are being performed in a laboratory which is certified to do so. The auditor will ascertain the name and address as well as the certification number of the facility performing the clinical tests. The question to be answered here is whether the laboratory is licensed under the Clinical Laboratory Improvement Act or is it accredited by some other body recognized by the Center for Disease Control. If the laboratory is not licensed, the auditor is instructed to determine whether the laboratory is making use of an acceptable proficiency testing program.

Usually, the auditor will also look at all the trip monitoring reports which are filled out by the medical monitor as a result of his on-site visits to the investigator. In examining these reports, the primary concern of the auditor is whether there is follow-up action on any deficiencies noted by the medical monitor in the report. For completed studies, the auditor also determines that a close-out monitoring report has been properly completed.

In reviewing CRFs at the site of the investigation, the auditor will normally select several CRFs at random and review a number of items against raw data. Possible items of concern to the auditor are patient number, clinical observations, verification of laboratory test results, legibility of case report entries, verification of concomitant therapy, type and severity of any adverse reactions, investigator and coinvestigator signatures, and validation of the actual doses of drug administered to the patient. If the study is ongoing, the auditor also determines whether the investigator is keeping the CRFs up to date or not.

C. Postaudit Activities

Following the audit, the auditor will usually meet with the investigator to discuss any major findings or findings that require immediate corrective action by the investigator. If there are no such findings, the investigator will be told so. Minor findings are usually not discussed with the investigator but, instead, are discussed with the medical study monitor first before corrective actions are taken. Once the internal and on-site audits have been completed, the auditor begins a series of meetings with the medical monitor to discuss the findings of the audit. No written report is issued until these findings have been discussed with the monitor and a reasonable course of corrective action has been determined, if necessary. These meetings with the medical monitor are held for two reasons. First, they help to ensure that the auditor's observations are accurate, and they help to identify facts that the auditor might not have been aware of in making his observations. Secondly, these meetings help to develop the corrective actions which should be taken, if necessary. Once the findings have been reviewed with the medical monitor, they are also discussed with other personnel having a vital interest in the findings. These might include personnel from the drug regulatory affairs department, drug packaging and shipment departments, and medical documentation center. At the completion of these meetings, a report is drafted and sent to the various parties for review and comment. If corrective actions have already been initiated by the responsible parties involved, they are asked to indicate what action has been taken and these are documented in the report. When all parties are satisfied with the audit report, it is finalized and issued by the quality assurance department to the management of clinical research and drug regulatory affairs.

If the findings of the report are serious and require extensive corrective actions, a follow-up audit is usually scheduled within 3 to 6 months, depending upon the length of the study. The entire audit process from the time the medical monitor is contacted before the initial audit to the time the final report is prepared will normally take 3 months. A competent auditor should be able to conduct at least three to four on-site audits per month. At any given time, he may have some 12-15 audits in various stages of progress. A successful audit plan must be based on a consideration for the feelings and concerns of the people who are being audited. Much effort should be made to avoid undercutting the relationship between the medical monitor and the clinical investigator. The main objective must be to work with both of them to improve the quality of clinical data while maintaining the efficiency of the systems used to obtain that data.

VII. CONDUCT OF CLINICAL TRIALS

Adequate control of clinical trials requires that good management practices be followed [14]. Most problems which arise in the conduct of clinical studies can usually be traced to a lack of planning and/or communication.

A. Planning

In planning a clinical trial, the monitor will often try to achieve too many objectives or answer too many questions in a single study. In trying to collect a large body of data from each individual patient to achieve his objectives, he will often end up with fragmented data from a large number of patients which cannot be interrelated or analyzed either clinically or statistically.

Study monitors often fail to recognize the limitations of the investigator's patient population, staff, and facility. Patient and drug exclusion criteria often are too exclusive, resulting in patients being entered into the study who do not meet the criteria—or patients not being entered into the study at all and the study simply lying dormant because of the exclusivity of the criteria. Frequently, the amount of data required by the protocol will overtax the investigator's staff or the staff of the institution where the study is being done. To this end, all required safety and efficacy parameters for a study should be discussed with each investigator in a realistic manner before the study starts to determine whether or not it is possible to perform the required measurement. It is often by the monitor's failure to recognize the limitations of an investigator's patient population, staff, and facilities in designing protocols that protocol violations are spawned.

Equally important in the overall planning of a clinical study is the basic design of the CRF used to transmit the data from the investigator's facility to the sponsor. It is imperative that the CRF reflect accurately all the requirements of the protocol. The CRF should be designed such that it aids the investigator in systematically providing the data required for a given patient as that patient enters the study and progresses to the completion of drug therapy. In designing a CRF it is important to make sure that it can stand on its own. All terminology and rating scales used in the CRF should be clear to the investigator. The investigator and his staff should not have to scurry back to the protocol to hunt for the definitions of the terms used in a CRF.

B. Communication

Effective communication is essential to the success of any clinical trial. To the clinical research auditor or monitor, effective communication has two aspects: (1) communicating auditing results to responsible management; and (2) communicating auditing results to the clinical investigator and helping the investigator establish corrective actions where required. Simply discussing corrective actions is not enough, however. One of the major problems encountered in this whole area of communications and corrective action is follow-up. An auditor or monitor will visit an investigator and determine that a particular problem exists. They usually will discuss the problem and potential solutions with the investigator. However, at this point the whole process falls down since they then assume that the problem will be taken care of by the clinical investigator. On the next visit, the monitor responsible for follow-up does not check or ask the investigator or the investigator's staff whether the corrective action agreed upon has been taken. Monitors should be required to review all previous monitoring or auditing reports prior to each monitoring visit to determine what follow-up is required from previous visits. Monitoring reports should note whether previous findings have been taken care of. A fundamental tenet of good auditing, effective communications, and (for that matter) good management is follow-up, yet this need is too often neglected or allowed to slip by in the conduct of clinical trials.

VIII. QUALITY ASSURANCE AND THE FINAL NDA PRODUCT

Just as the final package of medicament is the product for the marketing and manufacturing divisions of pharmaceutical companies [14], so the New Drug Application (NDA), when submitted to the FDA, can be viewed as the final product of the research division within a pharmaceutical corporation. Just as quality control, quality assurance, and manufacturing personnel will check the progress of a marketed product as it goes through the various manufacturing steps to a finished package for the consumer, so research quality assurance personnel must be intimately involved in the development of research products. They must also be involved in the final review and approval of the NDA product for compliance not only with federal regulations but with internal corporate standards of excellence before it is submitted to the FDA.

The need for an objective review by the quality assurance unit is obvious. Consider the FDA reviewer who may or may not be familiar with the particular drug being submitted. Data and the conclusions drawn from it may be very obvious to the medical monitor who carried out the clinical trials and wrote the clinical sections of the NDA, but they may not be all that obvious to the reviewer. Remember, the writer of the NDA product for compliance not only with federal regulations for several years. The FDA reviewer has not. The need for an internal objective review of all NDAs cannot be overemphasized. While the scientific aspects of the data presented to the FDA may be reviewed by pharmacologists, toxicologists, and medical personnel, the role of quality assurance in this review cannot be overlooked. Quality assurance personnel can provide an objective review of all final study reports, particularly from the point of internal consistency of the study report. Some of the questions to be answered by the quality assurance reviewer might include the following:

Are the individual toxicology, pharmacology, and clinical study reports internally consistent?
Is there consistency between the various sections of the NDA itself?
Does the data presented in the preclinical safety section of the NDA agree with or conflict with the clinical safety data?
Do the various graphs and tables presented in the NDA stand on their own, and do they agree with what is stated in the text?

IX. FDA INSPECTIONS

A. Plan for Inspections

Just as a plan is required for setting up and monitoring clinical trials, it is necessary also to establish a plan of action for government inspections [15]. The first element in this plan should be to establish personnel screening or security procedures for the particular facility. The second element in any plan is to establish a focal point for coordinating the FDA inspection. Based on its broad understanding of the day-to-day operations within a research function, the quality assurance unit is perhaps a logical focal point for coordinating FDA inspections. Basically, the responsibilities of the quality assurance unit in this role would include the following:

To coordinate all aspects of the inspection
To minimize the impact of the inspection on the time of the scientific personnel involved, whether they be preclinical or clinical personnel
To document in writing all details of the inspection proceedings
To follow up on inspectional findings and recommended corrective actions

B. Conduct of Inspections

The first step in any government inspection consists in determining what the nature of the inspection is. Is it a general inspection, or is it an inspection focused on a particular product which may be under consideration by the agency for approval? Having determined the answer to this question, the quality assurance unit should notify the appropriate scientific and administrative personnel concerning the nature of the inspection and should then arrange a meeting with the key scientific personnel who will be asked to participate in the inspection. At this meeting, there are a number of things which should be accomplished. First of all, certain ground rules should be established for the inspection. Of particular importance for discussion is the schedule of the senior scientific personnel involved in this study.

Once the inspection begins, it is imperative that throughout its various stages quality assurance personnel accompany the investigator and keep accurate, detailed records of what occurs. These should include the questions that were asked of personnel concerning data, standard operating procedures, and the actual conduct of the study. A list of the copies of the documents requested by the FDA investigator should also be maintained by the quality assurance personnel. Usually, quality assurance personnel will try to deal with questions concerned with standard operating procedures and facilities. However, discussions related to the scientific aspects of the study should be strictly reserved for the scientific personnel involved. In discussions with these scientific personnel, the quality assurance unit personnel should make sure that the FDA investigator does not overstep his authority by asking for information, i.e., financial records, to which he is not entitled.

C. Inspection Follow-up

Perhaps one of the most important aspects of any postinspection activity is the follow up that occurs. Once an inspector has discussed the findings of his inspection in an exit interview, a follow-up meeting should be held among corporate personnel involved in the inspection to discuss and evaluate the FDA investigator's findings. Once these findings have been evaluated, a decision should be made as to what course of action, if any, is to be taken by the corporation. Usually, the quality assurance unit is then charged by management to follow up on any corrective actions specified by the committee after its evaluation of the findings.

REFERENCES

1. *Federal Register 41*(225):51207 (Nov. 19, 1976).
2. *Federal Register 41*(225):51206-51230 (Nov. 19, 1976).
3. FDA Hearings on Good Laboratory Practices, Washington, D.C., Feb. 15-16, 1977. (Transcripts on file with the Hearing Clerk, U.S. Food and Drug Administration.)
4. *Federal Register 43*(247):59986-60025 (Dec. 22, 1978).
5. *Federal Register 42*(187):49612-49630 (Sept. 27, 1977).
6. Federal control of new drug testing is not adequately protecting human test subjects and the public. General Accounting Office report. U.S. Government Printing Office, Washington, D.C., July 1976.
7. *Federal Register 43*(153):35210-35236 (Aug. 8, 1978).
8. *Federal Register 43*(153):35186-35208 (Aug. 8, 1978).

9. *Federal Register 46*(17):8958-8980 (Jan. 27, 1981).
10. *Federal Register 46*(17):8942-8958 (Jan. 27, 1981).
11. Roger I. Justice, Regulatory affairs and clinical trials monitoring. Paper presented at the Fall Regional Meeting of the Associates of Clinical Pharmacology, Somerset, N.J., Sept. 26, 1980.
12. John Donahue, Qualifications and training of the biomedical research Q.A. professional. Paper presented at the American Society for Quality Control Symposium on Compliance and Beyond: Quality Assurance in Biomedical Research, Northbrook, Ill., Oct. 12-13, 1981.
13. John Donahue and Joseph Safaryn, Clinical quality assurance. *Pharm. Technol.* pp. 70-74 (Oct. 1981).
14. John Donahue, Drug Information Association Workshop on Clinical Aspects of Drug Development, Clinical Trial Management, Quality Assurance and Sufficiency of Data, Williamsburg, Va., Dec. 6-9, 1981.
15. John Donahue, Preparing for inspections. Paper presented at the International Meeting on Good Laboratory Practice, Rome, Italy, May 14-15, 1981.

19
ROLE OF CONTRACT RESEARCH ORGANIZATIONS

John D. Arnold and Dan M. Hayden
Quincy Research Center
Kansas City, Missouri

I. INTRODUCTION

In recent years, the contract research organization has emerged to support the pharmaceutical industry's clinical research effort in new drug development. Increased regulatory requirements and increased complexities of human clinical research have taxed industry resources and have broadened the role of the contract research organization and the clinical trials professional. Contract researchers often have the freedom to assume the initiative in the development of innovative new programs that speed clinical trials and expand the clinical approach to drug development.

Successful clinical research requires a clear view of how research goals may be defined and achieved, good judgments in the selection of investigators and research groups to obtain clinical data, and aggressive demands for research of the highest quality. The role of the contract research organization in this context includes provision of resources, experience, and expertise in conducting clinical trials. The contract organization must be skilled in marshaling diverse information systems to meet research objectives and must provide adequate quality assurance mechanisms addressing the following primary research needs:

1. Expedited design and execution of the research protocol
2. Protocol compliance
3. Accountability and data integrity
4. Prompt and accurate reporting
5. Confidentiality
6. Ethical practices in medical research
7. Data bases of populations and procedures

If the contract research organization can provide effective mechanisms to satisfy these needs, then it constitutes a unique resource for the drug sponsor.

The current chapter presents an overview of the role and responsibilities of the contract research organization in clinical research. Specific instances are identified in which it may not be cost effective for a drug company to set up and maintain its own research unit to provide clinical data. Considerations are raised regarding the role and responsibilities that should be assumed by the contract organization and should be expected by the sponsor company in meeting research needs.

II. THE CONTRACT RESEARCH ORGANIZATION

A. Definition and Scope

The contract research organization is a business enterprise whose primary roles in clinical research are to collect and defend experimental data and to provide ancillary research services. The contract organization may be university based or it may be a part of the free enterprise system. The contract unit may offer a single analytical or technical service, as might a clinical laboratory or a data processing unit. Some units may offer a wider array of services, including Phase I through Phase IV clinical testing of healthy volunteers and patients with medical conditions, multicenter data management and coordinating systems, and Investigative New Drug/New Drug Application (IND/NDA) preparation.

The contract research organization may have direct involvement in the generation of research data from the clinical setting. Alternatively, project management or data coordinating centers may have only indirect involvement in the data generation process. Regardless of the focus or extent of research services offered, the most prominent responsibility of the contract organization is to ensure the accuracy and accountability of its research data.

B. Structure, Operation, and Staffing

The principal investigator, associate investigators, and clinical trials professionals typically oversee the operations of the contract research organization. Investigators are often physicians with specialization in internal medicine or its branches and extensive clinical research experience. They may have a narrow range of research interests, or they may have experience with a wide array of chemical entities. Clinical trials professionals and technical staff may, depending on research services offered, come from diverse medical, academic, technical, and administrative backgrounds. These individuals complement and extend investigators' skills with expertise in project management and execution in all research aspects.

Operating characteristics of the contract research organization are diverse. Administrative systems and staffing policies provide support for the ongoing research effort by coordinating information resources, project logistics, and regulatory compliance. These efforts are often supported by local or national recruitment of consultants in many medical specialties and in professional and technical disciplines.

The contract laboratory usually has made advanced plans to anticipate research needs. Except in unusual circumstances, it is very difficult for a laboratory to carry out a new program for the first time with a new drug product. The contract laboratory should have established a track record with the particular project either from similar work done for other sponsors or by an in-house development program.

It can not be emphasized too strongly that any laboratory operating in unfamiliar territory has a considerable risk of failure. Many failures are attributable to miscalculations with respect to the available number of subjects; but even with an abundance of subjects, protocols and projects may impose unexpected difficulties in execution.

III. CONDUCTING CLINICAL RESEARCH

A. Limits to Research Resources

It is often not feasible for a company engaged in the manufacture and delivery of ethical pharmaceuticals to launch and maintain in-house research units to provide clinical data. A drug company may have few or widely differing research requirements: it may have few new chemical entities in development at any given time; it may have an insufficient flow of new entities; or it may have entities which differ greatly in their pharmacological class and expected activity.

The establishment of an in-house clinical testing unit under these conditions could constitute inefficient and costly use of resources. Even in the event that a company's economic position and/or research and development status permit consideration of an in-house unit, the establishment of such a unit might be undesirable for other reasons. An in-house unit may not have the right staff, it may not be able to recruit sufficient volunteers and patients, it may not have validated procedures and methods, and it may not have an adequate data base.

B. Alternatives

In some instances, a more practical and lower cost decision by industry management could be to select a contract research organization to conduct clinical testing or to manage a program or process the data. The contract unit of choice should have already successfully addressed the obstacles cited earlier, gained the requisite clinical research experience, and established a history of excellence in conducting and completing clinical projects of the types needed. Industry management may expect that the contract unit of choice will be able to provide complete accountability for data. If industry resources do not have to be diverted to the initiation of such units, these same resources may be more fruitfully brought to bear upon new drug development. Moreover, the availability of an outside contract organization permits smaller companies having only marginal resources to engage in research and development of new drug products.

The contract unit should have validated mechanisms already in place to deal with medical, ethical, and legal concerns related to research with human subjects. The contract unit may provide a buffer for the sponsor company for products that do not have good patent protection.

The contract organization should have a large population of subjects or a selection of experienced investigators so that subject recruitment can be as close as possible to project or protocol requirements. Finally, the contract unit may also be able to provide facilities for much larger research projects and for projects of longer duration. Such special facilities may handle large trials with many subjects or trials assessing safety of chronic, long-term, multiple dosing in cloistered volunteers.

C. Success of Clinical Research

The success of clinical research is the joint responsibility of the sponsoring company and the research organization. If a sponsor aggressively demands that the highest standards in research quality assurance and accountability be met by the research unit, then the production of quality data on new drugs may proceed in a straightforward fashion unimpeded by doubts concerning data integrity. The sponsor should define and specify, in advance, quality assurance and accountability standards that must be met for the project. Selection of a research or other contract unit should then be based on how competing units compare in meeting these standards, as well as on cost factors.

The standards that should be of concern to a sponsor in the selection of a contract research organization are discussed in the following subsections.

1. *Resources of the Contract Organization*

A prominent need in clinical research is to expedite the design and execution of the project protocol. Contract organizations are often consulted late in project planning when in fact they may have data and other information that would aid in protocol design. The efficiency with which research goals and mechanisms for meeting them are defined and written into a final protocol is of considerable importance. The contract researcher should provide specific assurances that mechanisms are in place to provide (a) sufficient experience and expertise, (b) volunteer availability, (c) staff availability, (d) valid procedures and associated documentation, and (e) standardized and certified clinical instruments and equipment, so as to allow the project to move along quickly.

a. Experience and Expertise

Does the contract unit have sufficient experience and expertise to keep the project moving at a good pace? With experience in conducting clinical trials, the contract research organization can provide guidance in the development and critique of protocols. Previous experience can often lead to improvements in the protocol that may not be apparent from existing literature or from ad hoc protocol development.

The experimental design, the measurement of clinical variables, and the size of volunteer samples can be established by inspection of results of previous trials. This information is often known to the contract organization. Variability may be computed from the unit's database, if available, and may then play a role in the selection of a final design and sample size for the project. Good design and sample size features of research projects are cost effective and can often enhance the interpretation of data.

The contract research organization can provide guidelines to the selection of variables, methods, and schedules of measurement that have been found clinically useful on similar projects and the reevaluation or modification of those that have not proven useful. Database characterization of the past behavior of prominent clinical variables may suggest alternate solutions in the selection of clinical variables. Variables that have been shown to be insensitive from previous experience may be deleted from the project or modified.

The methods and quality of physicians' assessments typically become keenly attuned to the clinical trials setting. A physician researcher's examinations and assessments may be quite different for clinical investigations of new drugs than for patient diagnostic services. Moreover, experimental treatment effects and adverse effects syndromes encountered during clinical trials of many new medications constitute a special area of medical practice and may not often be observed in the clinical setting or without special experience. The database of the physicians with experience with adverse drug effects and treatment effects of a variety of medications may distinctly enhance the interpretation of these events when they occur in clinical trials.

The contract research organization can often recommend better ways of handling and establishing data. For example, recent years have marked the advent of automated computer-based electrocardiographic analysis systems [1,2]. Such systems analyze digitized electrocardiogram (ECG) data. Programs are available to make measurements of the ECG waveform and apply a set of standard medical criteria to obtain rhythm and contour interpretations. Since the ECG has become a standard evaluation in the investigation of many new medicines, computer-based analysis and interpretation often resolve difficult problems of drug effect.

Upon completion of the protocol design, it is necessary for the contract research organization to have access to an Institutional Review Board (IRB) to carry out project review and approval. Submission of many protocols to an established IRB permits the contract unit to accumulate experience with the board and knowledge of its ethical concerns. Thus, protocol and informed consent requirements may be effectively refined before these documents are submitted for approval, and the risk of time-consuming rejection or modification of protocol and consent may be lessened.

b. Volunteer Availability

Does the contract unit have access to sufficient subjects or patients so as to be able to initiate and complete the project quickly? Clinical research may not, of course, proceed if a volunteer population is unavailable. It is necessary that the industry researcher assess the history and current capability of the contract unit to obtain acceptable volunteers from available populations. The contract unit of choice should be able to ensure (1) the effectiveness of volunteer recruitment and selection procedures, (2) the availability of a population for which a new drug is designed, and (3) the effectiveness of volunteer screening procedures in reducing contamination of research results from background "noise" in clinical variables.

(1) Recruitment and selection. Indicators of the effectiveness of recruiting procedures are (a) the size of the volunteer pool, (b) the proportion of volunteers who are selected repeatedly for project participation, (c) the consistency of recruiting efforts, and (d) the success rate of recruitment of new volunteer types. The contract researcher should maintain a database of individual subject participation histories and should be prepared to submit such information to the sponsor.

It is our experience that normal volunteers from different sources present different problems for the investigator. Some groups have better compliance rates than others. Some volunteer groups have different normal ranges for laboratory or other variables. It is also important that volunteers have a high degree of confidence in the research laboratory. This confidence is a very special asset to a contract laboratory and is often a hard-earned resource. The same problems of credibility and volunteer access apply not only to normal volunteers but also to patient populations.

(2) Population of interest. The objectives of a research project may be unattainable if the subject or patient sample does not represent the population group or subgroup for which the drug is designed. This is the fundamental issue in population selection; both the industry and contract researcher should aggressively pursue the more complete characterization of populations from which subject samples are to be drawn. For this reason, databases of the available populations are of great interest. In general, the larger the available populations, the more selective the recruitment process and, ultimately, the more satisfactory the results of a clinical trial.

In all clinical work, whether the group of interest is the rigidly defined "healthy" male or the patient with a specific medical condition, selection of research participants for inclusion in a project must be based upon minimizing the occurrence of out-of-range analyte results. This means that the contract organization needs to know the reference values for its populations and methods. The effect of out-of-range values or "background noise" in any project may be considerable. Contamination by background noise affects the interpretation of results. Noise also modifies the analysis of the role of the drug in producing the event. These problems are closely related to sample size requirements and the identification of positive findings for subsequent projects [3]. The central role of the organization's database cannot be emphasized too strongly. This has sweeping implications in terms of the ultimate cost of a study or series of studies, the speed of execution, and the confidence in the results.

It should be recognized that populations vary widely. Healthy young males often have higher rates of aberrant findings than do populations of middle age and older healthy males, for reasons that are quite obscure. It is not unusual for a laboratory to reject 60-90% of potential subjects in order to get very small coefficients of variation. Thus, the contract unit may incur substantial costs in the control of background noise to ensure that industry research is not undermined by subject selection that may degrade the study. This aspect of the process for selection of a contract organization is often overlooked in the project planning stage.

(3) Volunteer screening. Extensive activities of the contract research organization should be designed to minimize background noise and to provide acceptable samples of the group of interest. The volunteer selection process in clinical trials is a multifaceted set of procedures that provides for continuous assessment of a volunteer's acceptability and suitability for research projects. Recruiting procedures should include an in-depth interview of each volunteer to determine the volunteer's (a) fulfillment of protocol specifications; (b) interest in and availability for the project; (c) mental status, behavior, sociability, and compliance potential; and (d) substance use and abuse, e.g., tobacco, alcohol, and drugs of abuse. These criteria are used to assess the volunteer's emotional stability and, if required by the project, apparent capacity to tolerate the change in life-style likely to be imposed by the protocol.

In-processing procedures should include further assessments of the volunteer's medical, emotional, and medication history; preliminary review of organ systems and baseline screening tests; and physician evaluation of all screening results obtained to determine the volunteer's acceptability for the project. Once accepted, the volunteer's responses to treatment and procedures, compliance, and dependability are evaluated on the basis of past experience and the database for the appropriate population.

c. Staff Availability

Does the contract unit have access to sufficient staff to move the project along quickly? The success of clinical research is dependent upon technical and practical expertise in the laboratory or testing unit. Under internal pressure to meet project schedules, the industry researcher often must select an investigator, finalize a project protocol, obtain institutional review and approval, and initiate the project within a severely restricted time frame. Thus, it is particularly important that the contract unit have access to experienced, highly trained, and competent staff. Databases giving current performance effectiveness of the staff should be available. Periodic intra- and interobserver reliability checks can often estimate the degree of variability within or between projects that is attributable to variations in staff performance.

With the dramatic expansion in project administrative, management, regulatory, reporting, and documentation needs, particularly in the United States, the contract unit must also be able to ensure that these support areas will meet an audit review by the U.S. Food and Drug Administration (FDA).

d. Procedures and Operating Manuals
Does the contract research organization have validated research procedures that are fully described in written manuals? It is not uncommon in the typical experience of a contract unit that questions are raised about project results months or even years after project completion. Interpretation of such results may depend upon the precise mode in which certain methods and procedures have been performed. Documentation of the original procedure, procedure validation, periodic procedure audits and reliability checks, and current performance characteristics for procedures, instruments, and observers fulfills a prominent role in defending data and in creating an audit trail for a project.

The industry researcher must be assured that operating manuals for data management and manipulation will also facilitate the reconstruction of the contract researcher's project data files under audit conditions. Such manuals should allow the derivation of a complete data trail for any project by carefully documenting specifics of (1) data collection and storage, (2) movement and manipulation of data within the system, (3) data calculation or transformation, (4) data verification and editing, (5) statistical data analysis, and (6) data-reporting mechanisms. Data that stand alone unqualified by written procedures are difficult to interpret, often cannot be replicated, and are not defensible under audit conditions.

e. Instruments and Equipment
Does the contract research organization have records of standardization, certification, and repair of clinical instruments and equipment? Although implied in many regulatory actions and decisions, confirmation of equipment standardization and certification and also inspection of records of repair are rarely undertaken prior to initiation of a project. Nonetheless, industry and contract researchers are accountable for the consistent and reliable functioning of instruments and equipment. If a sphygmomanometer is off by 10 points, the impact on a study in hypertension can be substantial.

2. *Compliance*

What history of protocol and regulatory compliance does the contract unit have? There are many types of compliance deviations, ranging from minor deviations to major infractions, that actually affect the interpretation or acceptability of clinical research data. In the extreme case, project data may be of questionable worth due to noncompliance with ethical requirements of biomedical research, protocol inclusion/exclusion criteria, treatment regimens, methods and schedules of measurement, collection of biological specimens, standard procedures for management of adverse events, or project reporting requirements.

Evaluation of the experience of the research unit and the professional history of the principal investigator(s) should provide estimates of the ability of the contract organization to meet compliance requirements and of the worth and prevalence of systems designed to address these requirements.

3. *Accountability and Data Integrity*

Does the contract research organization have good accountability, audit, and general record-keeping systems to ensure data integrity? Clinical research typically generates large volumes of data, individual case reports, correspondence, and associated documents for each project. Project accountability, however, should begin well before the first volunteer is screened and may last for as long as 8-10 years, depending on a variety of industry and regulatory circumstances. Thus, to gain confidence that both immediate and future accountability needs will be met, it is of paramount importance that the industry researcher determine whether the contract research organization has current and projected stability in clinical research. Moreover, the industry researcher should receive assurances regarding the accountability for and preservation of project archive files. Selection or exclusion of a contract research organization may very well be based upon the quality of these mechanisms and record files.

It is becoming increasingly more common for a sponsor company to seek to reconstruct the results of earlier clinical work or to clarify results for products currently in development. An FDA audit may also occur long after project completion. The quality of these accountability mechanisms and record files will often have a considerable bearing on the success of the reviews.

Accountability mechanisms of the contract unit should include rigorous prior self-audit. Self-audit should focus initially on the setup and generation of primary data collection processes and records for an entire project. Next, the accuracy of descriptive information on these forms should be verified. As verification of descriptive information is completed, data collection activities begin; and experimental data verification should proceed on a daily basis to detect missing or illegible entries and data inconsistencies. During these self-audit activities by the contract research organization, the industry researcher should visit the site periodically to monitor the quality of the data collection and recording process. Increased industry vigilance is extremely important. Repeating studies because of flawed data or protocol execution or program design is extremely expensive.

Accountability mechanisms of the research unit should additionally include specific documentation of the location of record files. Ideally, the unit should have a description of its archive process to afford quick access to project information. It is inevitable that personnel changes will occur with time. What is not written down may be lost forever with personnel changes.

The contract unit should also be able to provide personnel experienced in audit conditions and requirements to assist in any project audit by the sponsor or the FDA. The availability of skilled personnel during an audit often has a significant and positive bearing on the outcome of the audit.

Systems ensuring data integrity should constitute a prominent feature of a contract unit's project accountability mechanisms. It has been suggested [4] that the research unit describe and report to the sponsor as to: (a) data handling and manipulation procedures, including records of purged, faulty, or erroneous data; and (b) quality assurance mechanisms and procedures. The establishment of higher reporting standards has also been called for [4], to include standards for quality control procedures, mechanisms for quality control monitoring, and specific procedures for ensuring data integrity.

4. Reporting

Does the contract research organization have prompt and accurate project reporting capabilities? Throughout all phases of clinical work in new drug development, excessive amounts of time are consumed in the reporting, analysis, and evaluation of project data and in the creation of written clinical or statistical reports. The contract unit should be able to greatly minimize time requirements for project reporting. Clinical summaries, for example, are increasingly expected within 30 days of project completion. Reasons for quick reporting may include: minimizing delays in additional work which depend on current results; an NDA filing deadline; ensuring that no untoward effects are being observed in subjects; and ensuring that future subjects are not put at needless risk.

Several multifaceted systems must operate harmoniously if reporting needs are to be met. Data collection, recording, and verification systems must be neatly interfaced with systems designed to provide individual case reports, statistical analysis, clinician evaluations, and medical writing. Considering the amount of data typically available, as well as the time constraints on reporting, computer support will be a definite asset. With satisfactory computer programming and database management, even stringent reporting requirements may realistically be met by the contract research organization. Some computer systems are now programmed to easily accept new protocol designs so that reprogramming is no longer needed for each new project or protocol.

5. Confidentiality

Does the contract research organization observe confidentiality and security requirements regarding proprietary information on new chemical entities in clinical research? Several procedures should be appropriate to ensure security and confidentiality. All employees, subcontractors, consultants, and IRB members should be requested to sign either an employment agreement which includes references to security and confidentiality requirements or letters stating that such requirements will be respected. Additionally, visitors badges, restricted access areas, the provision of discrete monitoring areas, and coded references to sponsors, protocols, or studies may enhance requisite standards of security and confidentiality.

6. *Ethics in Experimentation*

Does the contract research organization adhere to ethical requirements pertaining to the use of human subjects in biomedical research? The industry researcher should obtain assurances that informed consent will be obtained and that informed consent documents contain the 11 basic and 5 additional elements of informed consent, if appropriate. The FDA also requires that the informed consent document be worded in language that may be understood by the research participant. Good studies may be flawed if informed consent is not obtained; even though the validity of project data may remain unaffected, these ethical needs must be satisfied.

The established contract unit should have gained wide experience in implementing mechanisms to address difficult ethical considerations in biomedical research. The contract unit may often be able to identify lingering difficulties with ethical regulations currently in effect. It may be able to identify new difficulties which could not have been predicted but that, nonetheless, may have arisen since regulations were implemented. Moreover, the contract research unit may aid in the definition and resolution of new ethical issues in human experimentation.

IV. DATABASE CONSIDERATIONS

An additional new resource is database development for characterization of volunteer populations. The information potential available from database remains largely unexplored in clinical research, although its merits and methods appear widely accepted. Reference values for any given analyte may be known to vary with differences in demography and physiology of the subject as well as methods of observations. Moreover, circadian and seasonal fluctuations may be expected for certain variables. Yet actual database utilization in many instances may at best be limited to the comparison of research results with textbook reference values derived from small samples from an otherwise uncharacterized population. Comparison of research results with population- and procedure-specific reference values derived from databases makes much greater sense and much better science than comparisons with textbook values. There is little comfort or confidence in interpretation of sample data drawn from an uncharacterized population.

Much additional potential for database utilization stems directly from repeated sampling of the local population for clinical trials. Repeated sampling allows more complete population characterization and greater refinements in precision of population descriptions. Through repeated volunteer sampling and repeated exposure to many different new drugs, the contract researcher may be able to profile in advance the expected responses of its subject sample to a new drug. Moreover, database can detect when a random sample from the population is unrepresentative. More importantly, it can often identify specific instances in which control samples are clearly not representative of the unit's ordinary experience and may explain unusual findings in a particular protocol.

Database can gauge the incidence of background noise in a population for any given analyte. It is precisely the adequacy of control of background noise that affects the interpretation of research results. The cost of this exercise to the contract research organization may be considerable in terms of time and effort. However, the cost to the industry's clinical research goals of not controlling background noise can be even greater because of the effect such noise may have on collection and interpretation of data.

Finally, database can often confirm the reliability and validity of the experimental observations and the contract organization's own procedures. Without these, it may not be possible to defend the data.

V. CONCLUSIONS

The role of the contract research organization in clinical research encompasses the complete array of primary research needs of the pharmaceutical industry. The established contract unit can provide tested mechanisms for satisfying needs in areas of protocol development and critique, collection of quality data, validation of observations, provision or project audit trails, and the secure archiving of project data. With the development of effective research and administrative mechanisms, the contract organization constitutes a unique resource in clinical research. Of all the elements that must be evaluated in the selection of a contract research organization, the availability of subjects, validation procedures for observations, and databases of available populations may be most critical.

ACKNOWLEDGMENTS

Grateful acknowledgments are extended to the following for many excellent contributions to the manuscript: Arthur E. Berger, Sue W. Strickler, Anne DeSimone, Doug C. Moore, and Susan Kysela, for helpful comments during manuscript development. Grateful acknowledgments are also extended to the editor and to Marcel Dekker, Inc., for their firm support during manuscript development and preparation.

REFERENCES

1. R. Bonner and H. Schwetman, *Comput. Biomed. Res. 1:*367-386 (1968).
2. IBM Technical Report GH20-1249-1, *Health Care Support/Electrocardiogram (ECG) Analysis Program General Information Manual,* 2nd ed. International Business Machines Corp., White Plains, New York, 1974.
3. J. Lewis, *Trends Pharm. Sci.* pp. 93-94 (Apr. 1981).
4. C. Meinert, *Controlled Clinical Trials 1:*189-192 (1980).

20
DRUG REGULATORY AFFAIRS

Paul H. Roberts
Revlon Health Care Group
Tuckahoe, New York

I. INTRODUCTION

Most of the functions of a present-day Drug Regulatory Affairs (DRA) group are a direct outgrowth of the impact on pharmaceutical firms of the 1962 Harris-Kefauver amendments to the Federal Food, Drug, and Cosmetics Act. The regulations issued by the U.S. Food and Drug Administration (FDA) several years later as a means to implement the substantial changes in the act were written in a style considered almost unintelligible by those involved in pharmaceutical manufacturing, research, and marketing. The length of the regulations, the complexity, and the details covered left most firms unsure as to how to begin to comply. It *was* clear, however, that a substantially different type of communication and coordination system had to be set up within companies to cope with this entirely new situation. One or more individuals, usually from the existing company research organization, were assigned the task of interpreting the regulations and ensuring that the complex and extensive documentation submitted to the FDA by the company met the new requirements and accomplished the goal intended. As a quite natural consequence, written or verbal communications from the FDA were invariably referred to those newly assigned individuals, who in turn evaluated and interpreted the letters or telephone calls and arranged for responses. As the regulations became still more com-

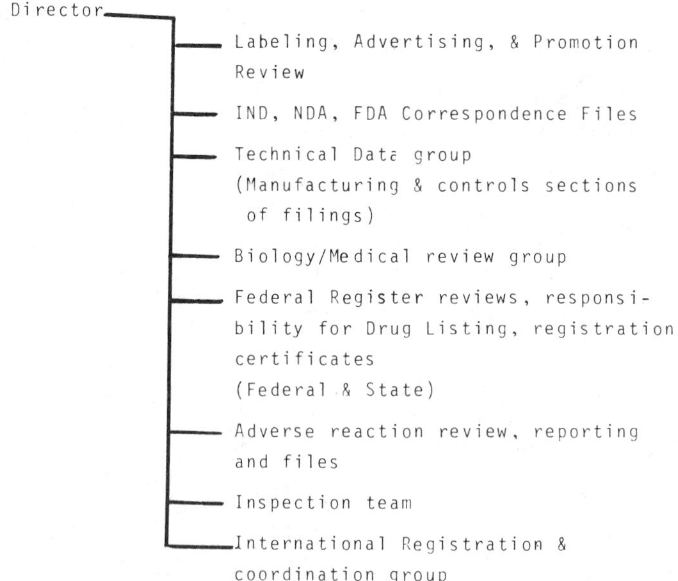

Figure 1 Organizational chart of a typical DRA group.

prehensive and complex, although no easier to understand or remember, the functions required of DRA grew.

Today, most DRA groups are organized in a manner similar to that depicted in the organizational chart presented here (Fig. 1). Depending on the size of the group, some functions may be combined or, in certain DRA groups, additional units may have been added for data processing/microfilm/microfiche, the writing of package inserts, and the technical writing required for manufacturing and control documents submitted to FDA. On the whole, the DRA group responsibilities are now fairly well stabilized, and the major functions are presented in Secs. II through VII as an introduction to the later discussion (Sec. VIII) of the role of DRA in the drug development process.

II. SUBMISSIONS TO THE U.S. FOOD AND DRUG ADMINISTRATION

A. INDs and NDAs

A notice of Claimed Exemption for an Investigational New Drug (IND) must be submitted to the FDA before a new drug may be evaluated in humans. Once a decision has been made to file an IND, the required information and data are assembled, put in the format specified by the regulations, reviewed, and submitted. Before a new drug can be marketed, a New Drug Application (NDA) or an abbreviated application (ANDA) must be filed with the FDA and approved by that agency. The following are other, supplemental types of submissions to FDA that are managed by DRA:

1. *Supplements*

After a new drug is approved by FDA, the firm may wish to make changes in the filing. New indications, a change in the dosage regimen switching from a capsule to a tablet formulation, or changing to a new manufacturing site are among the types of changes that require the preparation of documentation supporting the modification. The documents are submitted to FDA in a format specified in the regulations.

2. *IND Progress Reports*

A summary of data and information generated by the holder of an IND must be submitted to FDA annually. Failure to do so may result in withdrawal of the IND.

3. *NDA Annual Reports*

FDA regulations require the submission of an annual update of labeling and advertising (discussed next), unit sales information, and accumulated scientific data or literature reports.

B. Labeling and Advertising

The labeling (immediate container label, carton, and package insert) and advertising of prescription drugs are subject to extensive and detailed regulations. The DRA group, therefore, reviews the text of all such documents before release and maintains a permanent file of all approved labeling and advertising.

III. REGULATORY INSPECTIONS

Because a drug development unit is usually on the same location as a manufacturing unit, the site is subject to inspection by FDA for a variety of regulatory reasons. The FDA representative usually appears without giving prior notice and since the visit may affect one or many of the research and development (R&D) and manufacturing groups on the site, DRA meets with the FDA investigator upon his or her arrival and determines the purpose of the inspection. DRA answers the investigator's questions from its own files to the extent possible, alerts others within the R&D group or manufacturing area, and accompanies the investigator throughout.

The FDA regulations allow the investigator substantial but not unlimited powers to request and review records, obtain copies of documents, and personally inspect manufacturing and R&D sites. It is DRA's responsibility to know what lies within the FDA's jurisdiction. The R&D process can be inspected by FDA from several points of view. The FDA investigator may be assigned the review of a company's records on a specific IND or NDA, in which case a number of groups within R&D are directly involved. The Good Laboratories Practice (GLP) regulations define acceptable practices for conducting animal studies that are submitted to FDA in support of safety in humans. The conduct and record-keeping practices relating to clinical trials are subject to extensive FDA evaluation. Product complaints, whether from a marketed or investigational drug are occasionally the subject of FDA investigations. The production of marketed and investigational drugs are subject to the Good Manufacturing Practices (GMP) regulations.

In summary, and R&D/manufacturing organization is subject to extensive and detailed regulations and periodic FDA inspections from the time an IND is filed, and the regulatory impact continues as long as a product is marketed. The DRA group determines the subject of the investigation, remains with the investigator, keeps a running record of the findings, retains copies of all documents turned over to FDA, and participates in the "exit meeting" at which the investigator hands the firm a written list of adverse findings, if there are any. Subsequent to the meeting, DRA sees to it that a written response is sent to the FDA, replying to any adverse finding. Since DRA participates in all inspections, the group maintains a permanent record of all such visits.

Other inspection groups, such as the Environmental Protection Agency (EPA), state and local authorities, and groups representating foreign governments are handled in a similar manner.

IV. RECORDS

Since DRA submits the INDs, NDAs, supplements, and other correspondence to FDA and, in turn, receives almost all communications from FDA, the establishment and maintenance of a complete regulatory record system is an unavoidable consequence of the group's activities. These files have become, over time, the "official" regulatory records and are indispensable to the operation of an R&D/manufacturing/marketing organization. The regulatory record system is consulted frequently by all segments of the organization.

V. DISSEMINATION OF REGULATORY INFORMATION

Not only are the regulations governing R&D, manufacturing, and marketing complex and extensive, but they exist in a state of almost continuous minor and major modification. A daily government publication, the *Federal Register* is a means of officially notifying "those affected" of proposed or final changes in the FDA regulations and is a forum for displaying policy statements, official meeting announcements, and related matters. Unofficial trade publications, scientific symposia, state and local publications, and regulatory affairs meeting are additional sources of important information that often affects the drug development process. While many within an R&D organization scan the information, it is usually the responsibility of DRA to review all significant sources and disseminate the facts and probable impact within the company.

VI. MEETINGS WITH REGULATORY AGENCIES

It is usual that the same one or two DRA representatives attend all of the meetings with the regulatory agencies and they thus become very familiar with the "language" and style of the FDA. The FDA, in turn, comes to know the regulatory representatives of each company. This mutual familiarity is advantageous in that it helps ensure the accurate transfer of information and understanding. Mutual trust can be reinforced over time, and that is a significant advantage—but "special favors," if they are done at all, are extremely rare. The multilayer review process within FDA is an effective and proper deterrent to favoritism.

Face-to-face meetings with regulatory agency representatives are important in the drug development process. In addition to arranging and participating in the formal IND or pre-NDA program reviews held with FDA representatives, DRA arranges and holds frequent informal discussions with R&D representatives and the FDA on such diverse matters as the proper doses for long-term animal studies, the specifications and test methods of the formulated product, adverse reactions that have occurred during the investigational phase, and the wording of a proposed package insert. From a drug development point of view, the most significant meetings are those involving the question: "Is this program approvable?" An R&D organization may be quite prepared to defend its investigational program and results scientifically, but the policies, precedents, and regulations of the regulatory agency may dictate a significant change in the research plan. Thus, one or more meetings are arranged with FDA to review the proposed program, interim results, and the format for final presentation of the data. In a successful meeting of this type, the R&D and FDA positions are stated with sufficient clarity and detail so that a subsequent "memo of meeting" prepared by DRA and agreed to by the FDA is a commitment by both groups. Unfortunately, not all such meetings are completely successful, and subsequent changes are required in the research plan up to and occasionally beyond the date of filing the NDA.

VII. INTERNATIONAL REGISTRATION

Planning for worldwide research registration and marketing is an integral part of the R&D program of all multinational companies. The timing and sequence of U.S. and international programs vary so considerably with each project and with each research organization that a single, well-defined "usual" course of action cannot be described. Companies with worldwide research organizations may (and with increasing frequency are) beginning their clinical research programs outside the United States. When the U.S. clinical program is begun, the international investigations continue and, at a point determined for each project, the R&D organization and management may, after adequate review of the accumulated safety and efficacy data, elect to "release" the product for registration and marketing wherever government approval is possible. Thus, it is not unusual for a drug to be approved and available in many countries before FDA approval is obtained.

When a new drug under study in the United States is released for overseas registration by the R&D organization, the regulatory affairs group (usually in the United States) assembles the accumulated information in an agreed-upon format and forwards a copy of the entire package to the overseas regulatory affairs groups. They, in turn, evaluate the information and, if the data suffice, prepare the documents in an appropriate format and submit them to their governments.

VIII. ADDITIONAL RESPONSIBILITIES OF THE DRUG REGULATORY AFFAIRS GROUP

The detailed knowledge of the regulations required of DRA personnel leads, in some organizations, to their participation in mock inspections and audits of manufacturing, toxicology labs, clinical investigation programs and protocol review. The types of products marketed by a company may require regulatory affairs involvement in proprietary, over-the-counter (OTC) drugs, devices, cosmetics, veterinary drugs, or food additives. None of these potential involvements will be discussed here and the remainder of this chapter will be devoted to the close association of new drug development, FDA regulations, and DRA.

A. New Drug Development, Federal Regulations, and the Drug Regulatory Affairs Group

New drug development and regulation by the FDA are inseparable in the sense that the latter influences almost all stages of the former. To support this thesis, a brief diagram of the research process will be presented as an outline for the discussion (Fig. 2).

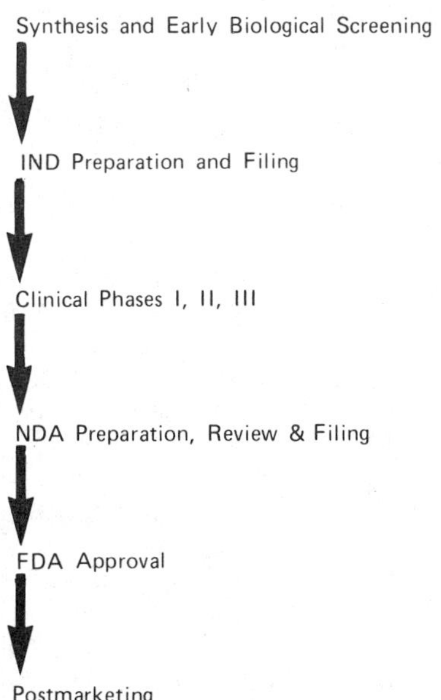

Figure 2 Diagram of the research process.

The synthesis and early biological screening phase is a basic research activity and is the only segment of the drug research process not significantly affected by FDA regulations. It also covers the shortest time span. This early phase begins when a chemical structure of interest is synthesized and tested in primary pharmacological and toxicological screens. If the results suggest serious consideration of the compound as a drug candidate for use in humans, it is subjected to additional and wider-ranging biological screening. The accumulated test results are reviewed by the basic research groups which produced the data and, if the collective opinion is still favorable, a comprehensive review/recommendation is issued to research management.

The favorable recommendation, in reality, proposes that the compound be tested in humans for safety and efficacy. In order to test the new drug in humans, the company or research organization must submit information to the FDA in the form of an IND.

The decision to proceed to an IND most often follows a formal internal R&D review of the chemistry of the compound, preliminary animal toxicology, and the pharmacology data which, collectively, leads to a recommendation by the research scientists that the compound be pursued as a potential new drug for the treatment or diagnosis of human disease. When the R&D organization and company management concur that an IND should be filed, the involvement of DRA begins in a formal way. Because the decision to file is based on preliminary internal reports, the various R&D groups continue their activities, now focusing on the accumulation of the data and information needed for the preparation of an IND. A list of the types of information required for an IND follows (and is shortened from the list available in the Code of Federal Regulations, Title 21, 312.1):

Descriptive name and chemical structure of the compound
List of components of the drug
Quantitative composition of the drug
Description of source; preparation of the new drug
Description of the new methods, facilities, and controls used for the manufacturing process
Preclinical summary
Previous marketing information (if any)
Informational material to be supplied to proposed investigators
The scientific training and experience expected of investigators
Name and curriculum vitae of each investigator and of the clinical monitor
An outline of the planned investigations in humans, including a description of the Institutional Review Board (IRB) and the names and addresses of each investigator
A commitment to inform the FDA if the investigations are discontinued and the reason therefore
An environmental impact statement, if requested by the FDA

Since the items listed above are not specific as to the information required, it is essential that the *purpose* and *scope* of the IND be clearly defined by R&D as soon as possible after a decision to file has been made. In a reasonably experienced R&D organization a formal definition of the project is an integral part of the approval to proceed to an IND and is never deferred. It is incumbent on DRA to be certain that the project is clearly defined because it is that group's responsibility to ensure that the diverse parts of a submission form a cohesive whole. The DRA group must know its own R&D organization and procedures sufficiently well to decide how much detail to include in the first formal internal DRA document—the outline of IND requirements. A specific outline for each IND is issued for wide distribution within R&D and, in its simplest form, consists of a list of the documented data expected from each of the groups contributing to the IND, together with a preliminary time schedule. The time schedule is not assigned by DRA but derives from the program schedule developed by R&D.

An early written definition of the purpose and scope of the IND is needed for the following reasons: (1) to establish a clear set of endpoints so that the R&D organization and DRA can plan a specific IND for a specific purpose; (2) to allow the groups to recognize when they have reached the initial goal; and (3) to act as a reference during the sometimes extended time period between a *decision* to file and the *actual* filing of an IND. Regarding the last point, it is surprisingly easy to lose sight of the intended goal as data are reported in the many project discussions that occur during the IND development phase. The temptation to follow interesting tangents suggested by preliminary research studies is almost irresistable. Parenthetically, if the tangents become literally irresistable, the project definition should be changed. Defining the *scope* and *intent* of an IND is important. An IND may have one of the following clinical study goals (and there may be others):

1. A single-dose drug metabolism study in normal volunteers plus a single (ascending) dose tolerance study to a defined dose limit
2. The studies in goal 1, above, plus single-dose and multiple-dose (for a few days) preliminary efficacy investigations in a small number of patients
3. The studies in goals 1 and 2, above, plus 30- to 90-day efficacy studies in approximately 100 patients
4. All of the above, plus studies in women of childbearing potential
5. Studies in infants and children

Each of these goals obligates the R&D organization and DRA to a different, although cumulative, set of IND requirements. If the initial intent of the IND is to allow studies through goal 3, above, the animal toxicology requirements are substantially greater than for goal 2. Similarly, if the stability requirements of the investigation drug formulation are different, then the formulation itself may have to differ. Although the overall intent of the R&D program is to develop a therapeutically useful and approvable product, the IND has limited goals initially; as these early goals are met, the scope and purpose of the IND change. The DRA group must keep track of the progress of the investigations through attendance and participation at R&D project meetings and R&D discussions. As the clinical investigation program becomes more comprehensive, changes in and additions to various parts of the IND are required. These changes must be anticipated well in advance and filed at an appropriate time so that, for example, the clinical program does not exceed the bounds of the available animal toxicology studies.

B. Pre-IND Activities of the Drug Regulatory Affairs Group

As the data leading to the preparation of the IND are being developed by the R&D organization, the DRA group undertakes a review of the regulatory files of the company and other sources of information that may provide useful guidelines for the preparation or the conduct of the investigational program.

The DRA group usually checks and reviews the following sources:

1. The company files, to determine if a similar or related drug has been the subject of an IND or NDA; FDA comments and evaluation as well as the firm's own reviews of earlier projects may well provide useful guidance.
2. After determining if chemically or pharmacologically related drugs are currently on the U.S. market, it is often worthwhile to request from FDA a copy of the "Summary Basis of Approval" (SBA) for one or more of these recently approved drugs. The SBA is a document prepared by FDA for each approvable NDA and is, as the name implies, a summarization of the data that the agency felt best supported safety, efficacy, and other regulatory requirements. One or more SBAs can provide very useful guidance for a research program and, at the least, will suggest questions that can be asked of FDA at the appropriate time.
3. Clinical and preclinical "guidelines" for various classes of drugs have been published in journals and elsewhere over the past 10 years; a number of the guidelines were produced by joint FDA-industry groups but most are, unfortunately, becoming out of date.
4. Position papers issued by FDA (but not necessarily published) are often specifically related to the class of compounds which

are the subject of the IND. These, again, may be obtained directly from the FDA division responsible for issuing the guidelines.

5. The *Federal Register* publishes regulations, guidelines, and comments 5 days a week on a wide range of topics. Very often the FDA has published on matters relating to the subject of the IND and a direct request to the FDA division involved will direct one to the proper documents.

Using these and other available sources as well as their collective experience, the DRA group provides as much guidance as possible for the preparation of the IND.

C. Pre- and/or Post-IND Meeting with the FDA

Often very useful and sometime essential, a meeting with FDA prior to filing the IND is not required by regulation and may be difficult to arrange. If the firm submitting the IND is really not certain of the FDA's position on, say, criteria for the clinical efficacy of a particular class of drugs or the doses to be used in animal toxicity trials and all available resources have been checked, they may wish to clarify these points before embarking on the investigational program. A misunderstanding or miscalculation of the FDA's opinion or guidelines can easily lead to significant delays and unexpected costs.

The FDA prefers not to react to seemingly hypothetical questions, that is, before an IND is filed. Therefore, a request for a pre-IND meeting should be made only if:

The FDA division to which the IND will be assigned is known.
The questions have been committed to writing and are specific.
The FDA division agrees verbally to the meeting; a consumer safety officer in the division will most often make that commitment after consulting with his or her superiors.

Assuming the FDA agrees to the pre-IND meeting, the firm will ordinarily be asked to submit a written agenda. The waiting time for such a meeting is usually 6-8 weeks after the written request and agenda have been received by the FDA. Because of the rather substantial delays involved in pre-IND meetings, many firms elect to file the IND and request a meeting almost immediately thereafter if there are important questions to be settled prior to embarking on significant animal or clinical investigational programs. In this case a written agenda can be submitted and the consumer safety officer called to arrange the meeting after the agenda has arrived at the FDA. A waiting period of 4-6 weeks should still be expected, but part of the working time may be simultaneous with the mandatory minimum 30-day review period required by the FDA after the IND is filed. Further, the FDA can respond "officially" to specific questions regarding the IND *after* it is filed.

DRA arranges for the meeting, mails the agenda to FDA, and accompanies the R&D group to the discussion.

D. How is the IND Assembled and by Whom?

As described above, there is a clear and definite format published in the FDA regulations. DRA has issued the list of requirements for the IND and acts as a focal point for the accumulated data and information. As each section of the IND is completed by the respective R&D department or division, it is forwarded to DRA. The regulatory affairs group reviews that section against the stated purpose and scope of the proposed IND and makes sure that it meets general regulatory requirements. Not all sections arrive in DRA simultaneously, and not infrequently a section is returned to the initiating group after DRA review has established that changes are needed. Thus, the content of the IND is altered during its development, but the changes needed are usually not substantial unless there has been a serious misunderstanding of the goal of the IND.

DRA is responsible for assembling the documents in the proper format, composing an appropriate cover letter to the FDA, and adding several documents that are not prepared by R&D. This "draft IND" is thoroughly reviewed in its final form by a combined R&D/DRA group to assure that it meets the regulatory requirements, that the various sections are consistent with each other, and that it fulfills the objectives established for the project.

E. Filing the IND

Someone from the DRA group prepares and signs the cover letter and submits the IND. The IND itself is an announcement to the FDA that the firm intends to evaluate a specific drug in humans. A mandatory 30-day waiting period after filing and before clinical trials can begin is imposed to allow the FDA to review the IND and to agree or disagree that the drug may be studied in humans. If 30 days pass without FDA comment, the clinical trials may begin. If the agency requires clarification or more information regarding the initial IND, a telephone call followed immediately by a written confirmation from FDA effectively stops the IND program until the questions are satisfactorily answered.

A rather extensive explanation has been given of the drug development process from inception to IND filing because the act of filing has committed the research organization to a comprehensive and long-term regulatory involvement. The type of documentation and the review process required from the R&D/DRA organization for the IND has established a format that is applied to the additional reports generated during the subsequent research process. The FDA opens a permanent file on the investigational drug at the time of IND filing, and thereafter the file is added to but no part of it is subsequently destroyed.

The DRA group also opens a permanent file at the time of IND submission. Copies of the IND and all subsequent communications to and from the FDA regarding the IND are retained in DRA permanently. The DRA filing system must be complete, up to date, and orderly because it becomes the single source of complete information regarding R&D/FDA interaction within the research organization. The FDA is fully aware that the DRA group has this function in almost all pharmaceutical firms, and therefore the dialogue between the firm and the agency is conducted through DRA representatives.

F. Clinical Phases I, II, and III

The evaluation of the drug for safety and efficacy in humans appears deceptively simple in outline form:

Phase I—human tolerance, metabolism and the determination of a safe dosage range
Phase II—initial efficacy trials in a limited number of patients
Phase III—assessment of safety and efficacy and optimum dosage schedule

During the time span covered by the clinical phases, a number of interrelated activities are taking place, and these activities influence and are influenced by the clinical investigation program. Synthesis of the drug substance is modified, improved, and set into its final pattern. Animal pharmacology studies extend the knowledge of the drug's action; long-term animal toxicology studies are begun and completed. The final dosage form is developed and studied extensively. Although the clinical investigation phases are neatly categorized as I, II, and III, there is often a considerable overlap of the phases during the execution of the evaluation program in humans. Nonetheless, most research organizations point toward a formal review of the entire program at the end of Phases I and II and before the extensive clinical studies usually required for Phase III are begun. The reasons for the review are several:

The Clinical Phase I and II studies have given a firm indication of therapeutic activity, the probable dosage of the drug, its metabolic fate, and preliminary human tolerance data.
The method of synthesis has been established.
The pharmaceutical dosage form is at or near its final development stage.
The long-term toxicology studies including carcinogenicity studies have begun or are about to begin.

The review documents for the Phase I/II review are prepared by each R&D group in the format intended for NDA submission. Following a formal review by the drug development group, a decision is made to proceed or not to proceed to an NDA. An affirmative decision means, in effect, that the R&D organization believes that it has sufficient proof of safety and efficacy to recommend the large-scale, long-term clinical and animal investigations required to obtain an approvable NDA. As the final detailed plans and protocols for the NDA are being prepared, quite frequently a formal meeting with the FDA is requested to review together the Phase I/II data, the animal pharmacology and toxicology results, the chemistry and formulation information, and the proposed Phase III clinical program and long-term animal toxicity trials.

Working closely with the various R&D groups, DRA prepares and submits an agenda for the Phase I/II meeting to the FDA, participates in the premeeting reviews of the presentations to be made, and joins actively in the resultant discussion with the FDA. If the R&D organization and the FDA staff have prepared sufficiently well for the meeting, a firm plan for the remaining studies leading to an NDA should result.

With an agreed upon Phase III plan available, the R&D organization schedules studies and investigations needed to complete the documentation for an NDA.

G. NDA Preparation, Review, and Filing

As the Phase III clinical studies and long-term animal toxicity studies proceed, there is a slow but definite shift of emphasis on the part of the R&D/DRA groups from "investigation" to "NDA preparation." During this transitional phase, and long before the NDA is actually submitted, the R&D project planning group and DRA meet frequently to agree on the final format for the NDA. The key and guiding document for the NDA format is the "Optional Expanded Summary" (OES), which is described in the Code of Federal Regulations. The OES subdivisions (outlined in Table 1) are consistent with the manner in which the project developed from IND through the Phase I/II review and to the NDA plan. Thus, reports prepared early in the investigation phases need not be rewritten but are added to and incorporated in more comprehensive summaries.

A second key document that completes the NDA outline (see Table 2) is a convenient means of subdividing the required sections and the complete reports underlying the OES.

A complete and detailed NDA outline (from which Table 1 and Table 2 are derived) can be found in the Code of Federal Regulations, Title 21.

Table 1 Optional Expanded Summary and Evaluation

Chemistry
Highlights of preclinical and clinical studies
Preclinical studies—summarized
Clinical studies—summarized
Clinical laboratory studies related to safety
Clinical laboratory studies related to efficacy
Summary of clinical literature
Overall results and conclusions
Annotated package circular

An integral and extremely significant part of DRA's responsibility with respect to preparation of the NDA is continued planned meetings with the FDA. Following the Phase I/II meeting and very much dependent upon the nature and complexity of the investigational program, a pre-NDA meeting may be warranted. In addition, special meetings with FDA chemists, pharmacologists, medical officers, and biostaticians are arranged to discuss and review selected sections of the emerging NDA. The pre-NDA meeting and other special meetings afford the R&D/DRA groups and FDA opportunities to reach agreement on the final format and content of each section of the application. The effort involved is worthwhile because significant modification of an NDA *after* submission may require the full efforts of the drug development group and DRA for many months, occasionally for several years.

At the beginning of the NDA preparation process, DRA establishes a filing system specific to the project and keyed to the agreed-upon final format for the NDA. When completed, the NDA is reviewed in its entirety by a combined R&D/DRA committee to be certain that the goals of the NDA have been met and the various sections are consistent with each other.

The DRA group assembles and delivers three copies of the NDA to FDA.

Table 2 Required Format of Reports Comprising the OES

Table of contents
Optional expanded summary
Manufacturing and controls; packaging
Preclinical reports
Adverse reaction reports
Case report forms
Bioavailability package

H. NDA Approval

Although the regulations require that the FDA "respond" to an NDA application within 180 days of the filing date, the average time between filing and approval is well over 2 years. The large, detailed filings required by the regulations, the multilevel reviews within FDA, and a chronic, seemingly irreducible backlog all contribute to the substantial delay.

The FDA has attempted to speed the review and approval process of drugs judged to be "important therapeutic gains." This has been done by giving higher priority to the "important" (Type A) drugs than to the "modest therapeutic gain" (Type B) drugs and "little or no therpeutic gain" (Type C) drugs. The assignment to a category is made by FDA, generally during the IND phase, and the classification can be changed (but with difficulty). The classification system has not reduced the backlog; it has simply accelerated the review and approval of some filings at the expense of others.

Another mechanism adopted by FDA to aid the approval process is to refer the proposed NDA to an established FDA Advisory Committee for review and recommendation. In recent years almost every NDA for a new clinical entity is subjected to Advisory Committee review. Most often, the NDA reaches the committee between 12 and 18 months after filing. Considerable preparation for the review is required of the sponsor of the NDA and of the FDA because both groups present or comment on the filed data. While the Advisory Committee system provides guidance and opinions useful to the FDA, the total review time is not thereby shortened.

During the FDA's review and approval process, DRA may receive written or telephoned questions about any aspect of the filing. The regulatory affairs group obtains the appropriate response from its own files or from R&D and responds, usually in writing, as promptly as is feasible. On occasion, the FDA requests a meeting to discuss a particular section of the filing.

The penultimate step in the NDA approval process is the receipt of an "approvable" letter from FDA. This communication describes the conditions, if any, that will be attached to forthcoming approval. Final approval is transmitted in written form: the FDA approves the application and, in particular, the printed package circular submitted by the company.

I. Postmarketing Studies: Phase IV

The manufacturer's responsibilities to the FDA do not end with the approval of the drug for marketing. Subsequent to approval, annual or more frequent periodic reports are submitted to FDA as required by the regulations. These reports include information on the quantity of the drug distributed, information concerning current clinical studies, reports of adverse effects associated with use of the drug, and copies of currently used advertising, promotion, and labeling. The periodic reports are submitted at specified intervals as long as the drug remains on the market.

The IND that allowed the investigations in humans to begin has continued in effect throughout the investigation and FDA review. Each clinical investigation commenced under the IND must be closed in a formal manner and the results of the study reported to FDA.

With increasing frequency, the FDA has insisted that formal postmarketing clinical investigations be undertaken as a condition of approval. These Phase IV studies are undertaken to evaluate long-term efficacy or to assess the occurrence rate of certain adverse effects of such low incidence that the clinical investigation population of 1,000 to 4,000 patients filed with the NDA may not provide sufficient information.

Phase IV studies required by the FDA may continue for many years after the drug is approved.

IX. CONCLUSION

The complexities of the drug regulatory process have highlighted the need for trained scientists, physicians, and administrative personnel in this specialized area.

The DRA group plays a pivotal role in the clinical research process. The rapport and relationships developed with the FDA are vital to the success of the clinical research process.

DRA personnel are the link between a pharmaceutical corporation and the FDA from the time an IND is filed through the NDA approval process and beyond. They interact with FDA on a continuing basis on behalf of the corporation on all regulatory matters.

The drug regulatory affairs function in its broadest sense is an essential ingredient in the clinical research process.

21
RELATIONSHIP OF MARKETING TO CLINICAL RESEARCH

Barry Strumwasser and Robert S. Cohen
Berlex Laboratories, Inc.
Wayne, New Jersey

I. THE ROLE OF MARKETING IN THE RESEARCH PROCESS

Until some 10 years ago, the prevailing attitude among scientists and corporate management was that the marketing group was charged with the commercialization of new drugs but could add little to the science and should, therefore, be relegated to a position of accepting whatever research and development (R&D) offered. Thus, marketing management did not become involved in the research process before Phase III studies at the earliest and sometimes only after the new drug application was filed.

This attitude prevails at some firms today. Yet, a number of pharmaceutical companies have begun fostering marketing's involvement much earlier in the research process. New product planning groups and strategic planners are becoming more prevalent on the business scene; and top management is seeking their advice and counsel on the disposition of valuable research assets.

While it has been a long time in coming, it seems clear that the age of the inventor in the pharmaceutical industry is giving way to the age of the professional planner.

The reasons for this evolution are obvious. There are inherent risks within pharmaceutical research. From executive management's perspective, if the research process is left unchecked, the firm's return on its research investment may not be optimized. Hence, we see the advent of the planner, who aids in formulating corporate decisions relative to deployment of resources for growth via emerging research. Economic constraints ultimately play a decisive role in the limitation of project funding as well as enhancing efficient, on-time development programs. Products emanating from these programs should not be inconsequential articles of commerce but, indeed, compounds which fill clinicians' needs and which will, in turn, help to ensure successful commerical ventures.

The cost of new drug discovery has skyrocketed from $50 million for each new entity in the 1970s to $70 million according to the latest estimates [1]. Moreover, the lag in new drug introductions, combined with the impact of patent expirations and generic intrusion on major brands, has served to put earnings pressure on most drug companies. At the same time, Drug Efficacy and Safety Information (DESI) review and the inherent threat of New Drug Application (NDA) revokation of products considered ineffective for given indications has put the sword of Damocles over products which provide retained earnings used to support research.

The net effect of all these issues is that pharmaceutical managements are taking keen interest in optimizing their returns on research investments and are employing professional planners to help reach these ends.

These planners are sometimes endowed with the title of corporate planners, strategic planners, business development planners, or simply new product planners. However, under any name, their charge remains basically the same, that is, to assist management in defining and reaching its strategic objectives within the resource limitations of the firm. Invariably, this process involves the organization, analysis, and planning of the firm's research, marketing, and financial activities.

Although one might get the impression that nonscientists' involvement in research planning represents meddling in science or that executive management is saying that research is too important to be managed by researchers, this is not the case. Rather, progressive management is inviting research to clearly define its objectives within those of the overall organization.

Throughout the process, it is marketing management's responsibility to reinforce both corporate and marketing objectives in order to ensure that those objectives are met. It is of paramount importance, however, that these objectives be achieved within the confines of good, responsible science.

Managers who seek to impose their wills on research without regard to science, scientists, and creativity are destined to fail in their mission. Balanced sensitivity and a clear understanding of the drug discovery process are required of marketing managers who interface with R&D. In order for nonscientific managers to be welcome into R&D, they must influence rather than demand, integrate rather than polarize, establish direction rather than lead. Therefore, the challenges faced in this process for both the scientists and marketing managers are best addressed by the matrix concept of management.

II. CLINICAL RESEARCH AND THE STRATEGIC PLAN

Strategic planning is the process of determining the major objectives of a corporation and the various strategies to provide for the acquisition, use, and disposition of resources to achieve these objectives. Such objectives include corporate missions which are generally long range in nature, defining the lines of businesses and/or product categories which the organization will pursue. Strategies are the means to deploy resources [2].

Figure 1 represents an example of a strategic planning network developed for a pharmaceutical company. Such networks allow for a strategy planning system.

The system, however, must function as an integrated network requiring the active involvement of key functional areas of the corporation, i.e., R&D, marketing, finance, operations, etc. The net result of the process is the strategic plan, which when adopted by top management is the blueprint for the future of the firm. Good strategic planning is the means for effecting innovation and change. Indeed, if the strategic planning system does not accomplish this, it is a failure [3].

The implementation of strategies implies the allocation of resources including funds, critical management talent, and technology. A corporation may seek to acquire additional products, expand product lines by the utilization of competences available within the organization, or go elsewhere for development of necessary resources. In essence, corporations will seek adaptation or integration to stay within their defined missions [3].

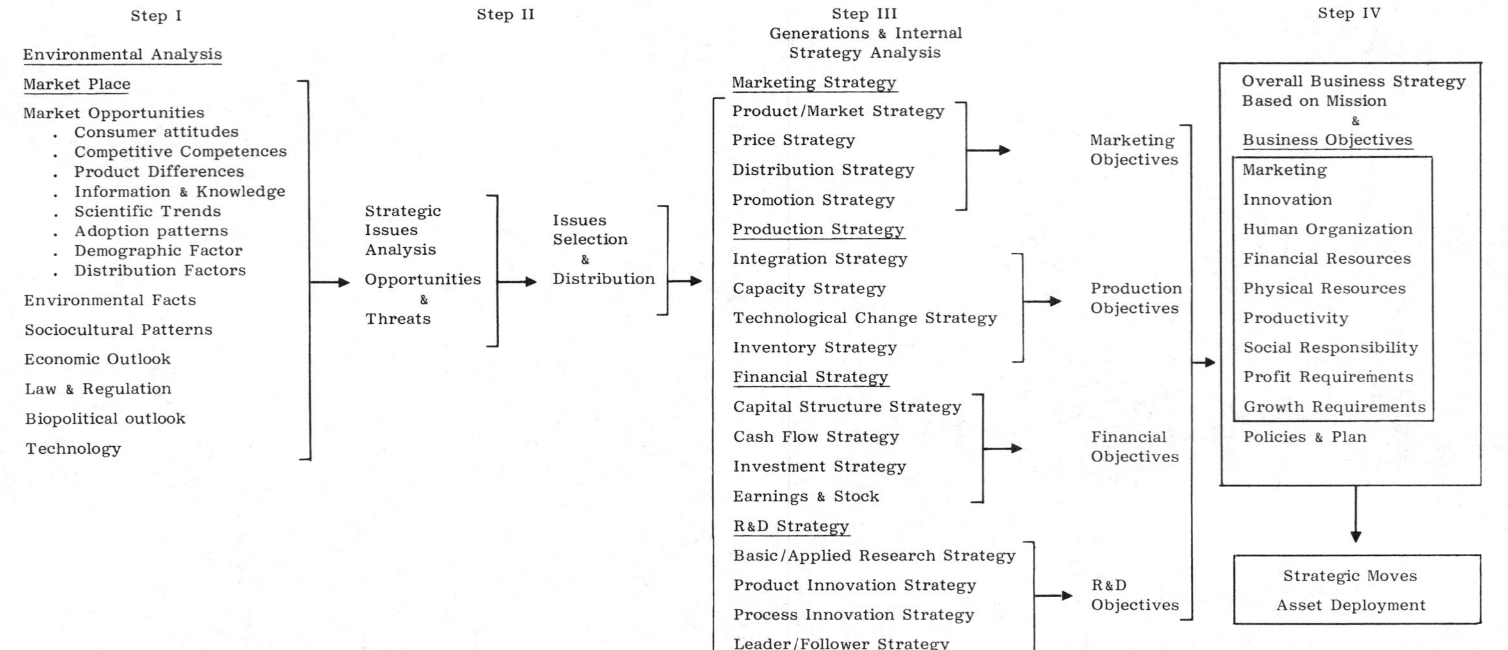

Figure 1 Example of a strategic planning network.

The business planners who have aided in the strategic planning process will then strive to achieve thorough implementation. Given this responsibility, the new product planner will seek involvement in the clinical research process to aid in the final development of all molecules, either developed by the organization's chemistry and screening groups or acquired through licensing arrangements. These planners will not only champion the product through clinical development but, frequently, are charged with the initial marketing of the brand. As such, clinical programs not only prove safety and efficacy of the compound but also contain the design factors which demonstrate unique sites and modes of action or other differentiating factors providing successful articles of commerce. These differentiating features should be incorporated into the cadre of assumptions leading to a preliminary forecast developed prior to filing of the Investigative New Drug (IND) application. During the course of the development program, features are then reassessed in order to reaffirm their original value conferred on the product, at times lowering or raising the revenue and earning forecast, or necessitating acceleration or cancellation of a program. In the latter case, if the program were continued without a particularly germane reason, the corporation could suffer a setback in earnings and/or return on investment. Additionally, this course could prove to be a conspicuous waste of valuable resources in talent and funds.

III. INFLUENCING THE RESEARCH PROCESS

The basic management influence is their expression of commitment to research by the definition of the implicit role of R&D in the overall mission of the firm. While many firms attempt to define the R&D role in public relations terms—"We make drugs as if your lives depended on it"—these commitments are meaningless unless they are backed up with clear, objective statements outlining the goals, structure, and function of the R&D commitment and supported by the financing required to reach those goals.

Therefore, some companies will clarify the R&D mission as one of development, whereas others will define it as discovery. Once the distinction is made, the proper expectations can be outlined and appropriate funding and staffing can ensue.

The common thread to both designs is the need for clinical research. Indeed, in the view of some managers, clinical research is a data-generating machine which provides the database required to prove efficacy and safety and, thereby, gain registration of the drug. This terse and primitive view frequently tends to ignore the basic marketing and corporate objectives of the firm.

Indeed, in our judgment, the primary function of clinical research is to establish all the qualities pertinent to the use of a drug, including those differential characteristics which endow it with value vs. other drugs within the same category.

Marketing's influence within this context is clearly one of support and guidance as the clinical drug candidate moves from Phase I to Phases III and IV. A cooperative relationship between marketing and clinical research benefits both disciplines, since only marketing can isolate the perceived needs of the medical community and only clinical R&D can develop compounds to meet those needs. Included among these issues are the following:

Efficacy and relative efficacy
Pharmacokinetics
Drug disposition and metabolism
Toxicity and relative toxicity

In today's environment, marketing drugs without some measure of differentiation which fulfills a therapeutic need is a formula for failure. Both sophisticated researchers and marketers are keenly aware of this axiom. Enlightened people in both disciplines recognize that by working together they will enhance the odds for both scientific and commercial success.

Unfortunately, in many companies the interdisciplinary prejudices continue to exist. These prejudices not only affect the marketing-R&D relationship but frequently are manifest in intramural R&D discord. In some instances, marketing can help resolve and even obviate discord by serving as the mediator, establishing better communication and timetables and reinforcing interdependency, thereby augmenting the basic tenets of research management.

The R&D-marketing discord usually revolves around some strategic discontinuity. Discontinuities generally take the form of timing differences to launch, failure to stay on the critical path, failure to receive or follow specific indications, changes in kinetic profiles, or abandonment of a mode of action or comparative trials.

Such discontinuity, therefore, invariably stems from failure in communications or follow-through. In our experience, there are few discontigous situations which cannot be resolved.

Basic to a successful relationship between R&D and marketing are mutual support and understanding. Clearly, good honest exchange and argument are vital to the clinical development process, but the underlying spirit must be one of empathy and mutual understanding of problems each discipline is facing.

The two basic principles on which the relationship is built are the following:

Marketing's awareness that clinical research is a complex process which requires time for the clarification of a drug's qualities
R&D's awareness that marketing must have drug differentiation and registration within a reasonable period of time to effect commercial success

Marketing and research remain two creative disciplines; and creative dissonance, given the above underlying principles, can only be healthy. The upshot of this partnership should lead to the partners becoming champions of the clinical candidate within their respective departments [4].

This mutuality of purpose is important since complacency is a constant danger to the vitality of any new product. Compounds which might be immune to complacency are few and far between, for example, propranolol and cimetidine. Most drugs introduced to the market are endowed with few differentiating characteristics. It is these drugs which can suffer the lethargy and complacency of both marketing and research management. To quote a colleague, "The 'just another ...' syndrome is the hobgoblin of the corporate mentality."

Compounds affected by this syndrome are often condemned to obscurity before they even reach the market. At some companies there are those within the structure who feed on destruction and denigration, initiating a process which, if left unchecked, will ultimately result in the disillusionment of key managers. Such disillusionment can result in either abandonment or withdrawal of adequate support for the clinical candidate.

On the other side of the coin lie the diehards—who cling tenaciously to those programs which are overdue for abandonment in the faint hope that the investment might somehow be recouped or that some obscure avenue remains to be explored.

In either case, R&D and marketing must come together to address these situations, eliminating the diehards and obviating the "just another" syndrome.

IV. MARKET EVALUATION

A substantial portion of the planner's time is devoted to generating marketing and financial assessments of new molecules. These documents are drafted prior to initiation of the clinical program, revisited during the clinical phase, and finally redrafted during development of the introductory marketing plan.

The sources of information are traditional marketing research audits published by Intercontinental Medical Statistics (IMS), including:

1. *National Prescription Audit* (NPA)—for demand trends
2. *National Disease & Therapeutic Index* (NDTI)—for usage patterns
3. *Drugstore and Hospital Audits*—for sales trends

These data provide an understanding of the size and trends of the market and the various products constituting the market. They provide a historic vantage point and some limited indication of future trends. In order to gain a better understanding of future developments, other sources of information are sought, such as the worldwide medical and scientific literature, new product patents, Wall Street analyses, and other syndicated sources including those of Stanford Research Institute (SRI). The whys and wherefores behind the IMS data are provided by sources within the R&D structure, through outside consultants, and through discussions with practicing physicians.

Research management and its operatives should provide critical information on emerging technologies, as well as existing trends within the field of clinical medicine. When these data are ultimately blended with marketing data, the planner can evaluate the product's potential and its fit within the needs of the medical community. At some companies the system is formalized under the concept of future scanning.

Given this information, concepts relevant to the product's potential are then tested with physicians, pharmacists, and (possibly) patients. These marketing tests should provide the commercial group and researchers with a feeling for the user's perception of the product with regard to:

Uniqueness
Differentiating characteristics
Awareness
Knowledge by the user
Potential satisfaction
Clinicians' needs

Finally, this evaluation should identify the advocates of this class of therapy, as well as the particular agent. This base of advocates can often form the springboard for success of a brand.

V. DEVELOPING THE DRUG TO MEET THE NEED

Having defined the value of various drug candidates to be developed, implementation of the program must be carried out in a manner that will provide physicians with a drug which will meet certain expectations.

We have previously outlined the overall objectives of the clinical program as being the demonstration of the compound's safety, efficacy, and differentiating characteristics between drugs in the market and dosage forms. Considerations regarding dosage forms should be given by the time Phase I studies are initiated.

Early clinical studies in Phases I and II might employ the most expedient dosage form possible, while exploring tolerance and early indications of efficacy. However, Phase III clinicals should employ the final finished dosage form planned for commercialization; the import of this statement rests on the necessity for planning development of this dosage form no later than initiation of Phase I trials. Inherent in this planning process are marketing considerations relative to ease of administration, flexibility of dosing, shape, colors, size, coatings, and costs. Beyond these factors are considerations for stability and feasibility of manufacture. Once these considerations have been resolved and pilot batches manufactured, optimization of the dosage form should begin. The market planners and R&D should work harmoniously during this process, ensuring that corporate and scientific goals are met.

Selection of investigators for Phase II and III studies should be made on the basis of assuring completion of such studies. Additionally, R&D and marketing should consider the need for investigators who can exert peer influence over clinicians in hospital and private practice settings. This could provide a fine basis upon which new drugs might accelerate through the adoption of innovation [5].

The planners should work closely with the corporation's clinical and regulatory affairs groups regarding the development of the program and its tie-in to final labeling. Given the knowledge that final labeling will impose constraints for introductory programs, special attention must be paid to studies which guide statements regarding preclinical data, clinical pharmacology, adverse reactions, side effects, concomitancy, and dosage regimens. All efforts should be made to define the full scope of the product's potential with regard to labeling parameters.

Phase III trials should include pivotal studies requested by marketing for purposes of comparative claims within promotion. Since such studies entail more research than that required for approval, there may be reticence on the part of R&D to initiate and complete substantial studies of this type. It should be noted that this part of the program may dictate success or failure of a product. Consequently, significant attention must be directed to such studies, necessitating a high level of discussions between the planners and clinicians.

VI. PRELAUNCH MARKET PLANNING

If the firm is progressing within the disciplines of strategic planning, the basic premises for a new product marketing plan have been laid down much earlier in the product's development. In essence, then, the marketing plan represents a tactical expression of the firm's overall strategy, where the new product fit within the firm's strategy has been defined years earlier.

In today's climate, all market planning is completed long before an NDA has been approved. Indeed, many new drugs are being marketed abroad while still awaiting approval in the United States. In these instances, it is sometimes possible to gain valuable insights into a new drug which may serve to expedite its adoption, once approved, in the United States.

Although the precise time when market planning commences varies from company to company, as a general rule it must commence no later than the end of Phase III.

The primary objectives of prelaunch market planning are as follows:

To carefully describe the ways in which the products fill a medical need

To precisely describe the revenue and earnings and other objectives a new drug will fulfill

To carefully define those strategies that will accomplish the foregoing expectations

To ascertain the level of marketing and other resources required to accomplish those strategies

To design a tactical plan for deployment of the firm's resources

While marketing personnel will usually spearhead the planning effort, the development of a successful new product plan requires the cooperation and involvement of other disciplines within the firm, including members of the medical department, production planners, sales management, financial planners, and distribution and trade relations staff. Hence, while the end product is usually titled "The Marketing Plan," in reality, it represents the interdisciplinary plan for the successful commercialization of a new drug.

A. Comprehensive Planning

Market planning styles vary from company to company. At some firms marketing plans are no more than promotional plans, whereas at other firms they are more comprehensive documents which live up to the objectives of market planning. In our judgment, anything less than comprehensive planning will not suffice. Therefore, any plan must, at least, address the following areas:

1. Market overview
    a. Medical trends
    b. Marketing research analysis
        (1) Prescription trends
        (2) Sales trends
        (3) Usage patterns
        (4) Attitudinal trends
        (5) Promotional trends
    c. Problems and opportunities
2. Strategic considerations and objectives
    a. Positioning
    b. Targeting
    c. Adoption
    d. Product mix

3. Tactical considerations
    a. Message objective
    b. Promotional concepts
    c. Promotion mix
        (1) Selling
        (2) Advertising
        (3) Direct mail
        (4) Professional education
        (5) Other media
    d. Distribution
    e. Measurement and control
4. Operational considerations
    a. Technologic requirements
    b. Capital requirements
    c. Manpower requirements
5. Financial considerations and objectives
    a. Sales forecasts, 3 years or more
        (1) Units
        (2) Dollars
    b. Impact on income
    c. Impact on balance sheet

It is not the mission of this chapter to get into the how-to's of marketing and business planning, but the above outline should be familiar to business executives at most companies. It serves to demonstrate several axioms (listed in the next section) which we believe are inherent to the practice of pharmaceutical marketing.

B. Planning Axioms

Here they are:

1. *The clinician's needs lead the marketing*—that is, unlike many other businesses, pharmaceuticals are very heavily dependent on the emergence of new knowledge. Furthermore, the product-buying decision is vested with the most sophisticated and well educated of all consumers—the physician. These two factors combine to the imperative that marketers must be conversant with the medical literature and current and future medical trends, if they are to successfully market pharmaceuticals.

2. *Database planning*—or the ability not only to review the marketing data for your product but to marry those data to the emerging trends in medicine. Instinctive marketing, in our judgment, will no longer suffice. If instinct is to be employed, it must be supported by quantified or qualified data since the risks are just too high to go on instinct alone.

3. *Integrative planning*—one can no longer consider marketing as one of many isolated disciplines which will be merged at the upper management echelons. Indeed, it is the planning process which should merge the disciplines. Through planning, discontinuities should be obviated earlier and better than they could be by top managers who are often detached and unfamiliar with the key issues.
4. *Planning precision and fiscal responsibility*—these are integral parts of the planning process. The cost of failure is simply too high to ignore. Therefore, the basic principles of precise planning and the financial effects resulting from those plans have impact at the product, division and sometimes corporate levels and should, therefore, be addressed clearly.

C. Prelaunch Marketing

Recently a new phenomenon is becoming apparent within pharmaceutical marketing, that is, prelaunch marketing. Perhaps the best way to describe this phenomenon is that it entails publicity emanating within the scientific and lay communities prior to new drug approval. Historically, publicity about new pharmaceuticals has been disseminated to the medical community via the worldwide medical and scientific press. Only recently have we noticed such announcements of new drugs in the mass media to the lay public.

While preapproval publicity represents a contentious issue which we neither condone nor deride, it is important to point out that public and professional relations, as they regard new products, is a factor in today's environment.

Among the various avenues for prelaunch marketing, the following are well accepted and considered "traditional" by companies, scientists, and regulators alike:

Scientific publications
Scientific exhibits
Scientific presentations at conventions

Within this context, investigators are informing the medical community about their results on scientific investigation of a new drug and such communications are offered in the spirit of adding to the body of medical knowledge. The manufacturer, however, might benefit (if the results and experiences are favorable) since such communication will frequently speed the new drug adoption process.

During the last few years, public interest in health care has made this usually dry scientific subject increasingly interesting and newsworthy. The lay press now attends major medical conventions. The news magazines' science and health departments are looking for good copy, as are the science and health care editors of the network news organizations.

Moreover, health care is big business and Wall Street has come to recognize that new pharmaceuticals and other health care technologies are predictive of tomorrow's stock values. Hence, the pressure on the normally conservative drug houses to discuss their new products is overwhelming, as is the temptation for top management to disclose new developments to the financial community prior to their approval by the regulators.

Given the nature of our society, freedom of speech, pressure from the media, and the need for information, combined with the traditional American view that challenges (disease or otherwise) are made to be overcome, we do not believe that one can successfully stifle information related to visceral health care issues.

Indeed, there is a positive side to this picture, where industry and government have worked together successfully to bring disease under control. A clear case in point is hypertension, where industry and government have cooperated to bring preventive medicine to more people by encouraging increased screening and earlier treatment with a view to preventing morbidity and mortality.

Yet the practice of prepromotion publicity is being carefully watched by the regulators. Recently several pharmaceutical companies were criticized for this practice, and surely new regulation will evolve to control prelaunch promotion activities.

VII. ORGANIZATION OF NEW PRODUCT PLANNING GROUPS

It is obvious that new product planning groups function best when the right individuals are employed. But precisely who is the right individual? Where should he or she operate within the organization? What skills should that individual possess?

In our view the new product planner is a generalist who understands both the business side of the firm and the scientific side—someone who might be best described as a scientific entrepreneur.

The background required for this position includes: a degree in biological science or pharmacy; a degree in business; and previous experience in marketing or other business disciplines.

Clearly, the new product manager should be proficient in planning and have a keen analytical sense and understanding of marketing research. He or she should have a working knowledge of finance. Underscoring these attributes, the new product planner must understand the research process from chemistry through clinical development and must be able to motivate, balance opportunity and risk, and be assertive in his/her convictions. In short the new product manager must be able to blend these skills to deal effectively with business and research management at all levels of the corporation.

Having the good fortune to have found such a person, we can now deal with the question of where he or she might best be positioned in the corporation. The options here are limited to R&D, within the corporate group, or within the marketing organization. As the decision depends on the size, organization structure, line of communication, and needs of the particular firm, it is difficult here to precisely define where new product planning might best be placed. However, given the longer-range objectives of new product planning, it is probably best to put some distance between new product planning and the in-line product marketing groups. By causing a distinction to exist between these groups, new product planners might better maintain the broad-balanced perspective required to deal with the strategic aims of the firm vs. the annual marketing objectives. To mix in-line marketing with new product planning might well result in the least favorable mix of objectives—where the long-range goals of new product planning are subsumed to the short-term, gap-filling needs of marketing [4,5].

For these reasons, we tend to favor new product planning as part of the business organization, either as a very distinct and separate part of marketing, clearly separate from in-line product groups, or as an arm of corporate planning. The reason for this preference is that new product planners should be in close proximity to those who are charged with deciding the firm's allocation of resources over the long term.

Therefore as a general rule, new product planning, wherever it is located, should have a short line of communication to the corporate president, chief executive officer, or executive committee. Given this reporting relationship, new product planners should fulfill their promise to the firm as a whole and not serve the functional needs of any given operating group. Moreover, such a relationship will allow for faster communication to the decision point in the organization, avoiding layers of management and organizational heirarchy.

REFERENCES

1. Discussion in *Scrip 555:* 8-9 (Jan. 12, 1981).
2. G. A. Sleiner, *Top Management Planning.* Macmillan, New York, 1969.
3. P. Lorange, *Corporate Planning: An Executive Viewpoint.* Prentice-Hall, Englewood Cliffs, N.J., 1980.
4. D. S. Hopkins, *The Conference Board: Options in New Product Organization.* Prentice-Hall, Englewood Cliffs, N.J., 1974.
5. P. Kotler, *Marketing Management: Analysis, Planning and Control,* 3rd ed. Prentice-Hall, Englewood Cliffs, N.J., 1976.

22

IMPACT OF THE PHARMACEUTICAL INDUSTRY ON HEALTH CARE AND HEALTH CARE ECONOMICS

Don C. Stark
Mead Johnson Pharmaceutical Division
Evansville, Indiana

Albert I. Wertheimer
College of Pharmacy
University of Minnesota
Minneapolis, Minnesota

I. INTRODUCTION

More than $240 billion is spent on health care services in the United States each year, i.e., nearly $1,000 per man, woman, and child in the country. About 8% of this sum, or approximately $20 billion, is spent on pharmaceuticals and pharmacy services. The amount spent on drugs and drug-related services includes: (a) sales of prescription-legend (prescribed) drugs; (b) over-the-counter (OTC) drugs; and (c) professional and dispensing fees charged by pharmacies. The pharmaceutical industry, then, is large, profitable, and economically important [1].

II. TRENDS IN HEALTH EXPENDITURES

The massive numbers just cited are significant, but they are considerably more meaningful when compared with the alternative avenues for expenditures or when the value of the benefits are computed. For example, let us examine the progression of cost increases in the health care sector over the years and the comparable percent of gross national product (GNP) figures. These data are seen in Table 1.

Before saying anything further, it is appropriate to examine comparative health costs as a percentage of GNP in other countries. Simanis and Coleman [2] provide this computation for us, as seen in Table 2. This table would seem to somewhat temper the frequent cries that health care expenditures in the United States are excessive. The proportion of the GNP devoted to health care in the United States is quite comparable to the figures seen in other industrialized countries.

Table 1 Cost of Illness in Dollars and Percentage of GNP

	National health expenditures	
Year	Amount in millions	Percentage of GNP
1929	$ 3,649	3.5
1935	2,936	4.0
1940	3,987	4.0
1950	12,662	4.5
1955	17,745	4.4
1960	26,895	5.3
1965	40,468	5.9
1970	71,573	7.3
1975	132,100	8.6
1979	212,200	9.0

Sources: For 1929 through 1970: *Historical Statistics of the United States*, bicentennial ed.: *Colonial Times to 1970*, Pt 1. U.S. Dept. of Commerce, Bureau of the Census, Washington, D.C., 1975, p. 74. For 1975 and 1979: *Health, United States 1980*, DHHS Publ. No. (PHS) 81-123 2. U.S. Dept. of Health and Human Services, Public Health Service, Office of Health Research, Statistics, and Technology, Dec. 1980, p. 206.

The *Statistical Abstract of the United States* [3] presents national health expenditures during the past two decades by type of service, as seen in Tables 3 and 4. It can be seen from these two tables that there have been increases in all areas of health care. Some areas, however, have increased more than others. For example, the data presented in Table 3 show an increase of 837% in hospital care expenditures between 1960 and 1979. Nursing home care expenditures increased an astounding 3,460% in the same time period. Expenditures for drugs and sundries, on the other hand, demonstrated relatively slow growth, posting a 359% increase in the 1960-1979 time period.

Let us examine the trend in drug expenditures a little more closely. Figure 1 illustrates that the proportion of all health care expenditures accounted for by pharmaceuticals has been declining over the past 50 years. As already mentioned, expenditures on pharmaceuticals currently contribute only about 8% of the total dollars spent on all health care needs. This is an especially interesting trend when one considers how much more effective the current drugs are relative to their earlier counterparts. This point will be discussed in more detail later in this chapter.

Table 2 Comparison of National Health Care Expenditures as a Percentage of Respective Gross National Products, 1975

Country	Health care expenditures as a percentage of GNP
Federal Republic of Germany	9.7
Sweden	8.7
Netherlands	8.6
United States	8.4
France	8.1
Canada	7.1
Australia	7.0
Finland	6.8
United Kingdom	5.6

Source: Simanis and Coleman [2].

Table 3 National Health Expenditures, by Object: 1960-1979

Object of expenditure	Expenditure (billion dollars)					Percentage			
	1960	1965	1970	1975	1979	1960	1970	1975	1979
Total	26.9	42.0	74.9	132.1	212.2	100.0	100.0	100.0	100.0
Spent for:									
Health services and supplies	25.2	38.6	69.6	123.8	202.3	93.7	92.9	93.7	95.3
Personal health care expenses	23.7	36.0	65.4	116.5	188.6	88.1	87.3	88.2	88.9
Hospital care	9.1	13.9	27.8	52.1	85.3	33.8	37.1	39.5	40.2
Physicians' services	5.7	8.5	14.3	24.9	40.6	21.2	19.1	18.9	19.1
Dentists' services	2.0	2.8	4.7	8.2	13.6	7.4	6.3	6.2	6.4
Other professional services[a]	0.9	1.0	1.6	2.6	4.7	3.3	2.1	2.0	2.2
Drugs and sundries	3.7	5.2	8.2	11.8	17.0	13.8	11.0	8.9	8.0
Eyeglasses and appliances[b]	0.8	1.2	1.9	3.0	4.4	3.0	2.6	2.3	2.1
Nursing home care	0.5	2.1	4.7	10.1	17.8	1.9	6.3	7.6	8.4
Other health services	1.1	1.3	2.1	3.7	5.2	4.1	2.7	2.8	2.4
Expense for prepayment and administration[c]	1.1	1.7	2.8	4.1	7.7	4.1	3.7	3.1	3.6
Government public health activities	0.4	0.8	1.4	3.2	6.0	1.5	1.9	2.4	2.9
Research	0.6	1.4	1.9	3.2	4.6	2.2	2.5	2.5	2.2
Construction	1.0	2.0	3.4	5.1	5.3	3.7	4.6	3.8	2.5

[a]Includes services of registered and practical nurses in private duty, visiting nurses, podiatrists, physical therapists, clinical psychologists, chiropractors, naturopaths, and Christian Science practitioners.
[b]Includes fees of optometrists and expenditures for hearing aids, orthopedic appliances, artificial limbs, crutches, wheelchairs, etc.
[c]Net cost of insurance and administrative expenses of federally financed health programs.
Source: *Statistical Abstract of the United States, 1980,* 101st ed. U.S. Dept. of Commerce, Bureau of the Census, Washington, D.C., 1980, p. 105.

Table 4 Health Services and Supplies—per Capita National Expenditures, by Object: 1960-1979 (in dollars, except percent)

Object of expenditure	1960	1965	1970	1975	1979
Total, national	137.00	194.91	333.89	569.42	899.03
Average annual percent change[a]	6.4	7.3	11.4	11.3	12.0
Hospital care	49.46	70.20	133.39	239.78	379.23
Physicians' services	30.92	42.84	68.81	114.66	180.41
Dentists' services	10.75	14.20	22.79	37.88	60.46
Other professional services[b]	4.69	5.22	7.65	12.04	20.83
Drugs and drug sundries	19.89	26.35	39.39	54.33	75.43
Eyeglasses and appliances[b]	4.22	6.12	9.24	13.72	19.34
Nursing home care	2.86	10.48	22.54	46.47	79.13
Other health services	6.02	6.60	9.87	16.98	23.02
Expenses for prepayment and administration[b]	5.93	8.78	13.39	19.05	34.31
Government public health activities	2.25	4.11	6.81	14.52	26.87

[a]Change from prior year shown; except for 1960, which is compared to 1955.
[b]See footnotes for corresponding objects in Table 3.
Source: Statistical Abstract of the United States, 1980, 101st ed. U.S. Dept. of Commerce, Bureau of the Census, Washington, D.C., 1980, p. 105.

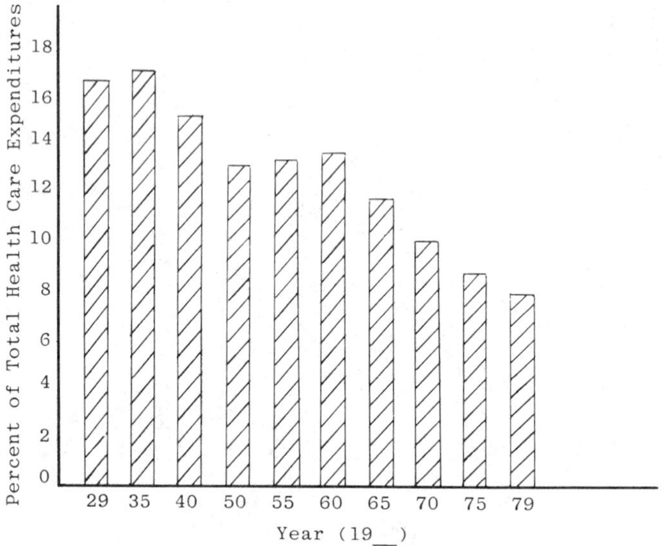

Figure 1 Proportion of total health care expenditures spent on drugs and drug sundries, United States, selected years, 1929-1979. (*Source:* For 1929-1979, adapted from *Historical Statistics of the United States,* bicentennial ed.: *Colonial Times to 1970,* Pt. 1. U.S. Dept. of Commerce, Bureau of the Census, Washington, D.C., 1975. For 1975-1979, adapted from *Statistical Abstract of the United States, 1980,* 101st ed. U.S. Dept. of Commerce, Bureau of the Census, Washington, D.C., 1980.)

III. MORBIDITY AND MORTALITY TRENDS

We will now leave these macro-figures and begin an examination of the component causes of morbidity and mortality during these same years. Perhaps Table 5 portrays these data better than any prose. In Table 5, we see the number of deaths per 100,000 population for selected years between 1900 and 1978. It is interesting to note the trends which become evident when the table is examined from this chronological perspective. The most obvious trend in these data is the decreasing importance of infections and other acute illnesses.

In Table 6, one may see a greater degree of specificity (for selected diseases) in examining the changes in mortality over time. It is noteworthy that the death rates due to several major chronic diseases are also declining (such as the various diseases of the cardiovascular system).

Table 5 Death Rate for Selected Causes: 1900-1978

Year	Tuberculosis, all forms	Syphilis and its sequelae[a]	Typhoid and paratyphoid fever	Scarlet fever and streptococcal sore throat	Hepatitis	Diphtheria	Whooping cough	Measles	Malignant neoplasms[b]	Diabetes mellitus	Major cardiovascular-renal disease	Influenza and pneumonia[c]	Gastritis, duodenitis, enteritis, and colitis[d]
1900	194.4	12.0	31.3	9.6	—	40.3	12.2	13.3	64.0	11.0	345.2	202.2	142.7
1905	179.9	13.8	22.4	6.8	—	23.5	8.9	7.4	73.4	14.1	384.0	169.3	118.4
1910	153.8	13.5	22.5	11.4	—	21.1	11.6	12.4	76.2	15.3	371.9	155.9	115.4
1915	140.1	17.7	11.8	3.6	—	15.2	8.2	5.2	80.7	17.6	383.5	145.9	67.5
1920	113.1	16.5	7.6	4.6	—	15.3	12.5	8.8	83.4	16.1	364.9	207.3	53.7
1925	84.8	17.3	7.8	2.7	—	7.8	6.7	2.3	92.0	16.8	391.5	121.7	38.6
1930	71.1	16.7	4.8	1.9	—	4.9	4.8	3.2	97.4	19.1	414.4	102.5	26.0
1935	55.1	15.4	2.8	2.1	—	3.1	3.7	3.1	108.2	22.3	431.2	104.2	14.1
1940	45.9	14.4	1.1	0.5	—	1.1	2.2	0.5	120.3	26.6	485.7	70.3	10.3
1945	39.9	10.6	0.4	0.2	—	1.2	1.3	0.2	134.0	26.5	508.2	51.6	8.7
1950	22.5	5.0	0.1	0.2	0.4	0.3	0.7	0.3	139.8	16.2	510.8	31.3	5.1
1955	9.1	2.3	(Z)[f]	0.1	0.5	0.1	0.3	0.2	146.5	15.5	506.0	27.1	4.7
1960[e]	6.1	1.6	(Z)	0.1	0.5	(Z)	0.1	0.2	149.2	16.7	521.8	37.3	4.4
1965	4.1	1.3	(Z)	(Z)	0.4	(Z)	(Z)	0.1	153.5	17.1	516.4	31.9	4.1
1970	2.6	0.2	(Z)	(Z)	0.5	(Z)	(Z)	(Z)	162.3	18.3	496.0	30.9	6.6
1975	1.6	0.1	(Z)	(Z)	0.3	(Z)	(Z)	(Z)	171.7	16.5	455.8	26.1	(ND)[f]
1978	1.3	0.1	(Z)	(Z)	0.2	(Z)	(Z)	(Z)	181.9	15.5	442.7	26.7	(ND)

[a] 1900-1920, excludes aneurysm of the aorta.
[b] Includes neoplasms of lymphatic and hematopoietic tissues.
[c] All years, excludes pneumonia of newborn; 1900-1920, excludes capillary bronchitis.
[d] All years, excludes diarrhea of newborn; 1900-1920, includes ulcer of duodenum.
[e] Denotes first year for which figures include Alaska and Hawaii.
[f] Z means less than 0.05. ND means that no comparable data are available.

Source: For 1900 through 1970: *Historical Statistics of the United States*, Bicentennial ed.: *Colonial Times to 1970*, Pt I. U.S. Dept. of Commerce, Bureau of the Census, Washington, D.C., 1975, p. 58. For 1975 through 1978: *Statistical Abstract of the United States, 1980,* 101st ed. U.S. Dept. of Commerce, Bureau of the Census, Washington, D.C., 1980, p. 78.

Table 6 Death Rates, 1960-1978, from Selected Causes

Cause of death	Death rates per 100,000 population			
	1960	1970	1975	1978
All causes	954.7	945.3	888.5	883.4
Major cardiovascular disease	515.1	496.0	455.8	442.7
Diseases of the heart	369.0	362.0	336.2	334.3
Active rheumatic fever and chronic rheumatic heart disease	10.3	7.3	6.1	6.1
Hypertensive heart disease[a]	37.0	7.4	5.2	4.7
Hypertension	7.1	4.1	3.0	2.5
Cerebrovascular diseases	108.0	101.9	91.1	80.5
Arteriosclerosis	20.0	15.6	13.6	13.3
Malignancies	149.2	162.8	171.7	181.9
Accidents	52.3	56.4	48.4	48.4
Respiratory diseases	ND[b]	15.2	12.0	10.0
Chronic and unqualified bronchitis	ND[b]	2.9	2.2	2.0
Emphysema	ND[b]	11.2	8.8	7.2
Asthma	3.0	1.1	0.9	0.9
Peptic ulcer	6.3	4.2	3.2	2.5
Infections of the kidney	4.3	4.0	2.1	1.4

[a]With or without renal disease.
[b]Comparable data not available.
Source: Statistical Abstract of the United States, 1980, 101st ed. U.S. Dept. of Commerce, Bureau of the Census, Washington, D.C., 1980, p. 78.

IV. THE ROLE OF THE PHARMACEUTICAL INDUSTRY

Moreover, when the role of the pharmaceutical industry is attached, even greater meaning in these figures becomes immediately apparent. As seen in Table 7 (a hybrid composed of data from several sources), a specific product of the pharmaceutical industry can be directly associated with the decreases in mortality shown. We will readily admit that it is incorrect to attribute the decline of these diseases *solely* to

Table 7 Changes in Mortality and Associated Pharmaceutical Industry Development

Disease	1940	1950	1960	1970	1940-1970 % decline	Type of pharmaceutical
Acute poliomyelitis	0.8	1.3	0.1	0.0	100.0	Vaccine
Diphtheria	1.1	0.3	0.0	0.0	100.0	Vaccine
Whooping cough	2.2	0.7	0.1	0.0	100.0	Vaccine
Measles	0.5	0.3	0.2	0.0	100.0	Vaccine
Typhoid and paratyphoid fever	1.1	0.1	0.0	0.0	100.0	Vaccine
Scarlet fever and streptococcal sore throat	1.8	0.2	0.1	0.0	100.0	Antibiotics
Syphilis and its sequelae	10.8	5.0	1.6	0.2	98.2	Antibiotics
Dysentery, all forms	2.0	0.6	0.2	0.1[a]	95.0	Antibiotics
Gastritis, duodenitis enteritis, colitis	10.3	5.1	4.4	0.6	94.2	Antibiotics
Tuberculosis, all forms	44.0	22.5	6.1	2.6	94.1	Antibiotics
Maternal deaths	6.1	2.0	0.9	0.4	93.4	Antibiotics
Hypertensive heart disease	—	56.6	37.0	4.1	92.7	Antihypertensives
Appendicitis	9.1	2.0	1.0	0.7	92.3	Antibiotics
Chronic and unspecified nephritis and renal sclerosis	28.0	16.4	6.7	3.7	86.8	Antihypertensives
Rheumatic fever and chronic rheumatic heart disease	22.4	14.8	10.3	7.3	67.4	Antibiotics
Meningitis	2.1	1.2	1.3	0.8	61.9	Antibiotics
Influenza and pneumonia (except newborn)	70.3	31.3	37.3	30.9	56.0	Antibiotics
Diabetes mellitus	26.6	16.2	16.7	18.3	32.2	Insulin

[a]In 1967.

Sources: The Economic and Social Contributions of the U.S. Multinational Pharmaceutical Industry. PMA, May 1976, p. 12. Cotton M. Lindsay (Ed.), *The Pharmaceutical Industry.* 1978, p. 48. *Historical Statistics of the United States,* bicentennial ed.: *Colonial Times to 1970,* Pt. 1. U.S. Dept. of Commerce, Bureau of the Census, Washington, D.C., 1975, p. 58. *Statistical Abstract of the United States, 1980,* 101st ed. U.S. Dept. of Commerce, Bureau of the Census, Washington, D.C., 1980, p. 78. *Vital Statistics of the United States.* U.S. Dept. of Health, Education and Welfare, Public Health Service, National Center for Health Statistics: 1950—I; 1960—II; 1975—II; 1977—II.

an innovation from the pharmaceutical industry. Other factors have, in most cases, also contributed to the lower death rates. Improvement in the quality of medical care in general may be one such factor. Rabin and Bush [4], for example, note that the decrease in the overall mortality rate in the nineteenth and early twentieth centuries were primarily due to improved environmental factors (i.e., better sanitation) and not due to medicines.

These caveats should not, however, detract from the role pharmaceuticals have played in the declining mortality rates illustrated in Table 7. Pharmaceuticals have been identified as contributing factors in the decline. In a recent paper, von Grebmer [5] suggests that the highest productivity increases in the health care industry have been due to advances in drug therapy.

A listing of the drugs introduced onto the U.S. market since 1975 is seen in Table 8, along with the date of introduction, generic name, name of developer, and drug category.

Due to its length, it cannot be reproduced in this chapter, but the serious student of this issue is referred to *Pharmacy Times 42*(3): 40 (Mar. 1976), in which the De Haen Organization list of drugs introduced in the United States from 1940 through 1975 is presented with date of introduction, therapeutic category, and other valuable data.

The process of examining the economics of health care delivery vis-ă-vis pharmaceuticals is made much clearer by a report from *Tile and Till* [6]:

> *The Revolution in Medical Care.* The research capacity of the pharmaceutical industry constitutes an invaluable national asset. In the last thirty years, this capacity to innovate has produced 90 percent of all new drugs. The results have helped revolutionize the treatment and prevention of disease.
>
> Yesterday no effective treatment was available for many diseases. Today these diseases are treated routinely and usually effectively:
>
> Addison's disease
> Bacterial endocarditis
> Choriocarcinoma
> Hodgkin's disease
> Hypertension (most forms)
> Meningitis (most forms)
> Mastoiditis
> Osteomyelitis (most forms)
> Parkinson's disease
> Peritonitis (most forms)
> Pneumonia
> Scarlet fever
> Schizophrenia (most forms)
> Septicemia (most forms)
> Typhoid fever
> Tuberculosis
> Venereal disease
> Whooping cough
>
> Measles, mumps, and poliomyelitis were common among children twenty years ago. Today they are, for all practical purposes, totally preventable.
>
> In 1950, the average stay in a mental hospital was 28 months; today it is eight months....

Table 8 Compilation of New Drugs, 1975-1979

Marketed	Trademark	Generic name	Manufacturer	Drug category
1975	Cylert	Pemoline	Abbott	CNS stimulant
1977	H-BIG	Hepatitis B immune globulin	Abbott	Vaccine
1977	Thypinone inj	Protirelin	Abbott	TSH analog
1978	Abbokinase	Urokinase	Abbott	Thrombolytic
1978	Depakene	Valproic acid	Abbott	Anticonvulsant
1979	Natacyn	Natamycin	Alcon	Antibiotic
1978	Bretylol	Bretylium tosylate	American Critical Care	Bronchodilator
1975	Calcimar	Calcitonin-salmon	Armour	Antihypercalcemic
1976	Duranest	Etidocaine HCl	Astra	Local anesthetic
1976	Peptavlon	Pentagastrin	Ayerst	Diagnostic aid
1976	Ticar	Ticarcillin disodium	Beecham	Antibiotic
1976	Amikin	Amikacin sulfate	Bristol	Antibiotic
1976	CEENU	Lomustine	Bristol	Antineoplastic
1977	BICNU	Carmustine	Bristol	Antineoplastic
1978	Platinol	Cisplatin	Bristol	Antineoplastic
1978	Stadol	Butorphanol tartrate	Bristol	Analgesic
1976	Asellacrin	Somatropin	Calbio	Growth hormone
1975	Lotrimin	Clotrimazole	Delbay	Antifungal
1976	Nalfon	Fenoprofen calcium	Dista	Anti-inflammatory
1975	DTIC-dome	Dacarbazine	Dome	Antineoplastic
1978	Tavist	Clemastine fumarate	Dorsey Labs	Antihistamine
1977	Lorelco	Probucol	Dow	Antihyperlipidemic
1975	Nebcin	Tobramycin sulfate	Eli Lilly	Antibiotic
1978	Dobutrex	Dobutamine HCl	Eli Lilly	Inotropic agent
1978	Mandol	Cefamandole nafate	Eli Lilly	Antibiotic
1979	Ceclor	Cefaclor	Eli Lilly	Antibiotic

1977	Azene	Clorazepate Monopotassium	Endo	Ataratic
1979	Nubain	Nalbuphine	Endo	Analgesic
1978	DDAVP	Desmopressin	Ferring	Antidiuretic
1977	Lioresal	Baclofen	Geigy	Muscle relaxant
1978	Lopressor	Metoprolol tartrate	Geigy	Beta blocker
1977	Topicort	Desoximetasone	Hoechst	Topical steroid
1977	Streptase	Streptokinase	Hoechst	Thrombolytic
1979	Cerubidine	Daunorubicin	Ives	Antineoplastic
1979	Surmontil	Trimipramine maleate	Ives	Antidepressant
1975	Loxitane	Loxapine succinate	Lederle	Antipsychotic
1976	Loxitane C	Loxapine HCl	Lederle	Antipsychotic
1979	Cyclocort	Amcinonide	Lederle	Topical steroid
1975	Ditropan	Oxybutynin chloride	Marion	Urinary antispasmodic
1976	Tolectin	Tolmetin sodium	McNeil	Anti-inflammatory
1975	Sinemet	Carbidopa and Levodopa	MSD	Antiparkinson
1977	Flexeril	Cyclobenazprine HCl	MSD	Muscle relaxant
1978	Clinoril	Sulindac	MSD	Anti-inflammatory
1978	Elspar	Asparaginase	MSD	Antineoplastic
1978	Mefoxin	Cefoxitin sodium	MSD	Antibiotic
1978	Timoptic	Timolol maleate	MSD	Glaucoma
1975	Renoquid	Sulfacytine	Parke-Davis	Antibiotic
1977	Vira-A	Vidarabine	Parke-Davis	Antiviral
1979	Estrovis	Quinestrol	Parke-Davis	Estrogen
1976	Minipress	Prazosin HCl	Pfizer	Antihypertensive
1976	Debrisan	Dextranomer	Pharmacia Lab.	Wound cleanser
1977	Duphalac	Lactulose	Philips Roxane	Laxative
1978	Didronel	Etidronate disodium	Procter & Gamble	Paget's disease
1978	Rimso-50	Dimethyl sulfoxide	Research Ind.	Anti-inflammatory
1979	Reglan	Metoclopramide	A. H. Robins	G.I. stimulant
1975	Clonopin	Clonazepam	Roche	Antihypertensive
1975	Solatene	β-Carotene	Roche	Antiphotosensitivity
1978	Rocaltrol	Calcitriol	Roche	Vitamin

Table 8 (continued)

Marketed	Trademark	Generic name	Manufacturer	Drug category
1978	Parlodel	Bromocriptine mesylate	Sandoz	Dopamine receptor agonist
1975	Diprosone cream	Betamethasone dipropionate	Schering	Topical steroid
1976	Vanceril	Beclomethasone dipropionate	Schering	Antiasthmatic
1977	Optimine	Azatadine maleate	Schering	Antihistamine
1977	Norpace	Disopyramide phosphate	Searle	Antiarrhythmic
1977	Tagamet	Cimetidine	SKF	Antiulcer
1979	Selacryn	Ticrynafen	SKF	Diuretic
1974	Halog cream	Halcinonide	Squibb	Topical steroid
1976	Kinevac	Sincalide	Squibb	Diagnostic aid
1979	Corgard	Nadolol	Squibb	Beta blocker
1976	Hibiclens	Chlorhexidine gluconate	Stuart	Antiseptic
1978	Nolvadex	Taxmoxifen citrate	Stuart	Antineoplastic
1976	Naprosyn	Naproxen	Syntex	Anti-inflammatory
1977	Colestid	Colestipol HCl	Upjohn	Antihyperlipidemic
1978	Florone	Diflorasone diacetate	Upjohn	Topical steroid
1979	Loniten	Minoxidil	Upjohn	Antihypertensive
1979	Prostin/15M	Carboprost tromethamine	Upjohn	Abortifacient
1975	Utibid	Oxolinic acid	Warner-Chilcott	Antibiotic
1977	Verstran	Prazepam	Warner-Chilcott	Ataratic
1978	Westcort cream	Hydrocortisone valerate	Westwood	Topical steroid
1975	Isopaque	Metrizoates	Winthrop	Diagnostic aid
1976	Danocrine	Danazol	Winthrop	Androgen
1978	Amipaque	Metrizamide	Winthrop	Diagnostic aid
1977	Ativan	Lorazepam	Wyeth	Ataratic
1979	Cyclapen	Cyclacillin	Wyeth	Antibiotic
1978	Duricef	Cefadroxil Monohydrate	Mead Johnson	Antibiotic

Source: Paul De Haen Information Systems, *De Haen New Drug Analysis, U.S.A.*, Vol. 16: *1975-1979—A Five-Year Survey*. Micromedex, Englewood, Colo., 1980.

New drugs have played the principal part in this "revolution," and the genesis of those new drugs was scientific research.

Research and Innovation. Lilly research and development involve 2,400 people, 700 with advanced scientific degrees ranging from cytogeneticists to biophysicists and including 66 physicians.

The yearly cost is more than $80 million. In addition, $68 million has been invested in laboratories and scientific equipment.

It takes more than people and money, however. There is a reservoir of insight into the complex life sciences and how to apply them to human disease problems. These competencies are not easily achieved by the simple collection of skills. They are the result of organizational learning and group expertise. This is related to internal atmosphere and the process of communication and integration as well as to the process of target definition, appraisal of results, and the coordination of resource use. A successful research organization depends on the skillful management of innovation, a skill that accumulates gradually over years and cannot be arbitrarily created in or transferred to another research group.

"Breakthrough" or "Me-Too"? Without equivocation, our policy is to direct major efforts toward "breakthrough" drugs that significantly improve the treatment of serious illnesses. In so doing, four basic conditions prevail:

1. The chances of success are least.
2. The effort and expense required are the greatest.
3. The potential benefit to society is the greatest.
4. The potential financial return to the company is sufficient to support such a policy.

These four conditions are compatible and productive. Society's greatest health needs are in areas such as heart disease, cancer, arthritis, and viral and infectious diseases. Lilly's major research efforts have been, are, and will be widely used and provide dramatic relief from major illnesses. That kind of usage would also provide potential economic return to Lilly. These simple but fundamental conditions maximize the continued productivity of our technologic and research capacity.

The Drug Discovery Process. Scientists in the pharmaceutical industry study and reject 8,000 compounds for every one that becomes a new medicine.

Breakthrough discoveries, of course, are even more rare and unpredictable. They are usually produced by chemically modifying compounds of known activity. Breakthroughs also come from new insights into the chemistry of living systems. In practice, the two approaches—chemical modification and more basic understanding of life processes—work together and constantly influence one another....

Much of the current criticism of molecular modification results from a misunderstanding of the scientific method. All science builds on existing information. It proceeds—usually one step at a time—from the known to the unknown. This basic process is the same whether the information concerns the characteristics of a subatomic particle or the biological properties of a known molecule. In this sense, or course, all science is a "me-too" activity.

In the case of drug research, the unknown is incredibly complex and unpredictable. Biological systems are enormously complicated and subject to individual variation. The possible molecular configurations for a drug are infinite. Moreover, drug action itself is almost always a mixed phenomenon involving both desirable and undesirable effects. Finally, even small changes in chemical structure can make a profound difference in the useful activity, toxic effects, or dependence properties of a drug. Yet, at present, there is no way to predict activity through mere knowledge of a compound's structure.

The search for new drugs, therefore, has a unique requirement for step-by-step procedures based, wherever possible, on proven information. Systematic modification of drugs of known activity is the one approach that best meets this requirement. It is a procedure that follows the dictates both of good science and of common sense.

The success of the method, however, is the real reason for its continuation. Chemical modification of the antihistamines, for example, led to the discovery of chlorpromazine, the first tranquilizer. Modification of chlorpromazine eventually produced other tranquilizers, the first antidepressants, and various drugs for the treatment of Parkinson's disease, motion sickness, and worm and skin infections. Similar efforts to improve sulfanilamide unexpectedly brought forth potent new diuretics, improved antibacterials, and the first oral drugs for the treatment of diabetes....

The Economics of Research. This Lilly discovery and evaluation process derives its strengths from many human factors...the scientific excellence of thousands of people and their desire to achieve... a reservoir of knowledge and skills built up over the years...deep concern for an individual's welfare during the testing process. Matters of the human mind and spirit are critical to the eventual success of Lilly research and innovation.

Matters of simple economics, however, are also critical. Massive efforts require massive funds—$340 million by Lilly in the last five years. Behind every commitment for such expenditures is the expectation of economic return from worthwhile accomplishment. The flow of funds from drugs discovered yesterday is today paying for the efforts to find tomorrow's drugs. The potential for appropriate funds and profits tomorrow makes today's investment worthwhile.

Table 9 Thirty Significant Pharmaceutical Advances, 1934-1963

1. Adrenocorticosteroids: primarily for "collagen diseases"
2. Basic penicillins, plus salts and ester: for antibacterial effects
3. Vaccines: for treatment of polio, measles, influenza, etc.
4. Synthetic anticoagulants: to prevent clots in veins and arteries
5. Streptomycin: important antibiotic
6. Isoniazid: for tuberculosis
7. Chlorpromazine and other tranquilizers
8. Hydantoins: for epilepsy and other convulsive states
9. Diphenhydramine and other antihistamines
10. Thiazides: diuretics
11. Sulfonamides: for respiratory and urinary tract infections
12. Tetracycline derivatives, salts, esters: important antibiotics
13. Rauwolfia and veratrum alkaloids: for hypertension and other problems
14. Meperidine: pain reliever
15. Chloramphenicol: potent broad spectrum antibiotic
16. Oral antidiabetic agents
17. Chloraquin compounds: antimalarials and antiamebics
18. Antithyroid agents
19. Immunoglobulins: temporary immunizing agent
20. Aminosalicylic acid, salts and esters: for tuberculosis
21. Isoproterenol: for bronchial asthma
22. Methantheline and other anticholinergics: for gastrointestinal conditions
23. Ganglionic blocking agents: for treating hypertension
24. Phenylephrine: to shrink nasal mucous membranes
25. Halogenated hydrocarbon anesthetics: for inhalation anesthesia
26. Surgical skeletal muscle relaxants
27. Organomercurial diuretics
28. Oral contraceptive agents
29. Trihexyphenidyl and other antiparkinsonism agents
30. Hypnotic barbiturates

Source: Report of the Commission on the Cost of Medical Care, Vol. III. Copyright 1964, American Medical Association, Chicago. As reported in Smith and Knapp [7].

> The ultimate question about research is whether it is worth the cost....
>
> The results of research have created today's health standards. The economic incentive to continue research will help determine the speed with which our remaining disease problems are solved."

Smith and Knapp [7] call our attention to a compilation prepared in 1964 by the Commission on the Cost of Medical Care of the American Medical Association, which is quite relevant to the topic at hand. The 30 significant pharmaceutical advances, 1934-1963, are seen in Table 9.

V. THE IMPACT OF PHARMACEUTICALS ON HEALTH CARE EXPENDITURES

The question which has not yet been addressed is: What has the net impact of pharmaceuticals been on health care expenditures? The answer, simply stated, is this: "As medical costs continue to skyrocket, the benefits of prescription medication become increasingly apparent as patients recognize the potential for savings by way of reduced or eliminated hospitalizations and/or physician visits" [8]. As was illustrated in Table 7, dramatic decreases in acute infectious diseases have occurred over the past several decades, at least partly the result of new drug therapies. Increases in the use of medicines in recent years are attributable primarily to the treatment of chronic diseases (hypertension, mental disorders, etc.). Even these chronic diseases have benefited from improvements in drug therapy [9]. A portion of the decreasing mortality rate for diabetes mellitus (shown in Table 5) and hypertension (shown in Table 6) is the result of improved drug treatments.

Two examples are used to illustrate these points in Figs. 2 and 3. Figure 2 shows the trend in health care expenditures (both in dollars and as a proportion of all health care expenditures) for treatment of patients in tuberculosis hospitals. Contributing to the dramatic decrease in tuberculosis hospital expenditures since the early 1950s have been improvements in sanitation and the introduction of three pharmaceutical products: streptomycin in 1946, isoniazid in 1951, and para-aminosalicylic acid in 1948 [9]. In Fig. 3, the trend in expenditures for psychiatric hospitals is illustrated. Although dollar expenditures have continuously risen, psychiatric hospitals have accounted for a decreasing proportion of total health care expenditures since the late 1950s. At least part of this decline must be attributed to the deinstitutionalization of psychiatric patients which is now possible because many mental disorders can be managed in the community setting by the use of pharmaceuticals [10]. The first major advances in this area were the introduction of reserpine and chlorpromazine in the early 1950s.

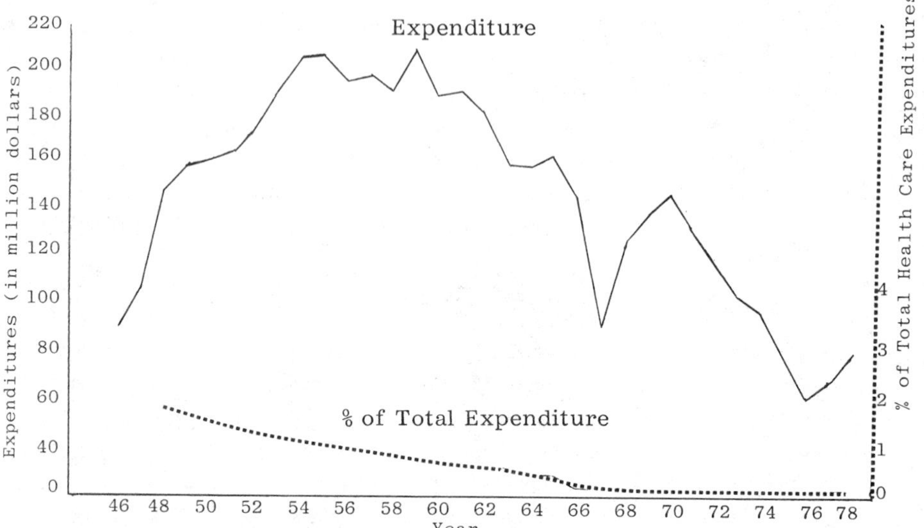

Figure 2 Health care expenditures for tuberculosis hospitals, 1946-1978 (dollars, and as a percentage of total health care expenditures, calculated as follows: the average daily census × hospital expenses per patient day × 365 days per year). (*Source:* From *Hospital Statistics of the United States*, bicentennial ed.: *Colonial Times to 1970*, Pt. 1. U.S. Dept. of Commerce, Bureau of the Census, 1975, Washington, D.C., p. 80; and *Statistical Abstract of the United States, 1980*, 101st ed. U.S. Dept. of Commerce, Bureau of the Census, Washington, D.C., 1980.)

Techniques analyzing benefit vs. cost and cost vs. effectiveness provide a useful framework for assessing various important economic issues involved in the use of pharmaceutical products. When carefully performed, they can be an important input in the evaluation of prevention and/or treatment strategies.

Sloan et al. [11], provide us with a means of measuring the impact of diseases. This construct is seen in Table 10. This type of analytical foundation provides us with a basis for performing cost-benefit studies.

A selected review of empirical benefit-cost and cost-effectiveness studies [12] found that behavior changes (such as reduction in smoking or consumption of alcoholic beverages; or the increased use of auto safety belts) could produce economic savings. Public health measures, such as fluoridation of drinking water, vaccines for rubella and polio-

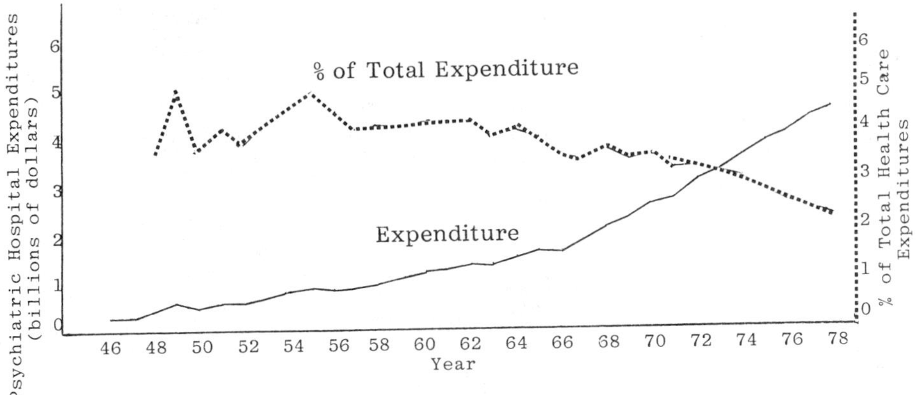

Figure 3 Health care expenditures for psychiatric hospitals, 1946-1978 (dollars, and as a percentage of total health care expenditures, calculated as follows: the average daily census × hospital expenses per patient day × 365 days per year). (*Source:* From *Historical Statistics of the United States*, bicentennial ed.: *Colonial Times to 1970*, Pt. 1. U.S. Dept. of Commerce, Bureau of the Census, Washington, D.C., 1975, p. 80; and *Statistical Abstract of the United States, 1980*, 101st ed. U.S. Dept. of Commerce, Bureau of the Census, Washington, D.C., 1980.)

myelitis, and salmonellosis surveillance were found to yield substantial economic benefits.

The benefit-cost relationships found in screening programs for phenylketonuria, congenital hypothyroidism, and spina bifida cystica also were favorable. Multiphasic screening programs for certain age groups also may result in benefits that exceed costs. In addition, there is preliminary evidence that the benefit-cost ratio of screening and treatment of hypertension and cancer of the colon and rectum could be favorable, at least for high-risk populations.

Aside from the inherent value of keeping people healthy, it has been found that there are many expenditures on prevention strategies that are economically sound investments. Many other prevention activities (for example, physical exercise or dietary changes) also may be economically sound, but the empirical evidence is as yet unavailable.

Recent studies have illustrated that significant cost savings will be possible by treating certain diseases with new drugs. A benefit-cost analysis of Tagamet (cimetidine, SKF) conducted by the Congressional Office of Technology Assessment (OTA) estimated that with use of this drug for duodenal ulcer 1977 national health care expendi-

Table 10 Potential Impacts of Infectious and Allergic Diseases

Category	Specific impacts
Demographic	Incidence and prevalence Mortality Fertility-fetal (wastage or damage) Migration
Health	Malnutrition Susceptibility to other illnesses Chronic disabilities Utilization of health care products Preventive health behavior
Child development	Intellectual achievement in children Physical and emotional development School attendance Ability to engage in athletics
Labor market	Type of work performed (including occupational choice) Work days lost Productivity loss on job Labor force participation and work hours Earnings
Household	Lost leisure time Lost time in performing household activities Household decisionmaking Stigma Personal appearance Pain and anxiety Family stability

Source: Sloan et al. [11].

tures could have been decreased by $645 million. This represents a gross savings of $679 million derived from increased productivity and decreased hospitalization and surgeon costs minus $34 million in drug costs [13]. The Center for Disease Control has estimated that Merck Sharp and Dohme's recently introduced hepatitis B vaccine will result in savings of $6 million per week ($327 million annually) [14].

Further drug treatment improvements may produce even greater savings in health care expenditures. Specific diseases for which improved pharmaceuticals hold a great potential to generate these benefits are hypertension, diabetes mellitus, depression, cancer, and several others. Kucherov and Trafford [15] suggest that the new calcium antagonist drugs (slow channel blockers), for example, may decrease the number of expensive coronary bypass surgeries.

Thus, it is very likely that the development of new drugs will continue to provide additional savings in health care costs. Additionally, the clinical research process in the pharmaceutical industry will continue to improve the currently available drug products. Even though new drugs are frequently more expensive than the old, Silverman and Lee [16] point out that new drugs can still result in cost savings. This is accomplished if the new drugs are "safe, more effective, and more rapid in their actions, reducing not only the risk of death from a disease but also the time and severity of pain and disability." [16].

A good example of just such a new pharmaceutical development is discussed in the aforementioned OTA report on use of Tagamet as a medical intervention. In the report, the OTA compares the cost of Tagamet for the treatment of duodenal ulcers to the use of antacids (the standard drug therapy prior to Tagamet). The study concludes that the treatment of duodenal ulcer with Tagamet will cost the typical patient no more, and possibly less, than treating the condition with antacids [13].

VI. CONCLUSION

In conclusion, it is obvious that drugs play an important role in health care—and a very cost-efficient one. As Fuchs says: "Drugs are the key to modern medicine. Surgery, radiotherapy, and diagnostic tests are all important; but the ability of health care providers to alter health outcomes—Dr. Walsh McDermott's 'decisive technology'—depends primarily on drugs. Six dollars are spent on hospitals and physicians for every dollar spent on drugs, but without drugs the effectiveness of hospitals and physicians would be enormously diminished." [17]. Without the clinical research process, pharmaceutical products could not be developed. Thus, the clinical research process is important not only to the pharmaceutical industry but also to the entire health care industry.

REFERENCES

1. Paul De Haen Information Systems, *De Haen New Drug Analysis, U.S.A.*, Vol. 16: *1975-1979—A Five-Year Survey*. Micromedex, Englewood, Colo., 1980.
2. J. G. Simanis and J. R. Coleman, Health expenditures in nine industrialized countries, 1960-1976. *Social Security Bull.* *43*(1): (Jan. 1980).

3. *Statistical Abstract of the United States, 1980,* 101st ed. U.S. Department of Commerce, Bureau of the Census, Washington, D.C., 1980, p. 105.
4. D. L. Rabin and P. J. Bush, The use of medicines: Historical trends and international comparisons. *Int. J. Health Serv. 4*(1):61 (1974).
5. K. von Grebmer, Competition in a structurally changing pharmaceutical market: Some health economic considerations. *Soc. Sci. Med.* 15C(2):77 (June 1981).
6. Creativity in pharmaceutical research. *Tile and Till 60*(2):19 (Fall 1974). Eli Lilly and Company, Indianapolis.
7. M. C. Smith and D. A. Knapp, *Pharmacy, Drugs and Medical Care.* Williams & Wilkins, Baltimore, 1979, p. 206.
8. How prescription drugs lower medical care costs. *Pharm. Times* p. 66 (Aug. 1978).
9. *Health, United States—1979.* U.S. Dept. of Health, Education, and Welfare, Public Health Service, Office of Health Research, Statistics and Technology, Washington, D.C., 1979, p. 154.
10. *Health, United States—1978.* U.S. Dept. of Health, Education, and Welfare, Public Health Service, Office of the Assistant Secretary for Health, Washington, D.C., 1978, p. 73.
11. F. A. Sloan, R. Khakoo, L. E. Cluff, and R. H. Waldman, The impact of infectious and allergic diseases on the quality of life. *Soc. Sci. Med. 13A:*473 (1979).
12. R. M. Scheffler and L. Paringer, A review of the economic evidence on prevention. *Med. Care 18*(5):473 (May 1980).
13. Tagamet patients spend more days on job than placebo patients, trials show. *F.D.C. Repts. 43*(45):7 (Nov. 9, 1981).
14. Merck Heptavax-B may generate $4.3 million in hospital costs savings per week, according to preliminary CDC estimate. *F.D.C. Repts. 43*(47):5 (Nov. 23, 1981).
15. A. Kucherov and A. Trafford, Coming: New kinds of drugs that could save your life. *U.S. News World Rept.* p. 53 (Oct. 5, 1981).
16. M. Silverman and P. R. Lee, *Pills, Profits and Politics.* Univ. of California Press, Berkeley, Calif., 1974, p. 183.
17. V. R. Fuchs, *Who Shall Live?* Basic Books, New York, 1974, p. 105.

BIBLIOGRAPHY

Bezold, Clement, *The Future of Pharmaceuticals.* Wiley, New York, 1981.

Christensen, Dale B., and Patricia J. Bush, Drug prescribing: Patterns, problems and proposals. *Soc. Sci. Med. 15A*(3):343 (May 1981).

Cocks, Douglas L., Economic competition in the ethical pharmaceutical industry. *Med. Marketing Media 14*(10):21 (Oct. 1979).

Cooper, Joseph D. (Ed.), *The Economics of Drug Innovation*, American University, Washington, D.C., 1969.

Cuatrecasas, Pedro, The downward trend in drug discovery. *Private Practice 12:*18 (Nov. 1980).

Goyan, Jere, The major vectors affecting health care in the 1980's. *Med. Marketing Media 15*(15):4 (May 1980).

Health, United States—1980. U.S. Dept. of Health and Human Services, Public Health Service, Hyattsville, Maryland, 1981.

Hughes, Richard L., Perspectives on changing healthcare delivery. *Med. Marketing Media 15*(8):30 (Aug. 1980).

James, Barrie G., A look into the future of the pharmaceutical industry. *Med. Marketing Media 13*(7):13 (July 1978).

Lasagna, Louis, The uncertain future of drug development. *Drug Intelligence Clin. Pharm. 13:*192 (Apr. 1979).

Lindsay, Cotton M., *The Pharmaceutical Industry.* Wiley, New York, 1978.

Mushkin, S. J., Health as an investment. *J. Pol. Econ. 70*(5):129 (1962).

Schaumann, Leif, Dynamics and opportunities in the 1980's. *Med. Marketing Media 15*(7):15 (July 1980).

Sorkin, Alan L., *Health Economics.* Heath (Lexington Books), Lexington, Mass., 1975.

Unhealthy costs of health care. *Business Week 2550:*58 (Sept. 4, 1978).

Weisbrod, Burton A., Research in health economics: A survey. *Int. J. Health Serv.* 5(4):643 (1975).

23

ROLE OF THE FDA IN THE CLINICAL RESEARCH PROCESS

Marion J. Finkel
U.S. Food and Drug Administration
Rockville, Maryland

I. INTRODUCTION

The U.S. Food and Drug Administration (FDA) assumed a major role in the clinical investigation of new drugs in 1962, when Congress passed the Kefauver-Harris amendments to the Food, Drug and Cosmetic Act (of 1938). These amendments provided that sponsors of investigational drug studies in humans notify the FDA of their proposed studies and that they submit data to support the safety of such research. In addition the amendments required that all marketed drugs be shown, by means of adequate and well-controlled studies, to be effective, except that those drugs first marketed prior to 1938 were exempt from that provision.

Before 1962 the FDA had no jurisdiction over investigational drug research. Its role with respect to new drugs was confined to review of New Drug Applications (NDAs) which were submitted by pharmaceutical firms and other sponsors to obtain permission for marketing. The FDA reviewed the NDAs for evidence of safety; the law prior to 1962 specified only that new drugs be safe to be marketed.

Since 1963 drug companies and individual clinical investigators have been required to submit a "Notice of Claimed Investigational Exemption" [referred to as an Investigational New Drug (IND) application] before embarking on research in humans with a new (nonmarketed) drug. An IND is also required when a marketed drug is studied for a new indication or in a dosage other than that encompassed by the labeling, provided that the investigator causes the drug to be shipped in interstate commerce for the purposes of the research. The FDA's authority usually extends only to drugs that move in interstate commerce, so that drugs prepared and confined wholly intrastate are theoretically exempt from FDA regulation; however, occasionally, when there is a particularly flagrant intrastate distribution of an unregulated drug, the FDA has exerted its jurisdiction over the drug by virtue of the fact that the raw materials were shipped interstate. INDs need not be submitted for marketed drugs used for unlabeled indications in the practice of medicine or for small clinical trials when the drugs are obtained from local sources; in the latter case, however, investigators will sometimes submit INDs either for whatever medicolegal protection such submission may afford or because their Institutional Review Board (IRB) requires it.

In addition to the FDA, the major participants in clinical investigation are drug company sponsors, clinical investigators, IRBs, and patients and other volunteers. This chapter is organized into sections that describe the FDA's role in relationship to the other participants.

II. THE FDA AND INDUSTRY

The FDA receives between 300 and 400 new INDs from pharmaceutical industry sponsors each year. Not all of these involve new chemical entities. Some consist of new dosage forms, new uses, or new combinations of marketed drugs or of generic drug products for which bioavailability or clinical studies are proposed. Among the INDs the bulk of the FDA's review time is devoted, quite naturally, to those involving new chemical entities.

Because the Kefauver-Harris amendments to the Food, Drug and Cosmetic Act were precipitated by the thalidomide disaster, and because the intent of Congress was to ensure the safety of investigational drug studies, the FDA's initial efforts in its review of INDs were devoted to determining whether the animal pharmacological and toxicological data supported the safety of the proposed clinical trials, whether patients would be adequately monitored for adverse effects, and whether the ensuing data demonstrated that it was safe for clinical trials to continue. Although drug companies are required throughout the investigational stage to submit copies of all clinical protocols prior to inception of the trials, the FDA's medical reviewers spent little time in determining whether the design of the studies was adequate to answer the question of effectiveness.

The FDA soon realized, however, that it was desirable to take a more active role in the drug development process when it was faced with NDAs for which the evidence of effectiveness was inadequate. Patients were being exposed to the rigors of clinical trials and the potential hazards of new chemicals for no useful purpose.

Thus began a series of steps by which the FDA became a major partner in drug research.

A. Adequate and Well-Controlled Studies Regulations

In 1970 the FDA published regulations describing the principles of adequate and well-controlled studies [1]. Although these regulations were promulgated primarily to facilitate the FDA's retrospective review for effectiveness of those drugs first marketed between 1938 and 1962, they served to underline the importance of the controlled clinical trial. The regulations list the four principle types of controls (placebo, active, no treatment, historical) without placing them in a hierarchy. Nevertheless, it is well-recognized that placebo and active controls generally provide the most rigorous proof of effectiveness.

B. Clinical Guidelines

Beginning in the early 1970s the FDA, working with the Pharmaceutical Manufacturers Association, the American Academy of Pediatrics, and its own advisory committees and consultants, developed guidelines for clinical investigation of a large number of drug classes as well as general guidelines applicable to all drug classes. To date 26 guidelines are available (Table 1). These guidelines describe important elements to consider in the design of studies to determine the safety and effectiveness of drugs in Phases I, II, and III of investigation. They are reviewed and updated on a biannual basis by the FDA's advisory committees. They are not considered to be graven in stone. Alternative design elements are accepted, and have been found, at times, to be more appropriate to the conditions under study.

Because the FDA accepts foreign data in support of safety and effectiveness of new drugs (provided, of course, that such data meet the standards of well-conducted and documented research), the FDA's clinical guidelines have had some impact on the type of studies conducted abroad. The guidelines have been circulated to foreign governments and have been translated into at least one foreign language. They have, in some instances, been considered by scientific bodies abroad in the design of their own clinical guidelines. Multinational drug companies and even wholly based foreign companies interested in marketing their drugs in the United States have consulted the guidelines in formulating their clinical trials.

C. Therapeutic Classification of Investigational Drugs

Since the mid-1970s the FDA has classified new molecular entities, new salts, new combinations, and new uses of marketed drugs undergoing investigation into categories based upon their potential therapeutic importance. Categories are also assigned to newly marketed drugs since the potential importance of an investigational drug may be found to increase or decrease as data are gathered from clinical trials. This classification serves the dual purpose of determining the degree of innovation in the industry from year to year and of intensifying the expenditure of FDA resources on drugs of major therapeutic importance. For the most part, within 6 months of submittal of an IND drugs are classified as of major (A), of modest or moderate (B), or of little or no apparent therapeutic gain (C), based upon their pharmacological properties and clinical data, if any, from studies conducted abroad. These classifications are reviewed semiannually and revised, when appropriate. Drugs assigned an A or B classification are handled in the same manner, although perhaps on occasion a little more effort may be directed to an A drug. The classifications are made by a team consisting of an FDA medical reviewer, a group leader, and a division

Table 1 The FDA's Clinical Guidelines

General considerations for the clinical evaluation of drugs
General considerations for the clinical evaluation of drugs in infants and children
Guidelines for the clinical evaluation of antidepressant drugs
Guidelines for the clinical evaluation of antianxiety drugs
Guidelines for the clinical evaluation of anti-infective drugs (systemic) (adults and children)
Guidelines for the clinical evaluation of anti-anginal drugs
Guidelines for the clinical evaluation of anti-arrhythmic drugs
Guidelines for the clinical evaluation of antidiarrheal drugs
Guidelines for the clinical evaluation of gastric secretory depressant drugs
Guidelines for the clinical evaluation of hypnotic drugs
Guidelines for the clinical evaluation of general anesthetics
Guidelines for the clinical evaluation of local anesthetics
Guidelines for the clinical evaluation of anti-inflammatory drugs (adults and children
Guidelines for the clinical evaluation of antacid drugs
Guidelines for the clinical evaluation of gastrointestinal motility-modifying drugs
Guidelines for the clinical evaluation of laxative drugs
Guidelines for the clinical evaluation of psychoactive drugs in infants and children
Guidelines for the clinical evaluation of bronchodilator drugs
Guidelines for the clinical evaluation of drugs to prevent, control and/or treat periodontal disease
Guidelines for the clinical evaluation of drugs to prevent dental caries
Guidelines for the clinical evaluation of analgesic drugs
Guidelines for the clinical evaluation of drugs used in the treatment of osteoporosis
Guidelines for the clinical evaluation of lipid-altering agents in adults and children
Guidelines for the clinical evaluation of antiepileptic drugs (adults and children)
Guidelines for the clinical evaluation of antineoplastic drugs
Guidelines for the clinical evaluation of radiopharmaceutical drugs

director. The sponsors are notified of A and B classifications and, if a drug is not so classified, a sponsor can request FDA to consider reclassification.

It should be noted that assignment of an A classification is limited to truly major therapeutic advances as well as to those drugs which may not necessarily be categorized as a major advance in treatment of disease but which, by virtue of their unique pharmacological properties, portend the development of improved molecules. The classifications are, or course, judgmental and are thus subject to differences of opinion on their assignment. Table 2 provides the number and percentage of new molecular entities that are the subject of active INDs and are assigned to A, B, and C categories.

The assignment of an A or B classification triggers a series of actions. The sponsor is invited to participate in an "end-of-Phase II" conference for those drugs, at which time the preclinical, clinical, and manufacturing data obtained to that point are discussed and the plans for Phase III clinical trials are described. Attending these conferences are clinicians, biostatisticians, pharmacologists, chemists, and—as needed—pharmacokineticists and microbiologists from the firm and the FDA. A member of the pertinent FDA advisory committee or another consultant is often present. The sponsor is also invited to submit detailed manufacturing controls information prior to submittal of an NDA in order to expedite review of such data and permit correction of any deficiencies. When the NDA is submitted, it is given a high-priority review by FDA and is scheduled, also, for review by an advisory committee. It has been estimated that these actions are capable of making important drugs available to the public some 1 or 2 years earlier than might have been anticipated.

D. Conferences and Other Communications

The FDA's involvement in clinical research is manifested not only by written communications to sponsors during the IND stage but also by meetings held throughout the process. Some sponsors, particularly of those drugs first studied abroad or of those drugs with unique pharmacological properties, will request a conference prior to submittal of an IND. Letters from the FDA during the IND phases may ask for clarification, recommend changes in clinical protocols, and in extreme cases require discontinuance of clinical trials on the grounds of potential serious hazard to patients. At the conclusion of end-of-Phase II conferences, letters are usually exchanged that summarize the under-

Table 2 Therapeutic Classification of New Molecular Entities in INDs: October 1981

Classification	No. (%)
A	34 (4)
B	120 (13)
C	770 (83)

standings reached at the conferences. (It should be noted that, time permitting, end-of-Phase II conferences are sometimes held, at the request of the sponsor, for drugs classified as C.) Many sponsors like to hold pre-NDA conferences to describe the data they have accumulated and to discuss the manner in which the data should be compiled to facilitate review by the FDA and, finally, at the time of NDA submittal, to explain the organization of the NDA. During FDA review of an NDA, conferences are held or letters sent to describe readily corrected deficiencies.

It is during the meetings held to discuss proposed protocols for the major controlled clinical trials that the FDA perhaps has its most direct input into clinical research. In addition to contribution by FDA scientists of their own ideas on appropriate design based upon experience and literature review, there inevitably occurs a cross-fertilization of ideas by virtue of the fact that the FDA is the repository for all clinical protocols and for the ensuing data on investigational drugs. Trade secret information is, of course, respected, but scientists do not feel constrained from sharing information on the principles of good clinical design that accrue to the benefit of the patients being studied and to the public awaiting new drugs.

E. Advisory Committee Review

The FDA's Bureau of Drugs currently has 15 advisory committees (Table 3), which meet two to four times a year. They consist of extra-

Table 3 Bureau of Drugs Advisory Committees

Anesthetic and life support drugs
Anti-infective and topical drugs
Arthritis drugs
Cardiovascular and renal drugs
Dermatologic drugs
Drug abuse drugs
Endocrinologic and metabolic drugs
Fertility and maternal health drugs
Gastrointestinal drugs
Oncologic drugs
Ophthalmic drugs
Peripheral and CNS drugs
Psychopharmacological drugs
Pulmonary-allergy drugs
Radiopharmaceutical drugs

mural experts in the designated drug classes and subspecialties and always contain at least one statistician. They are supplemented with consultants, as required by specific agenda items. The committees are asked to consider the data presented in NDAs classified as A or B and, sometimes, C and to advise as to whether there exist evidence of safety and substantial evidence of effectiveness based upon adequate and well-controlled studies. They are sometimes asked to review INDs, particularly for drugs with unique pharmacological properties and those with potential safety problems. The committees also review important issues relating to marketed drugs, e.g., new uses or newly discovered hazards, and on occasion they consider whether a drug should be removed from the market because of a serious hazard. Although the committees' role is advisory to the FDA, in virtually all cases their recommendations are accepted. The advisory committees not only supplement the FDA's own expertise but also bring an element of review independent of that of the federal government. Thus, when important decisions are reached by the FDA based on the committees' advice, the public knows that the matter has been considered from a variety of viewpoints.

F. Postmarketing Studies

The FDA's involvement in clinical research is not ended when a drug is approved for marketing. New uses often arise or can be predicted, and the FDA participates in the decisions on the appropriate action to be taken in those cases. At times, prior to approval of an NDA, an agreement is reached with the sponsor that postmarketing trials will be performed for uses or in patient populations not studied during the IND stages. For example, when a new beta blocker is approved for use in hypertension, agreement is reached that the sponsor will also perform studies in angina pectoris and cardiac arrhythmias to determine the effective doses, since it is anticipated that the drug will be utilized by practitioners for those diseases as well.

The performance of a postmarketing study to refine the incidence of certain adverse effects or to search for anticipated effects indigenous to a specific class of drugs is sometimes considered desirable by the FDA, and discussions are held with sponsors on the design of such studies.

The FDA provides financial support for several large ongoing epidemiological studies of adverse drug reactions and continually explores new methods for collection of information of that nature.

G. Benefit vs. Risk Considerations

Approval of an NDA is, in the end, based upon a decision that the benefits of a drug outweigh its risks. Prior to reaching such a conclusion, however, the FDA must ascertain that reasonable tests have been performed to elucidate the risks and that substantial evidence exists of effectiveness, derived from adequate and well-controlled studies. Ordinarily, effectiveness must be demonstrated by at least two independent studies. Substantial evidence of effectiveness means, however, that there is a preponderance of evidence pointing in the same direction. Two positive well-designed studies do not offset eight equally well-designed studies which are unable to distinguish between drug and placebo. Allowances are, of course, made for the difficulty of demonstrating drug effects (as, for example, in psychiatric illnesses) and for the size of the available patient population.

Most drugs that reach the NDA stage are eventually approved. Most firms are not interested in marketing drugs that are not at least as effective and safe as other available drugs, so that those with inferior characteristics are generally culled during the investigational stages. Occasionally, a drug reaches the NDA stage that is less effective and/or less safe than drugs for the same indication. Then it is necessary to determine whether there exists a patient population in which the benefits of the drug will outweigh its risks. In most cases the answer is yes, but such decisions are frequently made with the benefit of advice from the FDA's advisory committees.

III. THE FDA AND CLINICAL INVESTIGATORS

A. Responsibilities of Investigators

The FDA does not communicate directly with clinical investigators who perform studies for pharmaceutical company sponsors. All comments on protocol design and data are communicated to the sponsors. Investigators, be they independent or under company sponsorship, however, incur obligations under the Food, Drug and Cosmetic Act when they undertake such studies. They must be qualified to perform the studies (the qualifications are reviewed by FDA); must obtain the informed consent of patients and normal volunteers; must obtain approval of the study from an IRB prior to inception of the study; must maintain adequate records of the study; must dispose appropriately of unused drug; and must permit FDA investigators to audit their records.

B. The FDA's Clinical Bioresearch Monitoring Program

The FDA maintains a clinical bioresearch monitoring program the purpose of which is to monitor both sponsors and clinical investigators to determine whether they are fulfilling their responsibilities. In the past FDA performed inspections of clinical investigators on a random basis or "for cause." Most investigators were found to be generally fulfilling their obligations, with minor, easily corrected deficiencies.

These were brought to their attention as well as to the attention of the clinical pharmacology community. Since there are thousands of clinical investigators, the FDA's resources permitted inspection of only a small portion of the total. In recent years the FDA has found it to be more productive to audit major studies submitted in new drug applications. In occasional cases studies have been found to be so flawed as to jeopardize an entire NDA.

Many drug sponsors maintain their own sophisticated investigator monitoring programs, ensuring that their investigators understand all of their responsibilities and even auditing the data derived from their studies. Others have only rudimentary programs. A few years ago the FDA proposed detailed regulations that describe the responsibilities of both clinical investigators and sponsors of investigational drug studies [2]. These regulations have not been finalized as of this writing; the general principles underlying the proposed regulations are listed in Tables 4 and 5.

Table 4 Sponsor-Monitor Responsibilities

Sponsor:

1. Assure that investigator has obtained IRB approval
2. Establish written procedures for monitoring of investigators
3. Ascertain the adequacy of test facilities
4. Review case records and other data for safety parameters as soon as possible and no later than 10 working days after receipt
5. Review effectiveness data when an investigator has completed his study.

Monitor:

1. Visit each investigator prior to a study unless a visit has been made within the preceding 12 months
2. *Reasonably* assure that the investigator understands the protocol and his obligations under the regulations and that he is dealing with an investigational drug
3. *Reasonably* assure that the investigator has time to conduct the study and has an adequate number of subjects
4. Make periodic visits to the investigator to determine whether
   a. The facilities are still acceptable
   b. The investigator is fulfilling his obligations:
      (1) Adherence to protocol
      (2) Maintenance of records and reports
      (3) Submission of reports to IRBs and sponsor
5. Review investigator's source documents for case report information (voluntary)
6. Maintain record of findings

Table 5 Clinical Investigator Responsibilities

A. IRB
   1. Obtain IRB approval of the protocol and material to be used to obtain informed consent prior to initiating study or requesting the consent of subjects to participate
   2. Obtain IRB approval prior to alterations in the protocol which may increase the risk of research subjects
   3. Submit to the IRB annual progress reports and a report after termination of the study
   4. Report promptly to IRB (and sponsor of research) any serious not previously anticipated adverse effect or death reasonably regarded as associated with the test article; if such reports have come from the sponsor, they should be transmitted to the IRB

B. Test articles
   1. Maintain for use only by the investigator or by an individual responsible to him/her
   2. Maintain records of receipt and disposition of test article and dispose of unused material in the manner requested by the sponsor

C. Study
   1. Develop a written protocol
   2. Obtain informed consent

D. Records and reports
   1. In addition to reports to IRB, submit reports to sponsor
   2. Maintain detailed case records on each subject
   3. Retain all records for 2 years after the test article is approved for marketing by the FDA or 5 years after an NDA is submitted, whichever is shorter; if an NDA is not submitted, retain for 2 years after the *entire* investigation is completed or discontinued or an IND is withdrawn.

E. FDA inspection
   Permit an FDA investigator to inspect facilities and records for verification purposes and to copy such records, if necessary

C. Independent Investigators

In contrast to the lack of direct communication with clinical investigators performing studies under company sponsorship, the FDA communicates, when needed, with investigators who submit their own INDs. In addition to a requirement that a proposed study be shown to be relatively safe, FDA regulations require that the study be a reasonable one. This means that there must be a satisfactory rationale and that the design is of a nature that can be expected to provide useful information. In general, FDA does not attempt to intervene in small research projects conducted by individual investigators provided that the subjects' safety is assured. The FDA's reviewers may make recommendations for modification of the clinical protocols that may yield more appropriate information on effectiveness, but they do not insist that such recommendations be followed. When, however, a large number of investigators are performing studies on the same investigational drug, partly as research projects but mainly for the purposes of treatment, and no useful information on effectiveness can be expected because of the inadequacies of the protocols, FDA will at that point insist that the design of the studies be altered. Failure to do so can result in termination of the IND by FDA.

D. "Treatment" INDs

Most INDs from investigators are submitted for the purposes of conducting true research. Some, however, are so-called treatment INDs wherein one or a few patients are given an investigational drug, usually because the marketed drugs have been ineffective or are inappropriate for those patients. Some treatment INDs can be considered "Emergency" INDs in that the urgency of treatment necessitates a waiver by FDA from the 30-day period following submittal of an IND during which study cannot begin. (This interval gives the FDA time to review the IND.) Waivers are granted by telephone. In very urgent cases, waivers from submitting an IND prior to treatment are granted and the investigator need only submit his IND after the fact.

The FDA requires much less information for treatment INDs than for research INDs. In the former instance the information consists largely of the proposed circumstances of use.

As might be expected, treatment INDs generally yield little or no useful information on effectiveness of a drug but may provide interesting adverse reaction information; in one recent case, because of the large number of patients involved, they provided the bulk of the safety experience for NDA approval.

Sometimes a large number of treatment INDs will be submitted for the same drug. For those drugs that are used for a chronic disease, FDA will often develop a uniform protocol that will be sent to the IND sponsors. This maximizes the ability to obtain useful information on the safety of the drug.

Treatment INDs are not confined to independent investigators. Often a firm that is sponsoring an IND for an important new drug will receive numerous requests from physicians for shipments to treat individual patients. The difference between a company-sponsored treatment IND and one that is independently sponsored is that the company is obligated, under current regulations, to attempt to collect clinical data from the investigators on an annual basis. The FDA is considering a policy that would tailor the obligations of a company that sponsors a treatment IND to the situation surrounding the drug at the time it is distributed for treatment. That is, if much is already known about the safety and effectiveness of a drug, the investigators would merely be asked to report adverse reaction information but no attempt would be made to follow up on this request; where a drug is in the relatively early stages of investigation or where drug experience information from the investigators would be useful for the purposes of an NDA, a sponsor would try to obtain the desired information.

IV. THE FDA AND INSTITUTIONAL REVIEW BOARDS

The FDA has issued regulations that describe the standards that Institutional Review Boards (IRBs) are expected to meet in order to review clinical investigations regulated by the FDA or those that support applications for research or marketing permits for products regulated by the FDA, including drugs, medical devices, and biologicals for human use, food and color additives, and electronic products [3]. The regulations also state that all such research, whether or not performed in an institution, must be reviewed by an IRB on an initial and continuing basis. An exception is granted for emergency use of a test article, provided that such use is reported within 5 days to the IRB; subsequent use of that test article must be subject to IRB review. Certain types of research that involve no more than minimal risk, such as collection of blood samples or recording of data in adults using noninvasive techniques, or minor changes in approved research, are permitted to be reviewed under an expedited procedure consisting of review only by the chairperson or one or two members of the IRB.

Each IRB must contain at least five members of varying backgrounds (both scientific and nonscientific, e.g., law, religion, ethics). The IRBs are required to convene in order to review studies. They must develop written procedures that describe their functions and activities with respect to initial and continuing review of studies and the methods developed to ensure prompt reporting to IRBs by investigators of unanticipated risks and of changes in research protocols.

In order to approve proposed research, the IRBs must find that the risks to subjects are minimized by use of procedures consistent with sound research design and that do not unnecessarily expose subjects to risk; that risks to subjects are reasonable in relation to anticipated benefits, if any, to the subjects and the importance of the knowledge to be derived; that informed consent is obtained; and that privacy of subjects and confidentiality of data are maintained.

Because FDA regulations, as of January 1981, require that research in noninstitutionalized subjects be reviewed by an IRB, a number of review boards independent of institutions have been formed to take care of research for which established IRBs are unwilling to assume responsibility.

The FDA has conducted regional workshops for IRB members to acquaint them with their responsibilities under the regulations. The FDA also audits IRB performances and has found a great improvement in compliance in recent years.

V. THE FDA AND PATIENTS AND VOLUNTEERS

The FDA's regulations require that, prior to use of a test article, informed consent be obtained from the subjects of research regulated by the FDA [4]. Exceptions are granted in life-threatening situations in which it is not possible to communicate with the patient, there is insufficient time to obtain consent from the subject's legal representative,

Table 6 Elements of Informed Consent

A. Basic elements
   1. Purposes of the research; duration of participation; procedures to be followed, including any which are experimental
   2. Reasonable foreseeable risks or discomforts
   3. Reasonable expected benefits to the subjects or to others
   4. Alternative procedures or treatments, if any, that may be advantageous to the subject.
   5. Extent, if any, to which confidentiality of records will be maintained and notification that the FDA may inspect the records
   6. Where more than minimal risk is incurred, whether any compensation or medical treatments are available if injury occurs
   7. Whom to contact for answers to questions or in the event of injury

B. Additional elements, when appropriate
   1. Procedure may involve unforeseeable risks to the subject, embryo, or fetus
   2. Circumstances in which the subject's participation may be terminated by the investigator
   3. Costs which the subject may incur as a result of his/her participation
   4. Consequences of a decision by the subject to withdraw and procedures for orderly termination
   5. Any significant new findings which may bear on the subject's willingness to continue will be imparted
   6. The approximate number of subjects in the study

and no other approved therapy is likely to prove effective or as effective.

The elements of informed consent required by current regulations are listed in Table 6.

VI. PROPOSED CHANGES IN FDA REGULATIONS

The preceding pages have described the major interactions between the FDA and other participants in clinical research. Under consideration are changes in FDA regulations that, if implemented, would have a significant impact on certain aspects of investigational drug research conducted by pharmaceutical firms.

A. Foreign Clinical Data

As mentioned earlier, FDA accepts clinical data performed outside of the United States on the same basis as it accepts U.S.-derived data in support of INDs and NDAs, i.e., the studies shall have been performed by qualified investigators under an appropriate design, adequate information is available upon which to reach a conclusion on safety and effectiveness, and the studies are conducted under prevailing ethical principles (the Declaration of Helsinki or the principles of the country, whichever provides greater protection for the subjects). This has always been FDA policy, but it was not well recognized until the FDA published a regulation in 1975 that described the circumstances under which it would accept foreign-derived clinical data [5].

FDA regulations require that, except where a condition is rare in the United States, at least some studies must be performed in this country in order to permit approval of a drug for marketing. The nature and quantity of the U.S. studies depend, of course, on what is available from foreign research. In some cases one or two small studies may be all that is needed.

The scientific reasons advanced for requiring some U.S. studies include differences in diagnostic and therapeutic techniques; in patient population; and in bacterial sensitivity, in the case of anti-infectives. Obviously, not all of these factors would apply in each case, and in some they would not be applicable at all. Where they are not applicable the question has arisen as to why any studies need to be performed in the United States prior to marketing approval. The most important reason advanced is the fact that release of a drug in the U.S. market constitutes the largest incremental use of a drug in the world, not only by virtue of distribution in the United States but because many less developed nations approve a drug for marketing in their own countries based upon approval in the United States. Under those circumstances it is desirable that direct knowledge about the properties of a drug be gained from study by several experts well known to the research community in the United States.

Nevertheless, the question of whether a new drug can be approved for marketing in the United States based solely on foreign data—and, if so, under what circumstances—is currently under review by the FDA.

B. Modification of IND Requirements

Currently, pharmaceutical company sponsors must submit an IND prior to the first Phase I study. The INDs must contain detailed information on animal pharmacology and toxicology and the proposed Phase I clinical trials, as well as early information on manufacturing controls. As the studies progress, each new clinical protocol must be submitted prior to inception of the study.

Phase I studies (defined as pharmacological and safety studies in normal volunteers or patients and early dose-ranging effectiveness studies in patients) are the safest studies in the investigational process because they are small in size, the subjects are closely monitored, and the investigators usually adhere to well-defined and often-utilized protocols. It must be remembered too that, like all investigational studies, Phase I studies are reviewed by IRBs. It has been suggested that Phase I and perhaps early Phase II research might be expedited without harm to patients if one of two procedures were adopted: (1) an IND should not be submitted at all until Phase II; or (2), if one is to be submitted, it should contain only a brief summary of the safety data and a general outline of the proposed Phase I studies. Under this proposal, unless the FDA saw a serious safety reason for curtailment of the studies, they would be permitted to proceed.

The FDA is considering these suggestions, which not only have the potential for expediting research but also for permitting additional allocation of FDA resources to NDA review and thus expediting further the availability of important new drugs to the public.

REFERENCES

1. Code of Federal Regulations 21, Section 314.111(a)(5)(ii).
2. Obligations of clinical investigators of regulated articles. Code of Federal Regulations 21, Part 54 et al. (proposed in *Federal Register* Aug. 8, 1978, pp. 35210-35236); Obligations of sponsors and monitors of clinical investigations. Code of Federal Regulations 21, Part 52 et al. (proposed in *Federal Register* Sept. 27, 1977, pp. 49612-49630.

3. Institutional Review Boards. Code of Federal Regulations 21, Part 56.
4. Informed consent of human subjects. Code of Federal Regulations 21, Part 50.
5. Clinical data generated outside the United States and not subject to a "Notice of Claimed Investigational Exemption for a New Drug." Code of Federal Regulations 21, Section 312.20.

24
CLINICAL PHARMACY AND ITS RELATIONSHIP TO CLINICAL RESEARCH

William F. McGhan
University of Arizona
Tucson, Arizona

Glen L. Stimmel
University of Southern California,
Los Angeles, California

Gary M. Matoren
Clinical Research Practices and
Drug Regulatory Affairs,
Middletown, New York

I. INTRODUCTION TO CLINICAL PHARMACY

From a historical perspective, drug manufacturing was once primarily the responsibility of pharmacists. From roots, rhizomes, and minerals, early preparations were crafted with mortar, pestle, and percolators.

The technology of the twentieth century has revolutionized the production of "medicines," and the present-day pharmacist requires new and sophisticated skills to work in manufacturing or drug research along with many other professionals with diverse training and expertise [1].

Each new drug that is approved for marketing today is the result of a discovery and development process that requires up to 10 years of intensive investigation and review. An essential part of this drug development is the clinical phase of testing to assess safety and efficacy of the new chemical entity.

The clinically trained pharmacist is uniquely equipped to participate actively in clinical research. With an intimate knowledge of drugs and understanding drug effects on patients, pharmacists have a special appreciation for the importance of clinical research in the process of drug development and patient care [2].

A. Evolution of Clinical Pharmacy

Participation of pharmacists in clinical research has been a natural step in the evolution and development of clinical pharmacy as a discipline. The profession of pharmacy has undergone dramatic changes in the last several decades, particularly in the last 15 years. The emphasis in pharmacy practice has shifted: rather than being expert in formulation of products for patient use, today pharmacists must become expert in how drugs behave in patients. Product-oriented pharmacy practice is being replaced by patient-oriented practice. During the late 1960s and early 1970s, totally new roles in health care delivery were developed for pharmacists. Some pharmacists became active participants in drug therapy decisions, either as a drug information resource or by direct patient interaction and consultation with physicians. A new type of faculty emerged in schools of pharmacy during the early 1970s—the clinical pharmacy staff. The major effect of the new teaching approach was to further expand the clinical role of pharmacists and to train students to behave as clinical pharmacists, i.e., as specialists responsible for drug therapy. The final step in the evolution of clinical pharmacy as a discipline was the establishment of a research component. During the late 1970s and early 1980s, some clinical pharmacy practitioners and faculty emerged with added competence and training in the scientific method. This growing number of clinical pharmacists has assumed significant clinical research responsibilities. These individuals are actively involved as collaborators with physicians and principal investigators in clinical drug research.

B. Definition of Clinical Pharmacy

It is important to define what is meant by the term "clinical pharmacy." While some use the term to refer to the "patient orientation" sweeping the pharmacy profession as a whole, others use the term to refer to pharmacists with specialized knowledge and skills not common to all licensed pharmacists. While patient orientation should be the standard of practice for all pharmacists, clinical pharmacy best describes those pharmacists with "unique clinical skills." In the United States, there are now hundreds of clinical pharmacists with specific responsibilities that cannot be performed by the majority of pharmacists: drug therapy selection; dosage calculation based upon pharmacokinetic parameters; interpretation of laboratory tests and physical assessment used to evaluate drug response and adverse effects; and primary patient care [3,4]. Rather than being a divisive concept, recognition of these specialty practitioners points to the diversity and strength of the profession.

C. Training of Clinical Pharmacists

The Doctor of Pharmacy curriculum is intended to provide excellent training for clinical pharmacists. As stated in a June 1981 report of the Task Force on Pharm.D. Accreditation Standards, the Doctor of Pharmacy program should provide a core curriculum that is enriched with knowledge and clinical practice experiences significantly beyond that provided in baccalaureate programs [5]. More in-depth knowledge is to be gained in:

1. The pharmaceutical sciences—especially biopharmaceutics and pharmacokinetics
2. The biomedical sciences—pathology, biostatistics and research design
3. The social and behavioral sciences—especially patient communication and education
4. The clinical sciences and clinical practice—disease processes and diagnosis, clinical pharmacy practice, drug therapeutics, drug literature evaluation, clinical toxicology, and clinical pharmacokinetics

Each year for the last several years, approximately 7,000 pharmacists have been graduated in the United States, all of them well educated in the pharmaceutical sciences. They have been trained to serve as the drug expert on the health team and to maximize the effectiveness of the public's drug therapy. Although most students graduate after 5 years of training with a baccalaureate degree in pharmacy, there are now three schools in the United States that provide only a 6-year training sequence leading to the Pharm.D. degree, 26 schools that provide B.S. training with an option for a Pharm.D., and 14 more schools planning to have a Pharm.D. option [6,7].

Pharmacy students today thus receive significant training in patient care settings to develop clinical skills as well as knowledge. The clinical component of a Pharm.D. program provides a minimum of 1,500 hr of experience. For example, following clinical therapeutic lectures, case conferences and introductory clinical clerkship experiences,

Table 1 Final Year Clinical Clerkships: University of Southern California School of Pharmacy

Mandatory clerkships (18 weeks)	
Inpatient adult medicine	6 weeks
Outpatient adult medicine	6 weeks
Psychiatric hospital and clinic	6 weeks
Elective clerkships (16 weeks)	
Pediatrics	
Clinical pharmacokinetics	
Skilled nursing facility	
Oncology	
Cardiology	
Psychopharmacy	
Parenteral nutrition	
Medical subspecialties (various)	

Pharm.D. students at the University of Southern California spend their final year in full-time clinical clerkships (Table 1).

Paralleling the expansion of clinical training for pharmacy students has been the development of clinical pharmacy residency programs. The American Society of Hospital Pharmacists now has accreditation standards for two specialty residencies—clinical pharmacy and psychopharmacy [8,9]. In these programs, residents develop in-depth clinical knowledge and skills which include primary patient care responsibilities.

While many clinical pharmacists have developed clinical research skills on their own through collaboration with other investigators, specific research training programs have been developed during the past few years. Postdoctoral fellowship programs are now available in a variety of clinical areas including psychopharmacy, acute care medicine, pediatrics, oncology, clinical pharmacokinetics, and cardiology. Graduates of these programs are immediately assuming positions in industry and academia with clinical research responsibilities.

D. Clinical Pharmacists' Increasing Involvement in Research

The participation of clinical pharmacists in clinical research is only in its beginning phase. Clinical faculty in schools of pharmacy, expected to develop new roles in patient care during the 1970s, are now expected to greatly increase their clinical research activity. Schools of pharmacy can no longer afford to keep clinical faculty who provide only teaching and patient care. There are now many excellent clinicians in practice outside academia, many who are willing and interested in teaching students. For the 1980s, clinical research is the factor that will distinguish clinical pharmacy faculty from the clinical colleagues in private practice. For these reasons, clinical research by clinical pharmacy faculty will significantly expand. Additionally, many clinical pharmacists interested in research are going to industry and private research institutions. In all these settings, clinical research will increasingly involve the participation of clinical pharmacists.

The type of research conducted by clinical pharmacists falls into two major categories. Pharmacotherapeutics is the most obvious type, including studies of drug efficacy, adverse effects, interactions, and pharmacokinetics. Research on the impact of clinical pharmacy services is the other major type, including studies on the impact of cost on care, the quality of prescribing behaviors, and the effect of patient education on drug compliance.

One objective source of information concerning clinical research conducted by clinical pharmacists is an analysis of abstracts presented at annual meetings of the American College of Clinical Pharmacy. A total of 84 papers were presented at the 1980 and 1981 annual meetings [10,11]. The vast majority of papers concern pharmacokinetics, drug efficacy studies, and adverse effects/interactions of drugs (Table 2). Two-thirds of the papers involved collaborative research with physician coauthors. Geographic distribution of these researchers was widespread, with the 84 papers being presented by clinical pharmacists practicing in 22 states and in various parts of Canada.

Table 2 Abstracts of Papers Presented at the 1980 and 1981 Annual Meetings of the American College of Clinical Pharmacy

	1980	(N = 34)	1981	(N = 50)
Type of research	N	%	N	%
Pharmacokinetics	16	47	30	60
Drug efficacy	5	15	12	24
Adverse effects/interactions	8	23	6	12
Other	5	15	2	4
Collaboration:				
M.D. coauthor	21	62	37	74
Pharmacists only	13	38	13	26

E. Clinical Scientists in Pharmacy

As further evidence of the intense interest of the pharmacy profession in clinical research, the Millis Commission report entitled "Pharmacists for the Future" has recommended the training of "clinical scientists." This concept was derived from the models in medicine where the individual should be as competent at the bedside as he or she is in the research laboratory. There are different methods of training these clinical scientists including: (1) strengthening the pharmacy undergraduate curriculum to emphasize more clinical research along with clinical service; (2) taking clinical practitioners and giving them graduate research training; and (3) taking researchers and providing them with clinical training. Currently the University of Minnesota and the University of Utah have established training programs for clinical scientists. These clinical scientists may well make important contributions to the conduct of clinical research. It is also expected that the graduates of clinical scientist programs will become the instructors in pharmacy schools so that clinical research skills will more and more be a standard competency for all pharmacy graduates.

II. CLINICAL PHARMACY AND CLINICAL INVESTIGATIONS

A. Distribution of Investigational Drugs

Pharmacists' basic responsibilities in clinical investigations have in the past often been focused on the limited roles related to distribution of investigational drugs. The American Society of Hospital Pharmacists has developed "Principles in the Use of Investigational Drugs in Hospitals." This document states that "the pharmacy department is the appropriate area for the storage of investigational drugs, as it is for all other drugs. This [department] will also provide for the proper labeling and dispensing in accord with the investigator's written orders." The pharmacist is thus recognized by many as the appropriate person to be responsible for the dispensing, storage, handling, and dissemination of information regarding investigational new drugs [12]. The clinically trained pharmacist should be able to become involved far beyond just drug distribution and record keeping.

B. Clinical Pharmacy and Institutional Review Boards

Clinical trials can be viewed as an attempt to provide the public with medicinals for their benefit in a way that carefully attempts to elucidate the use of a drug to maximize its benefit/risk ratio.

Institutions in which clinical trials are conducted must establish Institutional Review Boards (IRBs) that have the function of reviewing the protocols of each new drug to be tested. The purpose of IRBs is to resolve questions about the proposed study that arise out of ethical considerations. They must, for example, weigh the risks and benefits of an experimental course of therapy.

Clinical pharmacists have new opportunities in serving on or working with IRBs. These boards are responsible for reviews of clinical studies for approval and for protection of human subjects' rights. Since July 1981, IRB approval has been needed for all clinical studies subject to U.S. Food and Drug Administration (FDA) regulations, whether conducted in physicians' private offices or in hospitals. The board must not only be able to review a proposed clinical study on its scientific merit but must also take into consideration sociological, ethical, and community issues. Each IRB must have at least five members and must include at least one nonscientific member (e.g., a lawyer or clergyman); one member must be included who is not affiliated with the institution (i.e., a community representative) [13].

C. Drug Research Monitors

Table 3 provides a description of the key participants in clinical investigations. Although clinical pharmacists can assume several different roles, the role of clinical monitor is being assumed by an increasing number of pharmacists. The clinical pharmacist is uniquely suited to the role of drug monitor for clinical trials because he has a combination of scientific knowledge of pharmaceutics, chemistry, and pharmacology as well as the ethical, administrative, and patient-oriented skills

Table 3 Key Participants in Clinical Investigations

Investigator—an individual who actually conducts a clinical investigation; often a physician, but a clinical pharmacist would be appropriate
Sponsor—a person who initiates a clinical investigation but who does not actually conduct the investigation
Monitor—a designated individual selected by a sponsor or control research firm to oversee the progress of a clinical investigation; clinical pharmacists are being increasingly utilized in this capacity
Sponsor/investigator—an individual who both initiates and conducts, alone or with others, a clinical investigation, i.e., the person under whose immediate direction the test drug is administered or dispensed to a subject; this can include a physician (clinical pharmacist) who is independently performing his/her own clinical study

gained during clinical training. These skills closely parallel those required of the drug monitor, whose duties include communication and liaison responsibility, dealing with and evaluating potential investigators, reviewing adequacy of research facilities, interpreting clinical laboratory results, and identifying and reporting adverse experiences.

In order to develop the parallels between the two concepts of clinical pharmacist and clinical monitor, a brief discussion of clinical investigational drug research is in order.

Clinical studies may be divided into two areas: clinical pharmacology and clinical research. The discipline of clinical pharmacology can be defined as the science that concerns itself with the description and determination of the mechanism of the effects of drugs in humans. It attempts to gain insight into drug action using appropriate scientific research methodology, biostatistics and medical "common sense." Clinical research, on the other hand, is the application of the experimental method to human therapeutics in an ethically suitable manner, for example, defining drug activity in human patients under the same conditions as would occur with usual use of the drug at the recommended dosage. This then forms the basis of what is termed"rational therapeutics."

The role that the clinical pharmacist plays in hospitals as the expert on drug therapy has been compared to the role of clinical pharmacologists. Some individuals have felt that clinical pharmacists have been very successful because they are more abundant (and less expensive) than clinical pharmacologists. As for maximizing the role of clinical pharmacists in clinical drug investigations, it should be pointed out that clinical pharmacists have many of the necessary skills and this could relieve some of the burdens placed on clinical pharmacologists [14].

Clinical monitors are expected to assure the quality of the drug investigation; they monitor patients' records, compliance, and reactions, following protocols (in essence, the plan and procedures for clinical trials)—all of which are familiar to the clinical pharmacist by way of experience and training. This parallel exists to such an extent that the content of pharmacy schools' courses agrees closely with the requirements for Investigative New Drug (IND) submissions and the description of the monitors' responsibilities, as prescribed by the Code of Federal Regulations.

D. Clinical Pharmacists as Investigators in All Testing Phases

It should by now be apparent that clinical pharmacy education and training provide many of the skills necessary to anyone who would succeed as a clinical drug investigator. The clinical pharmacist has important skills that are invaluable at all phases of clinical investigations, as described in Table 4.

Table 4 The Four Phases of Clinical Testing: Possible Role of Clinical Pharmacists

Phase I:	The cautious introduction of the new drug into normal, healthy subjects. These studies are intended to reveal information about the bioavailability, safe dosage range, toxicity, pharmacological actions, and preferred route of administration. If Phase I studies show an acceptable margin of safety, then Phase II studies may be initiated.
Phase II:	A limited number of patients with the target condition or disease are studied in these trials. The goal here is to discover a safe but effective dose of the investigational drug. This phase can be divided into two subphases. "Early" Phase II research deals in more detail with the observation of drug effects in normal subjects. "Late" Phase II research concerns itself with initial trial of the drug in the condition or disease for which the drug is indicated, searching for an effective dose. While it is true that Phase I and early Phase II studies have been primarily monitored by the clinical pharmacologist, the clinically trained pharmacist can play a major role in this area.
Phase III:	Utilizes a large number of patients with the target disease or conditions, frequently using the final dosage form. The objectives are to explore the safety and efficacy under expected conditions of use. It is realized with these larger sample sizes that problems with side effects and adverse drug reaction may appear at this stage of testing.
Phase IV:	"Market support" studies are conducted after the New Drug Application (NDA) has been approved to monitor continued safety and efficacy, to provide comparative studies, or to discover new indications for the drug. This postmarketing surveillance phase may continue for an extended period of time. Once a drug is marketed, clinically trained pharmacists in institutional and community settings can greatly assist by identifying and reporting side effects, adverse drug effects, and therapeutic success rates.

The job requirements of a clinical investigator are very similar to the skills required of a clinical pharmacist. The combination of a clinical pharmacist with a physician can often make an excellent investigating team. In such studies, the physician is generally responsible for making the diagnosis and initiating therapy, while the clinical pharmacist does the titratio s and monitors the patient's therapy under protocol. California law now allows pharmacists to modify the dose of regimen in institutional settings under protocol established with physicians and others. Pharmacy practice acts in the states of Washington and Michigan also allow pharmacists to become more involved in modifying drug regimens under cooperative agreements with physicians [15].

As the capabilities of clinical pharmacists are recognized and brought to the attention of those responsible for conducting clinical studies, clinical pharmacists undoubtedly will be recognized as a logical choice as clinical drug investigators along with physicians.

While it is true that Phase I and early Phase II traditionally have been monitored by the clinical pharmacologist, the clinical pharmacist can play a major role in this area.

Obviously, it is in Phase III, where adverse drug effects have the most likely chance to occur, that clinical pharmacists can serve in a very appropriate way. The clinically trained pharmacist's involvement in Phase IV is of special interest also. These pharmacists could play an important role in postmarketing surveillance by monitoring patients for the appearance of adverse drug effects that may occur; moreover, by counseling patients on proper use of new medications they might help to reduce patients' misuse of the products, thereby cutting down on drug therapy failures and decreasing negative side effects.

III. CONCLUSIONS

Clinical pharmacists are likely to be involved with all phases of clinical drug studies—as investigators, monitors, members of IRBs, dispensers of investigational medications, or study coordinators.

The clinical research of investigational drugs is a complicated process that requires sound clinical and pharmacological judgment, a solid grounding in medical ethics, and a good understanding of government regulations, economics, marketing, and organizational management. These skills are important assets found in the growing discipline of clinical pharmacy.

The clinical pharmacist by virtue of his education and training is well equipped to assume the role of a clinical monitor in the pharmaceutical industry.

Pharmacy education and training provide the skills necessary to anyone who would succeed as a monitor of clinical trials. Certainly, the skills gained as a drug monitor are invaluable in all phases of clinical investigation. It has been shown that the job description of a drug monitor and the qualifications of a pharmacist have a great deal in common from the legal and ethical points of view as well.

As this commonality is recognized and brought to the attention of those conducting clinical studies, pharmacists undoubtedly will be recognized as a logical (possibly preferred) choice as monitors of drug trials. While it is true that Phase I and early Phase II traditionally have been monitored by the clinical pharmacologist, the pharmacist drug monitor can play a major role in this area. Obviously, it is Phase III, where adverse drug effects have the best chance to occur, in which pharmacist drug monitors can have the most impact. The pharmacist's involvement in Phase IV is of special interest as well. Since prescription products sometimes become nonprescription drug products, the pharmacist's advice on their use, formulation, and application can be invaluable relative to the design of clinical studies in support of these products.

The increasing participation of pharmacists in professional societies dealing with clinical pharmacology implies the recognition that pharmacists can make major contributions to the field of general investigation. Pharmacists should continue, with time, to be more and more involved in this field.

REFERENCES

1. Anon., Industry supports professional development. *Am. Pharm.* [NS] *21*(4):(Apr. 1981).
2. N. A. Sceusa, R. A. Vukovich, S. Bolton, and G. Matoren, The pharmacist as a drug research monitor. *Am. Pharm.* [NS] *21*(4): (April 1981).
3. G. L. Stimmel, Special recognition for clinical pharmacy as a general specialty. *Am. Pharm.* [NS] *20:*6-11 (1980).
4. G. L. Stimmel, Clinical pharmacy and specialization. *Drug Intell. Clin. Pharm. 14:*540 (1980).
5. Report of the Task Force on Pharm.D. Accreditation Standards. American Association of Colleges of Pharmacy, Washington, D.C., June 1981.
6. J. F. Schlegel, Enrollment report on professional degree programs in pharmacy: Fall 1979. *Am. J. Pharm. Ed. 44:*177-192 (1980).
7. O. Kushner, Every new pharmacist a Doctor of Pharmacy? It may happen sooner than you think. *Am. Druggist 181*(2):13,83,97 (1980).
8. ASHP Accreditation Standard for Residency Training in Clinical Pharmacy. *Am. J. Hosp. Pharm. 37:*1223-1228 (1980).
9. ASHP Supplemental Standard and Learning Objectives for Residency Training in Psychiatric Pharmacy Practice. *Am. J. Hosp. Pharm. 37:*1232-1234 (1980).
10. Abstracts of papers from American College of Clinical Pharmacy: 1980 Annual Meeting. *Drug Intell. Clin. Pharm. 14:*632-639 (1980).
11. Abstracts of papers from American College of Clinical Pharmacy: 1981 Annual Meeting. *Drug Intell. Clin, Pharm. 15:*476-485 (1981).

12. H. M. Arbit, Regulatory aspects of Investigational New Drugs. *Am. J. Hosp. Pharm. 35:*81-85 (Jan. 1978).
13. C. L. Roehl, D. F. Miller, and T. S. Foster, Pharmacist involvement in institutional review of clinical trials. *Am. J. Hosp. Pharm. 38:*334-339 (Mar. 1981).
14. R. R. Miller, An overview of clinical pharmacy and clinical pharmacology. *J. Clin. Pharmacol. 21*(5-6):238-240 (May/June 1981).
15. G. L. Stimmel and W. F. McGhan, The pharmacist as prescriber of drug therapy: The USC pilot project. *Drug Intell. Clin. Pharm. 15:*665-672 (Sept. 1981).

25
CAREER OPPORTUNITIES IN INDUSTRIAL CLINICAL RESEARCH

Martin C. Sampson
Sampson, Neill & Wilkins Inc.
Upper Montclair, New Jersey

I. PERSPECTIVES

In recent decades, career opportunities in industrial clinical research and related areas have become increasingly numerous and varied. This change has resulted partly from the expanding impact of new scientific technologies upon development of new drugs and partly from the expanding development of new, nondrug modalities, such as various medical devices and diagnostic tests [1-3]. Contributing to these developments have been varying degrees of governmental requirements in the United States and other countries as well [61]. These requirements are that marketed drugs, devices, and diagnostic agents be proved not only safe but also effective [4-6]. The industry's response to these requirements has been to create the in-house regulatory affairs unit, which has become critically important in facilitating the governmental approval of new drugs, devices, and diagnostics [7-9].

Incorporated into industrial clinical research in the last several decades have been many sophisticated disciplines, including management by objectives, continuous applications of biostatistics, science information systems, and the ever-enlarging roles of computers in day-to-day operations (10-17,63].

Public attention to industrial toxicology and to factors influencing such disorders as cancer and hypertension has led to increasing utilization of epidemiology and the requirement, accordingly, that appropriately trained epidemiologists be sought for evaluation of clinical experience data [18]. Also, there have been growing concerns relating to environmental pollution and to exposure of workers to potentially toxic substances in the workplace. Stemming from this has been the emergence of organizations devoted to contract toxicology research.

A reflection of how times have changed is the fact that as recently as 30 years ago, when I entered the industry (to stay for 16 enjoyable years), physicians in the industry were often affiliated with pharmaceutical companies on only a part-time basis and were often considered to be "internal medical consultants." They were thought of as clinicians, not managers, and—more often than not—"necessary window-dressing" [19]. Thirty years ago statisticians and computer specialists were little involved in clinical research activity; in fact, clinical research was generally of a noncontrolled type, with testimonials abounding as to favorable drug actions.

II. DEFINITIONS

A. Opportunities Available in the Total Field Encompassed by Industrial Clinical Research and Related Opportunities

Opportunities available include those for Clinical Research Directors (with M.D. or Ph.D. degrees, or both); Clinical Research Monitors (with M.D. and/or Ph.D. degrees); Clinical Research Associates (with Ph.D., M.S., B.Ph. and/or B.S. degrees); Statisticians (with Ph.D., M.S. and/or B.S. degrees); Science Information Specialists (with Ph.D., M.S. and/or B.S. degrees); Clinical Pharmacists (with Ph.D., M.Ph. and/or B.Ph. degrees); Computer Specialists (with Ph.D., M.S. and/or B.S. degrees); Clinical Data Analysts (with M.S. and/or B.S. degrees); Epidemiologists (with Ph.D. and/or M.S. degrees); Medical Writers and Editors (with M.S. and/or B.S. degrees); Regulatory Affairs Specialists (with M.D., Ph.D., J.D., M.S., B.Ph. and/or B.S. degrees); and Administrative Managers (with M.D., Ph.D., M.S., B.Ph., and/or B.S. degrees) [10,11,64-67].

B. Industrial Organizations Where Clinical Research and Related Opportunities Are Available

These include diversified health care corporations, pharmaceutical and chemical companies, companies with proprietary and over-the-counter medicinal products, toiletry and cosmetic companies, medical device companies, diagnostic product companies, clinical laboratory organizations, contract toxicology and contract medical research organizations, and medical-academic-industrial complexes.

III. MOTIVATIONS AND APTITUDES

A. Some General Remarks

In general, those who enter clinical research careers have a basic interest in the sciences and obtain gratification from dealing with complex problems that can be resolved through mathematics or a logical system or reasoning. In addition to the basic scientific set of aptitudes, those in clinical research positions also may have other appropriate aptitudes. For example, the ability to relate well to other people (e.g., outside clinical investigators) in a liaison function is important for those who wish to become Clinical Research Associates. (Clinical Research Associates function as extensions of and assistants to Clinical Research Monitors, in the development and administration of clinical research studies.)

Motivations and aptitudes may vary, depending upon the stage of clinical research in which the individual is involved. Thus, commonly in clinical pharmacology (Phase I and early Phase II new drug research studies in humans), the Clinical Research Monitors will have a basic research orientation and will enjoy interrelating with the preclinical laboratory sciences.

In broad-scale clinical investigation studies (Phase III), the goal is to develop a sufficient volume of controlled clinical data to define the clinical efficacy and safety of a new drug compound (or a new medical device or diagnostic agent). This, in turn, leads to development of a New Drug Application (NDA). Hence, the Clinical Research Monitors (M.D. or Ph.D. degrees) who handle such studies will be therapeutically oriented and will enjoy seeing the statistical results of their efforts. These Clinical Research Monitors will enjoy problem solving and intricate organization of information and also be able to handle many projects in parallel at the same time.

In clinical development studies (Phase IV), which take place once a drug has been approved for marketing, the Clinical Research Monitors will develop more extensive clinical experience with the new drug for approved indications and, at times, for new indications. These Clinical Research Monitors will be pragmatic in their outlooks and will primarily enjoy seeing the prescription-use end results of their individual efforts.

B. Physicians Entering Industrial Clinical Research

Physicians who enter clinical research in industry may have a variety of motivations, both positive and negative[10,11,19-26].

On the positive side of motivations, physicians entering industrial clinical research are attracted by the opportunities to participate in the clinical development of important new therapies, even breakthroughs at times. A key advantage is that required funding and administrative services are readily available. Also important are the opportunities to contribute to the well-being of large groups of people through contributing to development of important therapies. To many physicians, the opportunities to develop administrative, communication, and clinical research skills are key advantages. To others, the opportunities to utilize special knowledge or abilities such as in various medical subspecialties, biostatistics, epidemiology, or medical writing are enticing. Underpinning these opportunities are the industry stanchions of financial security and adequate time to develop a balanced way of life.

On the negative side of motivations, physicians may find that medical practice no longer provides challenge. They may be burdened by high overhead costs of practice, high malpractice insurance costs, the requirements of third-party payments, and continual public scrutiny of the medical profession. They may be weary of the morbidity and death encountered in everyday practice. They may be enduring a mediocre or poor quality of life in terms of time for family and recreation. For those in medical academia, there may be concern about the shrinking of funding for academic medicine and academic clinical research (with some notable recent exceptions related to development of industry/academic complexes).

The aptitudes of physicians who enter industry, by virtue of the selection process involving those who attend medical school and go through the rigors of medical training, are often those required of successful top industry leaders: the ability to think broadly and deal with various problems innovatively yet logically, rationally, and dispassionately; the ability to handle many problems concomitantly and effectively; the ability to be individualistic and have the courage of one's convictions in the face of traditionalism; the ability to integrate dissimilar concepts and data; and the ability to lead and direct other members of the health team.

Future physicians are selected by medical schools partly because of self-reliance, and they are then taught to be even more self-reliant. If physician graduates have excellent academic records (including specialty training) and can also operate effectively in group or team settings, and if they have natural qualities of credibility and leadership and the ability to plan and to follow through as well, they will be the ones who most likely become Directors of Clinical Research or Vice Presidents of Clinical Research in the industry.

C. Ph.D. Scientists Entering Industrial Clinical Research

Ph.D. scientists entering industrial clinical research will be those whose scientific curiosity extends beyond basic preclinical research in animals. Such Ph.D. scientists often have had a long-standing interest in medical therapeutics and have the advantage of being able to apply a knowledge of the basic medical sciences, particularly pharmacology, toxicology, biochemistry, and physiology, to therapeutic perspectives. A number of outstanding Directors or Vice Presidents of Clinical Research in the industry have either Ph.D. degrees or both M.D. and Ph.D. degrees [10,11]. Those with both degrees are among the best qualified in the pharmaceutical industry.

As with physicians, the aptitudes of Ph.D. scientists who rise administratively in industry will include the ability to operate effectively in a team and to plan and follow through, as well as the ability to exercise leadership. The Ph.D. scientist who rises to become Vice President of Clinical Research or Scientific Affairs is likely to be an individual with scientific vision and imagination coupled with the other aptitudes described.

D. Others Entering Industrial Clinical Research in Support Functions

Those who become members of support functions in industrial clinical research—Statisticians, Science Information Specialists, Clinical Pharmacists, Computer Specialists, Clinical Data Analysts, Epidemiologists, Regulatory Affairs Specialists, Medical Writers and Editors, and Administrative Managers—will have the strong desire to be members of teams of expert professionals and to benefit from all of the financial and technical support available in industry [10,11]. In addition, as presented subsequently, those who enter such functions at staff levels will value the opportunities for administrative advancement within their respective fields.

Those who enter industrial clinical research support functions will often have become disenchanted with the bureaucratic, political, and/or budgetary environments of academia or government. They will recognize the industry's ability to provide sophisticated administrative functions that will enable them to achieve their goals most expeditiously and effectively.

The aptitudes of those who provide clinical research support functions will vary, of course, with the function. Thus, *Clinical Research Associates* will be those with an interest in science and medicine who also have the excellent ability to communicate with leaders in clinical research. They will be very much people-oriented. The same will hold true of *Administrative Managers,* who in addition will have strong aptitudes in planning, organizing, and following through.

On the other hand, those who go into *Statistics, Science Information,* or *Data Management* and *Data Analysis* will be mathematically inclined and meticulous in the handling of detailed information. At the same time, they need not be exceptions in their ability to deal with people. Those who become managers, nonetheless, will have developed these appropriate administrative talents regardless of the functional responsibility.

Other support functions will require other aptitudes. Thus, those who enter *Regulatory Affairs* should have the ability to focus on details yet at the same time be able to handle projects that encompass inputs from all areas of an industrial organization. Such inputs will come frm the Preclinical and Clinical Research, New Drug Development, Legal Manufacturing, Quality Assurance, Advertising, and Marketing Departments.

Those who enter *Medical Writing* or *Editing* in industrial clinical research must obviously have good command of the written language.

In addition, they should have a real interest in science and medicine and should have the ability to grasp complex scientific subjects and reduce them to simple written forms.

As clinical research has become more complex and diversified, it has become more and more necessary to have people who are skilled in the administration of clinical research. This holds true not only in the development of protocols and overseeing of clinical programs but also in planning, targeting of objectives, budgeting, and related functions. Hence, Managers or Directors of *Clinical Research Administration* have become increasingly important, particularly in the pharmaceutical industry. Such Managers are innately good planners and relate well to others at all levels. They have the ability to conceptualize and also to keep track of details, and to delegate to others. Their backgrounds may be in the biological sciences, with Bachelor's, Master's or Ph.D. degrees, or in pharmacy, or other related disciplines.

IV. EDUCATIONAL AND EXPERIENTIAL REQUIREMENTS

A. Clinical Research Directors in Industry

Clinical Research Directors will hold M.D. or Ph.D. degrees, or both. They will have had substantial prior clinical research experience, either in industry or academia.

Prior experience in industry will be favored over prior experience solely in academia, however [10,11,19,21]. The main reason for this is the crucial experience, which one can gain only in industry, relating to successful development and submission of a New Drug Application (NDA) to the U.S. Food and Drug Administration and the final approval by the FDA for the marketing of a new drug, device, or diagnostic product.

The educational backgrounds for Clinical Research Directors will be the same as for Clinical Research Monitors (as described in the next subsection). Ordinarily, such Directors will have started in industry as Clinical Research Monitors and with appropriate experience and demonstration of appropriate abilities will have been promoted to the directorship positions.

Clinical Research Directors will have mastered the science and art of multicentered, double-blind clinical research methodology and data handling. They will comprehend both the assets and limitations of medical statistics. They will have had successful prior administrative achievement involving the supervision of Clinical Research Monitors, support staffs, programs and projects. They will have had successful experience in planning and in management by objectives. They will have developed in-depth knowledge of FDA requirements for new drugs and/or new devices and diagnostic products. They will have frequently made presentations to the FDA, in concert with their own company's Regulatory Affairs and Legal Departments.

B. Clinical Research Monitors in Industry

Clinical Research Monitors will hold M.D. or Ph.D. degrees, or both. They may or may not have had prior industry experience. At times they will have had clinical research experience in academia. Most often, such Monitors will not have had industry experience but will be selected for their intellectual and personality qualities—particularly high intelligence, scientific and medical curiosity, an eagerness to learn and achieve, ability to communicate well, and a high-energy level with high output [10,11,20,22].

Undergraduate education in college should ideally include concentrated studies in one or more of the following: biochemistry, pharmacology, toxicology, molecular biology, pharmaceutics, physiology, bioengineering, or biostatistics. For those entering the earliest phases of clinical research (Phases I and II), achievement of a Ph.D. or Master's degree in clinical pharmacology is ideal. In case one has an M.D. degree, then subsequent achievement of a one-year Fellowship in Clinical Pharmacology is worthwhile, as is prior or subsequent achievement of a Master's or Ph.D. degree in any of the other aforementioned basic sciences [27,28].

If a physician enters the later stages of clinical research (Phase III, encompassing broad-scale controlled clinical research studies leading to development and submissions of NDAs to the FDA; and Phase IV, encompassing postmarketing clinical research studies for approved claims and/or new indications), a background in internal medicine is ideal; also worthwhile are backgrounds in family practice or pediatrics, with Specialty Board Certification [10,11].

Physicians who enter clinical research with internal medicine or other specialties will be sought after if they have internal medicine subspecialty training in cardiovascular diseases, immunology/allergy, infectious diseases, oncology, endocrinology, hematology, rheumatology, or gastroenterology.

Physicians who are specialized in psychiatry will also find careers in industrial clinical research, particularly if they have had some experience or training in neuropharmacology or psychopharmacology.

Some physicians with more discrete specialties, e.g., ophthalmology or dermatology, will occasionally have fine opportunities in industry—with those companies whose products match these specialties.

Physicians who have had training and experience in nuclear medicine, radiology, computer sciences, and/or engineering prior to or after their medical training may be of special interest to medical diagnostic and device companies for the clinical research on their new products.

Industry will also value physicians who are subspecialized in epidemiology or biostatistics, e.g., through achievement of a Master or Doctor of Public Health degree.

There are some physicians and Ph.D. scientists subspecialized in medical writing and medical editing. They, too, will find appropriate opportunities in industrial clinical research.

C. Clinical Research Associates in Industry

Clinical Research Associates (CRAs may have Ph.D. degree training in basic sciences or pharmacy training but more likely will have a Master's or Bachelor's degree-level education. Ideally, such education will be in the basic sciences, particularly biology, chemistry, pharmacology, and/or physiology [10,11].

Some CRAs are registered pharmacists and have had some special training in clinical pharmacy. Others, on occasion, may have had their training as registered nurses or physician assistants and then become affiliated with industrial organizations.

In many cases, CRAs have originally been with a pharmaceutical company's professional sales staff and in that capacity have shown a particular success in understanding and transmitting scientific and clinical research information, even though they may have had no basic science education. In other cases, CRAs may have started with a pharmaceutical company's (or device/diagnostic company's) basic research division. In such cases these CRAs would have had basic science education and subsequently would have shown a particular ability to understand clinical research data and the relationship of such data to basic science information. These will also have shown the ability to relate to people in general and to communicate well.

D. Statisticians in Industrial Clinical Research

Statisticians who have their career paths in industrial clinical research may have Bachelor's degree-level training in mathematics, statistics, or biostatistics or may have Master's degree and Ph.D. degree-level training as well [10,11]. In industrial clinical research, the higher level positions, e.g., for Department Director or Head of Biostatistics, are almost always filled by those with Ph.D. degrees. A Statistician may enter industry directly from college or may enter after several years of academic experience.

As the need for sophisticated, controlled clinical studies has increased in the past several years, so has the need increased for sophisticated Biostatisticians with understanding of clinical frames of reference [13,14,29-32]. Hence, Biostatisticians should not only have basic understanding of life sciences and/or biological sciences but also have competent knowledge or experimental clinical research design. In this regard, it is important for Statisticians to understand the difference between "statistical significance" and "clinical significance." A particular finding may be statistically significant at a particular confidence interval (p value) but not be clinically significant, and vice versa [33].

E. Science Information Specialists in Industrial Clinical Research

Science information Specialists will have educational backgrounds ranging from the Bachelor's degree to the Ph.D. level. The explosion of information in the medical sciences in recent decades makes the Science Information Specialist's role increasingly important in providing background information to clinical research operations within industry [15, 34,35].

In recent decades, the ability to store huge volumes of data and to rapidly locate and retrieve specific documents has become increasingly translated into utilitarian effectiveness, with many industry applications. Science Information Specialists have eagerly adapted and developed sophisticated data storage/retrieval systems and have enjoyed the fruits of this age of the computer.

The health industries have departments of Science Information within various companies. Those qualified to head up such departments will have had not only significant education (usually a Ph.D. degree relating to the biological or information sciences) but also significant experience, including administrative experience.

F. Clinical Pharmacists in Industrial Clinical Research

Clinical Pharmacists will have obtained the Bachelor of Pharmacy degree and the Registered Pharmacist license initially, and then they will have had more or less additional training in which the knowledge and techniques of pharmacy will have been correlated with clinical pharmacology and drug therapeutics [10,11,28,36].

Clinical Pharmacists may have had experience in clinical pharmacology or clinical research in academia, with close participation in studies within large hospitals and medical centers. Such specialists may also have received Master's degree or Doctorate level training in pharmaceutical sciences, clinical pharmacology, and/or health administration.

Clinical Pharmacists may go directly into industry at the time of completion of their education or they may come into industry later, after experience in clinical research. In addition to clinical research monitoring, such Clinical Pharmacists may enter industry in support areas, e.g., Regulatory Affairs or Medical Administration.

G. Computer Specialists and Clinical Data Analysts in Industrial Clinical Research

Computer Programmers will have appropriate training and experience in programming. While education may be solely at a technical school level, often there will be a Bachelor-level college degree in mathematics or computer science, with varying degrees of experience thereafter.

Those who supervise computer operation or programming analysts—or who become Clinical Systems Analysts or Programming Analysts—will have achieved a Bachelor's degree in mathematics, computer science, physical science, or engineering. In addition, such candidates may have achieved Master's degree training in one of the biological sciences or in business administration. The latter specialists will have knowledge of programming languages such as FORTRAN and will be familiar with statistical software and with multitasking operating systems, data base concepts, and project management [10,11,16,34,37,38]. In the device/diagnostic area such candidates, in addition, will often have knowledge of minicomputer and microprocessor systems [17].

Clinical Data Analysts will ordinarily have a Bachelor's degree in one of the biological or physical sciences. In addition, Master's degree education will be an advantage in future career progress.

Data Analysts become more important in industry as they gain experience in clinical-trial data management, data computerization, and data tabulation and presentation methods [39,40]. Further, an understanding of statistical analysis approaches and FDA regulations concerning "Good Clinical Practices" (GCP), double-blind controlled protocols, and adverse reaction reporting will be especially valuable.

H. Epidemiologists in Industrial Clinical Research

Epidemiologists will usually have a Master's degree or Ph.D. degree in epidemiology. These degrees may be in public health or preventive medicine and may be achieved subsequent to an M.D. degree. Such candidates may have experience in academia or in public health following their education, or they may go directly into industry upon completion of the appropriate education [10,11,18].

Clinical Epidemiologists will act as bridges between Biostatisticians and Clinical Data Managers and Clinical Research Monitors [41]. Epidemiologists should have been prepared by their education to have in-depth understanding of both the statistical approaches to clinical data and the therapeutic approaches.

Those in this field will usually concentrate their efforts upon the adverse reaction implications of various clinical data, whether related to effects of drugs or effects of environmental impacts. The proper evaluation of possible adverse reactions from new drugs will have a significant impact upon the claims that a company decides to make for a new drug in its NDA to the FDA and in the package insert of information that it proposes to be provided to the medical profession.

Epidemiologists may also play a valuable role in determining the cost savings of a new type of drug therapy. For example, they might calculate how many patients with a particular medical disorder have been spared hospitalization by the advent of the new drug therapy.

I. Regulatory Affairs Specialists in Industrial Clinical Research

Regulatory Affairs Specialists may have a variety of educational backgrounds—say, a Bachelor's degree in liberal arts or biological sciences, or an education in pharmacy or law, or Ph.D. degrees in sciences or mathematics, or an education in medicine [10,11].

Regulatory Affairs functions in industrial companies are usually headed by managers with M.D. or Ph.D. degrees, but not always. Such Managers may, alternatively, have J.D. degrees or may have Bachelor's or Master's degree levels of education.

Those who start in Regulatory Affairs functions in industry may begin in various Regulatory sections, e.g., Regulatory Compliance, Regulatory Surveillance, Clinical Documents Control, Adverse Drug Reaction Reporting, Medical Advertising and Medical Promotion Review, and/or coordination of U.S. with international regulatory activities [4-9,42-45].

A major responsibility under Regulatory Compliance in pharmaceutical companies involves the preparation and filing of Investigational New Drug (IND) applications and final NDAs. (The latter, when approved, define the parameters of what can and cannot be claimed about respective products for marketing purposes.) There also are periodic filings with the FDA of adverse drug reaction reports. In companies developing and manufacturing device and diagnostic products, there are different types of product-experience data to be submitted to the FDA, particularly a Premarketing Approval Application. In all industrial companies involved with health products, postmarketing surveillance of experience with the new drug or device or diagnostic product is becoming increasingly important. Thus, both the FDA and the National Academy of Sciences have recently given concerted attention to this area [7,46,47].

J. Medical Writers and Editors in Industrial Clinical Research

Medical Writers and Editors will have a background of at least a Bachelor's degree in college, usually in English or in one of the related communiation areas. Often there are individuals with Master's degrees and even Ph.D. degrees in Medical Writing or Editing among those who enter clinical research in industry. Some of these people also have training in the sciences, with either Ph.D. or M.D. degrees.

Those who have training and experience in medical writing and also have had successful administrative experience may head up Departments of Medical Communication or Medical Editing within industrial companies. There are opportunities for Medical Editors whose entire experience is basic-science oriented and puristic, on the one hand; and opportunities also for those whose entire experience is pragmatic and market oriented, on the other [10,11].

K. Administrative Managers in Industrial Clinical Research

Those who become Administrative Managers in industrial clinical research will have a variety of educational backgrounds, as we saw with those who enter Regulatory Affairs functions. Such Managers may have their initial training in pharmacy or in business administration (with emphasis upon planning). They may, on the other hand, not have had any formal training in management or administration but rather in the sciences or medicine, along with considerable subsequent industry experience and development of administrative skills and track records during that experience [10,11].

The Administrative Manager will usually report to the Vice President of Clinical Research and will be charged with keeping the admistrative wheels rolling for the various clinical research functions. Hence, such a Manager is usually responsible for budgeting; project planning; reporting and logging of planning by objectives; coordination of clinical research support services; monitoring of office space and equipment; and maintenance of personnel and related activities [34,48-50].

In some companies Administrative Managers are being given responsibilities for clinical quality assurance, i.e., assurance that clinical monitoring of studies in the particular Clinical Research Department be kept at highest quality constantly, in conformity with preset standards. Those who have this function are likely to have M.D. or Ph.D. background plus considerable clinical research experience prior to development of the clinical research administrative function [51].

Those who become Administrative Managers in Clinical Research Departments may have arrived at such responsibilities from a number of different directions. Usually, but not invariably, such Managers will have had appropriate experience in clinical research in industry—in any one or more of the various clinical research functions described in this chapter. In some cases, such Managers may not have had prior industry experience but may have had significant prior such experience in large hospitals, in academic medical research centers, or in contract medical research organizations.

V. CAREER POTENTIALS

A. Some General Remarks

In the past three decades, the need for the specialists described herein who contribute to industrial clinical research has far outstripped the growth of the industries in which such clinical research is an essential function [10,11,14,15,19,22,24,28,31,34,36,38]. This stems from the progressively greater requirement over recent decades for more and more clinical research activity and documentation, relating both to efficacy and safety of new prescription drugs, over-the-counter drugs, consumer health products, toiletries and cosmetics, dental products, and devices and diagnostic procedures [1,3,4,6,12,13,15-18, 32,37,46,50].

Thus, as earlier observed, when I first entered the pharmaceutical industry 30 years ago there were far fewer full-time physicians in clinical research monitoring or in pharmaceutical medical activities than is the case today. Ph.D. scientists, who were always full-time, were generally in charge of clinical research, but such research tended to be incompletely controlled by today's standards.

Thirty years ago Clinical Research Associates did not prevail as they do today. There were medical liaison representatives, whose forte was not knowledge of clinical trials but rather the ability to relate to outside investigators and to obtain a favorable hearing for their company's new products (either prior or subsequent to marketing) under the then prevailing conditions of clinical trials. Thirty years ago there were some Biometricians with industrial organizations, but they were scarce. They were primarily involved with basic laboratory research but could be borrowed for occasional clinical research needs.

Thirty years ago Medical Information Science was beginning to be developed in various industrial companies and some companies had substantial numbers of such scientists organized into separate departments—but those companies were the exceptions. Today they are the rule. Likewise, Medical Computer Programmers and Medical Data Analysts have come into their own only in the last decades, with such separate units now usually reporting to the Vice President for Clinical Research or the Medical Director.

Finally, 30 years ago federal regulatory controls on new drugs, devices, and diagnostic products were far fewer than is the case today. Then, an NDA would often be approved in 90-120 days. Now it is often a matter of years.

Career opportunities for various professionals in industrial clinical research can be understood to a better extent from the interrelationships of the various titles for each function—assuming each title adequately relfects each function. This is illustrated in the Appendix, which summarizes hierarchal relationships of various functions via their respective titles [10,11]. A prototype composite is presented.

In the Appendix we may note a variety and a stratification of directorships in Clinical Research, Clinical Monitoring, and related areas. These techniques of title stratification help industrial organizations attract the services of the maximum number of best qualified professionals within their budgetary means. Of course, the term"Director" always means a degree of administrative supervisory responsibility, depending upon company size and the magnitude of its clinical research program. Hence, a Clinical Research Director in a consumer products company or a biodevice or medical diagnostics company will usually have less administrative supervisory responsibility than a Clinical Research Director in a pharmaceutical company.

B. Advancement Potentials for Clinical Research and Related Specialists

Those involved in clinical research in industry, particularly the *Clinical Research Monitors*, may progress either up the scientific ladder or up the administrative ladder [10,11].

Those Ph.D. scientists and physicians who have outstanding ability as Clinical Monitors and are individual performers can find satisfying and productive careers in this context.

Physicians who ascend the administrative ladder can expect eventually to qualify for positions such as Vice President, Clinical Research, or Vice President, Medical Director. If such physicians have basic science background in addition to the medical background, particularly with a Ph.D. degree in an appropriate subspecialty (as detailed earlier) and some background in animal research, they may be qualified eventually to fill positions such as Vice President of Research and Development [10,11].

On the other hand, once they have reached the vice presidential level, if such physicians are market-oriented and have exceptional leadership qualities they may on occasion become General Managers or even Presidents of pharmaceutical and other health industry companies [10,11].

Ph.D. scientists who begin as Clinical Monitors may well achieve directorships—particularly Directorships of Clinical Pharmacology, because of the interrelation of clinical pharmacology with preclinical basic sciences. Ph.D. scientists may also become responsible for all clinical research activities, from first studies in humans to the development and conclusion of NDAs and continuing clinical studies after marketing [10,11]. Such responsibilities carry commensurate titles, such as Director of Developmental Therapeutics or Director of Scientific Affairs.

Occasionally, a Ph.D. scientist with clinical research experience who also has a marketing interest and a Master of Business Administration or equivalent degree will gravitate into pharmaceutical marketing and achieve a responsibility such as Director, Scientific Marketing [10,11].

Ph.D. scientists with both basic research and clinical research background and administrative abilities may well qualify to become Vice President of Research or of Research and Development. Those with the broadest backgrounds and the best leadership qualities will also qualify to become General Managers and even Presidents of industrial organizations, both pharmaceutical companies and device or diagnostic companies [10,11].

Those who become *Clinical Research Associates* in industry will for most of their careers be engaged in a good deal of travel around the United States. They may have their headquarters either in the corporate headquarters or in regional offices in different parts of the country.

Should CRAs show aptitudes in management and leadership, they then will have opportunity to become heads of Clinical Research Associates Departments or units [10,11]. With such career progress they will be responsible for coordination of projects (including participation in and planning of clinical trials, writing of protocols, and writing of clinical study reports) and for the management of the staff of CRAs. The CRA work needs good logistical management because often the CRAs are engaged in providing service to more than one clinical monitor and priorities must be continually reset and adhered to.

Eventually, a manager of Clinical Research Associates may qualify to become a general Administrative Manager with a Clinical Research Department.

Statisticians who enter industrial clinical research may broaden their responsibilities to include preclinical as well as clinical research. They may also start in the basic science area and then advance into the clinical area.

In clinical research, Statisticians will usually start at a staff level. Initially they will demonstrate their abilities in problem solving and in adaption of biostatistics to the development of clinical research protocols and to the proper statistical evaluation of the results of such studies. They will have opportunities for creative use of their knowledge, e.g., in refinement of techniques of sequential analysis or in techniques of retrospective case-control study.

With demonstration of the ability to manage projects and poeple, the Biostatistician will eventually have the opportunity to become head of a biostatistics department [10,11]. From there, he or she may consequently receive increased administrative responsibility—for example, in the area of clinical research planning or in the area of total clinical data management, where the biostatistical function is closely coordinated with clinical data transfer functions; or there may be a melding of biostatistical responsibilities with science information and clinical data management responsibilities, providing additional advancement potential for the Statistician-Manager [10,11].

Biostatisticians with administrative abilities may also move into Regulatory Affairs or into the area of general Clinical Research Administration.

Science Information Specialists who have good education in the biomedical sciences and who have administrative abilities will have potentials for advancement to become heads of Science Information Departments. Such Science Information Specialists will acquire knowledge of computer applications, including programming, in the course of their industry experience. In order to progress they also should demonstrate ability to communicate well, both in writing and orally.

Eventually, Science Information Specialists who advance in industry may find that their operations will become parts of larger groupings of functions that will include Biostatistics, Clinical Data Management, Science Information Services, and Computer Data Processing [10,11,63].

Clinical Pharmacists may progress in industry within the functions of Clinical Research Associates, Regulatory Affairs, or Administrative Management (see the discussions devoted to these respective functions).

With Ph.D.-level training, Clinical Pharmacists may become Clinical Monitors and advance administratively as previously described. They may also be able to apply their knowledge and experience in areas outside of clinical research, e.g., pharmaceutical product development or pharmaceutical pilot plant production [10,11].

With the increasing utilization of multicenter studies, involving sophisticated protocols and objective as well as subjective measurements of therapeutic effectiveness, the role of the *Computer Specialist* has become more and more significant in the clinical research area of industry. There may be Departments of Computer Programming and *Clinical Systems and Clinical Data Analysts*, with appropriate directorships and staff and subsidiary positions as well [10,11].

More and more, particularly in sophisticated pharmaceutical companies, there is beginning coalescence of the functions of *Data Management*, i.e., Science Information, Computer Programming, Data Analysts, and Biostatistics [10,11]. The few such groupings that already exist are headed by a Director who is administratively responsible to the Director or Vice President of Clinical Research (or the equivalent) and who has a dotted-line relationship with similar such corporate-level functions (particularly Science Information and Data Management). Corporate Data Management structures will vary from one company to the next, but the end result is generally the same, namely, that those responsible for clinical research need to have the ability to utilize such functions on first-priority and accountability bases.

Furthermore, eventual advancement potentials exist for Computer Specialist Managers and Clinical Data Managers to head up their functions on corporation-wide bases.

Clinical Epidemiologists may progress, depending upon their education and experience, as Clinical Research Monitors, Clinical Data Managers, Regulatory Affairs Specialists, or Medical Administrators. With the special knowledge and skills that such epidemiologists can bring to evaluation of possible adverse drug reactions, they may find particular career potentials in the postmarketing medical area—which involves not only the continuing clinical research of drugs once they have been marketed but also the tracking, surveillance and evaluation of adverse reaction reports possibly involving such drugs [10,11]. The latter will include not only U.S. experience but international experience. Such experience will ultimately provide the best understanding of a particular drug's benefit/potential side effect ratio.

Clinical Epidemiologists may also move to areas of a company outside of the medical area, particularly marketing research.

Regulatory Affairs Specialists may initially progress administratively within Regulatory Affairs Departments to become Managers or Directors of the following functions: Regulatory Surveillance; Regulatory Compliance; Adverse Drug Reaction Reporting; Product Labeling; Clinical Documents Control; Medical Advertising Review; or, U.S./International Regulatory Coordination.

A section manager within a Drug Regulatory Affairs Departments may eventually become Director of Drug Regulatory Affairs and then advance to Vice President of Regulatory Affairs. A Director of Regulatory Affairs or Vice President of Regulatory Affairs may report to the Director or Vice President of Clinical Research, the Vice President of Scientific Affairs, the Executive Vice President, or the President of a company [10,11].

A Regulatory Affairs executive may become Director of Corporate Compliance, or with outstanding administrative skills, may enter corporate planning and/or general management [10,11].

Current potentials for career growth and advancement in the field of Regulatory Affairs are expanding, particularly in the medical device/diagnostic industries [10,11].

Medical Writers and Editors may apply their skills as science communicators or eventually as administrators of departments of other writers. They may concentrate in the basic science area, developing summaries of scientific information and investigational-use brochures for first trials of a new drug, device, or diagnostic test in humans. Or they may be more involved in clinical sciences, helping to write summaries of clinical research studies for inside or outside investigators [10,11]. Such reports will often be submitted for publication in reputable medical journals.

Medical Writers/Editors may contribute to the development of new product package information brochures for pharmaceutical products (known as "labeling") prior to FDA approval and market introduction. Such writers may eventually become Regulatory Affairs staff specialists [10,11].

Once a product is on the market, Medical Writers may assist in developing summaries of studies for scientific exhibits at major medical conventions. They may also contribute to the development of symposia for special segments of the medical profession with proceedings later to be published and utilized for product promotion or corporate good will. Medical writers who enjoy product promotion may eventually become advertising copywriters.

Advancement potentials for *Administrative Managers* exist either within a Clinical Research Department or Medical Affairs Department or in a Research and Development function within an industrial company, pharmaceutical or otherwise. Administrative Managers in Clinical Research may become Directors or Senior Directors of Administration. If they have M.D. or Ph.D. degrees, they may become Directors of Clinical Research or Vice Presidents of such function. They may also become Directors of Clinical Quality Assurance, as described in Sec. IV.K.

Since Administrative Managers of Clinical Research are necessarily involved with Functions of planning, budgeting, and personnel, they may be able to apply these talents on broader bases within a particular corporation—with progression into Corporate Research and Development, Corporate Planning, and/or Corporate General Management [10, 11].

C. Ancillary Benefits in Industry

In industry, in all of the areas described in the foregoing discussion, the individual will have opportunity for continuing education and development of his/her particular expertise. Such new expertise may come from educational courses outside of regular work hours. For example, a physician Clinical Monitor might take an evening course in clinical pharmacology or might embark on a 3-year evening program to obtain a Master of Business Administration degree. In either case the cost of such education is usually borne by the industrial organization.

A physician who wishes to keep his/her hands in clinical medicine can often spend one-half day per week in a nonremunerative affiliation with a local university or medical center. He or she, as well as other professionals in the industry, are often enabled to take one or two refresher courses per year in areas of their special knowledge, at company expense. In addition, professionals in various areas of clinical research will likely attend one to several national scientific meetings per year pertaining to their areas of professional expertise and/or areas of company product development.

In some companies, management development programs are in place for Ph.D. scientists and physician Clinical Research Monitors who are seen to have administrative potential. Such development programs, particularly for physicians, are relatively new and proving to be worthwhile [10,11].

Individuals in industrial clinical research will often enter other pathways. For example, one with knowledge of foreign languages and with some travel experience in Europe or Latin America may eventually decide to make his or her career in the international division of a given company. Such positions require a good deal of international travel and absences from home.

There is no rule against movement of professionals from one area to another. Hence, a Computer Specialist or Science Information Specialist who starts in preclinical research might make the transition into clinical research. A Ph.D. Clinical Monitor in drugs might find that he or she can do just as well with medical devices. A physician who starts as a Clinical Monitor and develops a particular interest in FDA regulatory requirement may, after a while, decide to concentrate totally in the Regulatory Affairs area.

V. CONCLUSIONS: A LOOK INTO THE FUTURE

The future looks bright for those who are planning to enter industrial clinical research or who are already in this exciting and challenging field [52]. The value of sophisticated clinical research programs has been borne out in the last several years by FDA approval and subsequent successful marketing of drugs varying from those that control severe infections to those that effect healing of peptic ulcers within 2-3 weeks to those that decrease the incidence of death from recurrent heart attacks.

In the foregoing instances and in many others, sophisticated, controlled clinical research has paid off in new benefits for the afflicted and expanded business for the parent pharmaceutical companies. This has also been true for new medical device and diagnostic products and, on a smaller scale, for proprietary over-the-counter drug products, health and beauty products, and dental products.

Furthermore, advances in biotechnology are revealing new horizons in health industries [1,53-56]. For example, genetic engineering will probably lead to the mass production of previously rare or unavailable proteins of different types, e.g., growth hormone, interferon, and monoclonal antibodies, for treatment of various illnesses. Another example is the currently rapid growth of immunodiagnostic and immunotherapeutic research.

Overriding the new biotechnology era are the impacts of microelectronics and microcomputerization—both in performance of research (e.g., computer-assisted design of new therapeutic chemical molecules) and in development of diagnostic and therapeutic devices (e.g., the newest generations of blood microanalyzers and cardiac pacemakers) [17].

Therefore, we can look forward to an expansion of research efforts on the part of knowledgeable industrial organizations that are involved in the health field, representing commitments of tens of millions of dollars annually by each such organization [62]. Such commitments will probably lead to even more growth of research-intensive organizations, often via mergers and acquisitions.

Various industrial organizations whose main business has not been in health products, particularly chemical and petrochemical companies, will enter or expand their footholds in the health field. Examples of these, noted in recent public announcements, include Dow Chemical, Monsanto, Olin, and Shell Oil [57,58].

While in-house corporate clinical research functions will correspondingly expand as a result of all of the foregoing developments, we also can foresee a continuing role for outside contract medical research organizations. Such should become especially useful when highly specialized clinical research efforts are required.

We also can probably anticipate an expanding relationship between industry and academia. For example, plans have been publicized for possible industrial/academic collaboration between Hoechst AG Pharmaceutical Company and Massachusetts General Hospital [59] and between Technicon Instruments and Massachusetts Institute of Technology.

Such industry-academic collaborations will undoubtedly be mutually beneficial. On the industry side, there will be access to some of the best brainpower available to solve certain research problems—leading ultimately to profitable advances in diagnostic and therapeutic technologies. Such benefits will accrue not only to the pharmaceutical industry but also to related consumer health industries and the device/diagnostics industries. While specific industrial companies may put tens of millions of dollars into research at particular institutions, they will as a result receive certain advantages in patentability or priority of marketing. The advantages to the academic medical institutions and related scientific institutions of collaboration with industry will be primarily in the area of funding—a critical factor in these times when government funding for research continues to diminish, year after year—and also in the area of beneficial collaboration among scientists of different disciplines and superior abilities, with such scientists present in both academia and industry. Academic institutions will also benefit from royalty agreements with industrial companies relating to profits from marketing of new products, and such royalties should help the academic institutions survive and prosper [57-60].

Last but not least, the general public should benefit from the new industrial/academic collaborations because of the likelihood of acceleration of development of new drug breakthroughs and new technology breakthroughs in the fields of science and medicine.

We also can look forward to more management attention to management development programs for those who are in virtually all areas of clinical research, from physicians to data analysts. All who may express such interest and who demonstrate management potential will likely be given greater systematic attention in the future than has been the case in the past—with fuller opportunities for developing their administrative potentials.

The final result will be a growing supply of qualified and experienced Managers of Clinical Research and related functions. Out of this pool of talent will come future corporate officers in pharmaceutical, consumer health product, medical device/diagnostics, dental, and animal health companies—in key positions of Medical Affairs, Scientific Affairs, Research and Development, Regulatory Affairs, Administration, and, ultimately General Management.

APPENDIX: HIERARCHAL RELATIONSHIPS AMONG VARIOUS CLINICAL RESEARCH AND RELATED FUNCTIONS, AND THEIR RESPECTIVE TITLES (COMPOSITE, PROTOTYPE ORGANIZATIONAL STRUCTURES)

**Vice President, Clinical Research*
(*or* Vice President, Medical Director, or Vice President, Medical Affairs)

†*Executive Director, Clinical Research (Phases I, II, and III)*

Senior Director, Clinical Pharmacology (Phases I and early II)

Associate Directors, Clinical Pharmacology

Assistant Directors, Clinical Pharmacology

Clinical Pharmacology Staff (Monitors)

Group Directors, Clinical Research or Clinical Investigation (Phases II and II)/Cardiovascular, Anti-inflammatory, Anti-infective, etc., by Therapeutic Areas

Directors, Clinical Research *or* Clinical Investigation/by Therapeutic Areas

Associate Directors, Clinical Research *or* Clinical Investigation/by Therapeutic Areas

---

*The Vice President, Clinical Research (or Vice President, Medical Director, or Vice President, Medical Affairs) will usually report to one of the following: President; Executive Vice President; Senior Vice President, Scientific Affairs; or Vice President, Research and Development

†These functions (by title) signify those reporting directly to the Vice President, Clinical Research (or Vice President, Medical Director, or Vice President, Medical Affairs)

Assistant Directors, Clinical Research *or* Clinical Investigation/by Therapeutic Areas

Clinical Research *or* Clinical Investigation Staff (Monitors)/by Therapeutic Areas

†*Senior Director, Clinical Data Management*

Director, Clinical Data Analysis

Assistant Director, Clinical Data Analysis

Clinical Data Analysts

Director, Clinical Science Information Services

Assistant Director, Clinical Information

Senior Clinical Information Scientists

Clinical Information Scientists

Case Report Designers

Director, Clinical Systems and Programming

Assistant Director, Clinical Systems Analysts

Senior Clinical Systems Analysts

Supervisor, Clinical Programming

Senior Clinical Programmers

Assistant Director, Electronic Clinical Data Processing

Computer Operators

Director, Biostatistics

Assistant Director, Biostatistics

Senior Biostatisticians
Biostatisticians

Manager, Biostatistical Services

Clinical Epidemiologist

†*Director, Medical Administration*

Associate Director, Clinical Project Planning and Tracking

Manager, Project Documentation

Associate Director, Clinical Quality Assurance

Staff

Manager, Clinical Research Associates

Senior Clinical Research Associates

Clinical Research Associates

Associate Director, Personnel and Budgeting

Assistant Director, Personnel

Assistant Director, Budgeting

Staff

‡*Executive Director, Clinical Development (Phase IV) and Medical Services*

Senior Director, Clinical Development (Phase IV)

Associate Directors, Clinical Development

Clinical Development Staff (Monitors)

Senior Director, Medical Services

Director, Medical/Marketing Liaison

Assistant Director, Medical/Marketing Liaison

Director, Medical Communications

Associate Director, Professional Inquiries

Supervisor, Medical Correspondence

Staff

Associate Director, Medical Writing and Editing

Senior Medical Writers

Staff Writers

Associate Director, Medical Education

Senior Director, Clinical Safety Evaluation

---

‡The Executive Director, Clinical Development and Medical Services may report to the Vice President, Clinical Research (or Vice President, Medical Director, or Vice President, Medical Affairs)—or, alternatively, to a top marketing executive, e.g., Executive Vice President of Marketing. In the latter case, the Executive Director, Clinical Development and Medical Services may achieve vice presidential status. In the latter case, also, the function of Clinical Safety Evaluation will likely be divided between the Clinical Research and Marketing areas, depending upon the stage of individual drug development.

- Associate Directors, Clinical Safety Evaluation
  - Assistant Directors, Clinical Safety Evaluation

¶*Executive Director, Regulatory Affairs*

- Director, Regulatory Compliance
  - Associate Director, Research and Clinical Compliance
    - Staff
  - Associate Director, Product Labeling (Claims)
    - Assistant Director, Document Control
      - Manager, Document Submissions (INDs and NDAs)
        - Staff
  - Associate Director, Adverse Drug Reaction Reporting
    - Staff
  - Associate Director, Medical Advertising and Promotion Review
    - Assistant Director, Medical Advertising and Promotion Review
- Director, Regulatory Surveillance
  - Associate Director, Regulatory Surveillance
    - Assistant Director, Regulatory Surveillance
- Director, U.S./International Regulatory Coordination
  - Assistant Director, U.S./International Regulatory Coordination

---

¶The Executive Director, Regulatory Affairs may report to the Vice President, Clinical Research (or Vice President, Medical Director or Vice President, Medical Affairs)—or, alternatively, to the Vice President, Scientific Affairs, Executive Vice President, or President. In the latter case, the Executive Director, Regulatory Affairs may achieve Vice Presidential status.

REFERENCES

1. M. Waldholz, Pharmaceutical firms prepare to introduce new "wonder drugs." *The Wall Street Journal 15*:1 (1982).
2. Pharmaceutical Manufacturers Association, Science and Technology Division, *Medical Device Industry Profile.* Pharmaceutical Manufacturers Association, Washington, D.C., 1982.
3. D. Ramroth (Ed.), *Getting New Products to the Market.* Health Industry Manufacturers Association Report 79-3. Hyatt Regency O'Hare, Chicago, 1979.
4. A. Hayes, Food and drug regulation after 75 years. *JAMA 246*: 1223 (1981).
5. U.S. Food and Drug Administration, U.S. Dept. of Health and Human Services, Obligations of clinical investigators of regulated articles. *Federal Register 43*:153 (1978).
6. F. Coulson and A. Kolbye (Eds.), *Regulatory Toxicology and Pharmacology*, Vol. 1. Academic Press, New York, 1981.
7. J. Ballin, Regulation and development of new drugs. *JAMA 247*(21):2995 (1982).
8. W. Wardell and L. Lasagna, *Regulation and Drug Development.* American Enterprise Institute for Public Policy Research, Washinton, D.C., 1975.
9. W. Wardell (Ed.), *Controlling the Use of Therapeutic Drugs: An International Comparison.* American Enterprise Institute for Public Policy Research, Washington, D.C., 1978.
10. Candidate Data Files. Sampson, Neill, and Wilkins, Inc., Upper Montclair, N.J., 1968-1983.
11. Client Data Files. Sampson, Neill, and Wilkins, Inc., Upper Montclair, N.J., 1968-1983.
12. L. Friedman, C. Furberg, and D. DeMets, *Fundamentals of Clinical Trials.* Wright-PSG, Littleton, Mass., 1982.
13. J. Boissel, Today's challenges for statisticians and designs. *Controlled Clin. Trials 1*:333 (1981).
14. E. Gehan, The training of statisticians for cooperative clinical trials: A working statisticians's viewpoint. *Biometrics 41*:326 (1978).
15. E. Grochla and N. Szyperski (Eds.), *Information Systems and Organization Structure.* DeGruyter, Walter, Hawthorne, N.Y., 1975.
16. V. Sondak, H. Schwartz, and N. Sondak, *Computers and Medicine.* Artech House, Dedham, Mass., 1979.
17. J. Meindl, Microelectronics and computers in medicine. *Science 215*(12):792 (1982).
18. J. Paul, *Clinical Epidemiology*, 2nd ed. Univ. of Chicago Press, Chicago, 1976.
19. C. Lyght, The role of the Medical Director in the pharmaceutical industry. *Curr. Ther. Res. 17*:4 (1975).

20. W. Abrams, Doctors seek good life in executive ranks. *The New York Times* p. 1 (Nov. 28, 1976).
21. L. Adair, The physician executive: What he brings to the device/diagnostics industry and why he is attracted to it. *Surg. Business 40*(2):28 (1977).
22. H. Sherwood, Would you be happy in a grey flannel suit? *Am. Med. News* pp. 9-10 (Apr. 25, 1977).
23. J. Burnham, American medicine's Golden age: What happened to it? *Science 215*:1474 (1982).
24. G. Levey, D. Lehotay, and M. Dugas, The development of a Physician-Investigator training program. *N. Engl. J. Med. 305*(15):887 (1981).
25. M. Sampson, How can engineers gain respect (as Contrasted with Physicians)? *Mech. Eng. 97*:95 (1975).
26. E. Heller, *Physicians in Management: Why Are They There?* MIT Press, Cambridge, Mass., 1982.
27. J. Bickel, Editorial response on expansion of M.D.-Ph.D. Clinical Investigator programs. *N. Engl. J. Med. 305*(12):704 (1981).
28. Questions and answers: The Pharm.D. degree. *JAMA 248*(1):93 (1982).
29. D. Sackett and M. Gent, Controversy in counting and attributing events in clinical trials. *N. Engl. J. Med. 301*:1410 (1979).
30. M. Zelen, A new design for randomized clinical trials. *N. Engl. J. Med. 300*:22 (1979).
31. N. Breslow, Perspectives on the statistician's role in cooperative clinical research. *Cancer 41*:326 (1978).
32. C. Klint, The conduct and principles of randomized clinical trials. *Controlled Clin. Trials 1*:283 (1981).
33. W. Westlake, Design and statistical evaluation of bioequivalence studies in man. In *Principles and Perspectives in Drug Bioavailability* (J. Blanchard, R. Sawchuck, and B. Brodie, Eds.) Karger, Basel, Switzerland, 1979, pp. 192-210.
34. B. Vetter, *Supply and Demand for Scientists and Engineers*, 2nd ed. Scientific Manpower Commission, Washington, D.C., 1982.
35. D. Nelkin, Intellectual property: The control of scientific information. *Science 216*:704 (1982).
36. T. Moulding, The unrealized potential of the medication monitor. *Clin, Pharmacol. Ther. 25*:2 (1979).
37. B. Williams, *Computer Aids to Clinical Decisions*, Vol. 1. CRC Press, Boca Raton, Fla., 1982.
38. *Data Processing Occupations*, Job Guide 2R-E. New Jersey Department of Labor and Industry, Washington, D.C., 1980.
39. E. Lee, *Statistical Methods for Survival Data Analysis*. Lifetime Learning Publ., Belmont, Calif., 1980.
40. D. Knuth, The art of computer programming. *Seminumerical Algorithms 2*:104 (1981).

41. E. Johl, F. Christensen, and N. Tygstrup, The epidemiology of the gastrointestinal randomized clinical trial. *N. Engl. J. Med. 296*:20 (1977).
42. A. Ferguson (Ed.), *Attacking Regulatory Problems: An Agenda for Research in the 1980's*. Ballinger, Cambridge, Mass., 1981.
43. R. Haynes, D. Taylor, and D. Sackett (Eds.), *Compliance in Health Care*. Johns Hopkins Univ. Press, Baltimore, 1979.
44. C. Hammer (Ed.), *Drug Development*. CRC Press, Boca Raton, Fla., 1982.
45. W. Curran, Reasonableness and randomization in clinical trials: Fundamental law and governmental regulation. *N. Engl. J. Med. 300*:22 (1979).
46. W. Wardel, Postmarketing surveillance of new drugs: I. Review of objectives and methodology. *J. Clin. Pharmacol. 19*(2&3):85 (1979).
47. W. Wardell, Can improved postmarketing surveillance permit earlier drug approval? *Drug Ther*. pp. 143-153 (Feb. 1982).
48. J. Rudman, *Administrative Manager*, C-1754. Career Examination Services, Washington, D.C., 1977.
49. D. Yoder and H. Heneman (Eds.), *Administration and Organization*, LC74:8047. Bureau of National Affairs, Washington, D.C., 1977.
50. L. Dyer (Ed.), *Careers in Organizations: Individual Planning and Organizational Development*. New York School of Industrial Relations, New York, 1976.
51. G. Knatterud, Methods of quality control and of continuous audit procedures for controlled clinical trials. *Controlled Clin. Trials 1*:327 (1981).
52. H. Lahon, R. Rondel, and C. Kratochvil, *Pharmaceutical Medicine: The Future*. Proceedings of the Third International Meeting of Pharmaceutical Physicians, sponsored by the International Federation of Associations of Pharmaceutical Physicians, Brussels, Oct. 1978.
53. D. Fredrickson, Venice is not sinking (the water is rising): Some views on biomedical research. *JAMA 247*(21):2995 (1982).
54. E. Roberts, R. Levy, S. Finkelstein, J. Moskowitz, and E. Sondik, *Biomedical Innovation*. MIT Press, Cambridge, Mass., 1981.
55. J. Paul (Ed.), *Genetic Engineering Applications for Industry*. Noyes Data Corp., Park Ridge, M.J., 1981.
56. D. McKerrow, Funding the next generation of molecular biologists. *Pharm. Executive* pp. 28-29 (Dec. 1981).
57. B. Culliton, The academic-industrial complex. *Science 216*:960 (1982).
58. B. Culliton, Monsanto gives Washington U. $23.5 million. *Science 216*:1295 (1982).
59. B. Culliton, The Hoechst Department at Mass. General. *Science 216*:1200 (1982).

60. P. Gray, M.I.T. wants closer ties with business. *The New York Times*, p. 1 (Sept. 27, 1981).
61. D. Altman, R. Greene, and H. Sapolsky, *Health Planning and Regulation: The Decision-Making Process.* Health Administration Press, Ann Arbor, Mich., 1981.
62. *Industrial Research Laboratories*, 15th ed. Bowker (Jacques Catell Press), New York, 1979.
63. F. Stitt, *Clinical research data management and analysis: A physician's view.* Paper presented at the Drug Information Association and Pharmaceutical Special Interest Group 1978 Workshop on Clinical Data Processing and Analysis, Arlington, Va., Feb. 27, 1978.
64. *200 Ways to Put Your Talent to Work in the Health Care Field*, No. 400M. National Health Counsel, New York, 1977.
65. *Health Careers in New Jersey.* New Jersey Department of Labor and Industry, division of Planning and Research, Washington, D.C., 1976.
66. J. Nassif, *Handbook of Health Careers: A guide to Employment Opportunities.* Human Sciences Press, New York, 1980.
67. W. Wilkins, Executive recruiting: An outside view. *Med. Marketing & Media 12*:46 (1977).

26

RECENT TRENDS: THE IMPACT OF THE ENVIRONMENT ON RESEARCH STRATEGIES AND PRODUCTIVITY

Richard E. Faust
Hoffmann-La Roche Inc.
Nutley, New Jersey

I. INTRODUCTION

Pharmaceutical innovation and drug development have been influenced by many environmental constraints over a long period of time and these have had a cumulative, pervasive and negative effect. Many of the constraints are interrelated and additive in their influence on the research process and the corporation. For example, all of the regulations, such as Good Manufacturing Practices, that impose demands on the production function will influence how quickly and efficiently the corporation is able to utilize the output of its R&D operations. Also, all of the regulatory and pricing policies that make it difficult for marketing to compete successfully will influence research to the extent that, without marketing success, funds will not be available to reinvest in new drug discovery and development. All of these and related environmental pressures place burdens on the corporation and make it increasingly difficult for senior management to direct more and more resources to the costly R&D function.

II. IMPACT ON DRUG DISCOVERY AND INNOVATION

The complexity of modern biomedical research and the focus on more challenging diseases, such as cancer, cardiovascular disorders, and arthritis, account in part for the slowdown in new drug flow. However, the regulatory environment since 1962 has also had a significant negative impact on drug innovation and development. The effects of these forces and patterns over the past two decades on research operational strategies and outputs may be summarized.

A. New Product Introductions Down

The average number of new products and new chemical entities introduced annually between 1961 and 1970 was 149 and 20, respectively, compared with 356 and 43, respectively, over the 1951-60 decade. The United States rate of pharmaceutical innovation remained low during the 1970s [1]. (Hopefully, however, the upturn in new drug approvals seen during the past couple of years will continue.)

B. Drug Development Cost/Time Up

The time to develop a new drug today can extend to 13 years and cost over $70 million compared to pre-1962 estimates of approval time of 1 to 2 years and costs of less than $7 million [2].

C. Diminished Patent Protection

Because of the longer development cycles the effective patent life after New Drug Application (NDA) approval has declined from over 15 years to the current 9.5 years [3].

D. Less Return on Research Investments

The estimated rate of return on R&D investments has declined from 21.7% for drugs introduced in 1958 to 10.8% for drugs introduced in 1978 [4]. A recent study indicates that during the period 1967-1976, only 25% of a cohort of new chemical entities (NCEs) achieved higher-than-average returns on investment. The remaining 75% failed to generate cash flows adequate enough to recover the cost of the R&D investment [5].

E. Shift of Research Overseas

There has been a shift of research investments and efforts overseas within major U.S. pharmaceutical firms. From less than 10% of company spending in the late 1960s, today nearly 20% of the approximately $2,000 million is spent overseas [6].

F. Changes in Research Emphasis

As a result of the many pressures on research, there have been a number of changes in research strategies and emphasis. Because of the high cost of moving compounds through the development cycle, most firms have fewer such agents in the pipeline. There is a tendency to be more selective and concentrate on those high-usage therapeutic categories having significant sales potential. Often, fewer funds are available for exploratory research, as resources are directed to drug development and efforts to extend the life cycle of existing products through the creation of new dosage forms and/or broadened indications

G. Less Flexibility and Serendipity

A creative and productive research environment is favored by an open and flexible approach that encourages change and experimentation and that is characterized by a rational cost/benefit and risk/benefit orientation. As forces generate pressures on research investments, we often see the evolution of a more structured and regimented research organization and one that is "analyzed" by financial, marketing, and corporate planning personnel who are concerned with research productivity.

H. Increased Technology Transfer Through Licensing and Collaborative Programs

Increasingly, firms are seeking to complement and strengthen internal research through collaborative programs with university research groups or other companies. For example, this has been an important strategy for many firms moving into the recombinant-DNA field. The licensing of products from other companies represents a key growth strategy for many firms, even those that are considered highly research-intensive. The magnitude of this activity in the world arena is reflected in the fact that nearly 300 products were the subject of licensing or joint development agreements in 1980 [7].

I. Problems of the "Static" Research Organization

In addition to the multitude of changes in research strategies and output already noted, the various environmental forces that impact negatively on the corporation and bring about reduced sales and income growth (or actual declines) also may reduce commitments to R&D, or at least influence them adversely. When this happens and we have a relatively static or shrinking research organization, a number of negative operational patterns may emerge which present challenges to research management and affect directly esprit de corps within the scientific staff [8]. These challenges resulting from environmental constraints on the firm and the research process have not always been fully recognized and communicated and include, for example, those associated with the evolution of a middle management "bulge," or trend toward a "production" mentality, and attempts to reorganize as the only solution.

III. ACCELERATING THE DRUG APPROVAL PROCESS

A number of proposals have been put forth to speed up the drug approval process. Testifying before the Congressional Science and Technology Committee, Dr. McMahon (Chairman of the Congressional New Drug Review Commission) offered the following recommendations to accelerate the approval process and make drugs available to the medical profession earlier: "(a) Minimize FDA involvement in Phase I-II by reducing the IND Phase I-II requirements to a 'Notice of Initiation of Studies.' (b) Require (as now) a Form 1639 to be submitted should toxicity or serious adverse reactions occur. (c) Require only a 'Notice of Cessation' to be submitted to FDA when a drug fails in Phase I-II. (d) Require only a certified summary of preclinical and clinical data when a drug is proposed for Phase III studies—with a Phase III program outline to result in approximately 500 controlled patients for

NDA submission. (e) Require only a Certified Summary for NDAs. (f) Require NDA decision within 180 days. (g) Require postmarketing surveillance for new drugs [9].

In a talk at the 1981 Annual Meeting of the Pharmaceutical Manufacturers Association, Health and Human Services (HHS) Secretary Schweiker reaffirmed the need to encourage innovation through research and speeding up the drug approval process. He outlined the following eight approaches, some related to those proposed by McMahon, to be pursued either administratively or through legislative channels: (a) loosening FDA regulation during the earliest stages of drug research; (b) allowing local institutional review and approval of preliminary drug testing; (c) reducing the paperwork burden by providing for greater use of summaries by the FDA in making its drug approval decisions, requiring the submission of raw case data only upon request; (d) providing for a treatment Investigational New Drug application (IND) under which the primary purpose of drug use is to treat patients with a serious illness not satisfactorily treatable with alternative therapy; (e) accepting foreign clinical studies to the same extent as domestic clinical studies in Phase II and III under an IND and for approving an NDA; (f) establishing a procedure for the review and resolution of scientific disputes by public advisory committees; (g) making the drug reviewers more accountable in giving the drug sponsors prompt notice regarding review of their application; and (h) requiring the reviews to be conducted concurrently within the statutory time frame [10].

In a letter to President Reagan, Congressman Scheuer of the House Committee on Science and Technology reaffirmed the need to streamline the drug approval process in the United States [11]. He pointed out that 13 of 14 drugs classified by the FDA as important and approved between July 1975 and February 1978 had been available in other industrialized countries from 2 months to 12 years earlier. In the letter he made the following specific recommendations: "(a) That the FDA fulfill the intent of Congress to make medicines promptly available. (b) That the FDA fulfill its legislative mandate by clearing drugs within the statutory 180 days. (c) That the FDA itself meet the requirements it sets in its own regulations and stop arbitrarily shifting standards. (d) That the FDA stop overreaching, particularly in its intrusions into pharmacologic investigations, clinical research and medical practice. (e) That the FDA be stripped of its *sole* power to grant special exemptions from drug use restrictions for individuals and small groups on an emergency basis under the compassionate drug approval provisions. (f) That the FDA stop adding new layers or bureaucratic requirements." Congressman Scheuer noted, furthermore, that the introduction of a postmarketing surveillance system to speed a new medicine into use should be linked to drug clearance earlier and not be just an add-on to the current extensively long approval process.

A number of these proposals have been embodied in the proposed NDA regulations published in the October 19, 1982 issue of the *Federal Register* and are estimated to reduce the cost of NDAs by $2.5 million and reduce NDA approval time from 27 to 21 months for most submissions. The thirteen proposed changes may be outlined briefly:

1. Streamlined format for applications
   - Overall summary (50-200 pages)
   - Section summaries (clinical, pharmacology, chemistry, statistics, biopharmaceutics, microbiology)
   - Parallel review within FDA
2. Minimized supporting information
   - No case report forms submitted routinely
   - Tabulations of patient information
3. Fewer supplements to approved applications
   - Certain information
     - No longer required
     - Included in annual reports
     - Not subject to agency preclearance
4. Reduced recordkeeping and reporting requirements on approved applications
5. Time frames for filing and FDA review of applications clarified
   - FDA action within 60 days
     - Application "acceptable" or not
   - FDA action within 180 days
     - Approval letter
     - Approvable letter
     - Not approved
   - Extension for major amendments limited to review time only
6. Meaning of action letters clarified
7. Expedited approval based on draft labeling
   - Approval letter (rather than approvable letter) issued if labeling deficient only in editorial or minor respects
8. Expedited hearing procedures on refusals to approve an application
   - Specific time frames for issuance of notice of opportunity for hearing and notice of hearing
9. Improved procedure for resolving scientific disputes
   - To be implemented 30 days from 10/19/82
10. Policy on acceptance and use of foreign data clarified
    - May rely solely on foreign studies that meet U.S. standards
11. New safety update reports for pending applications
    - Required every four months following NDA submission and receipt of approvable letter
12. Strengthened postmarketing surveillance for approved drugs
    - 15-day "alert reports" for serious adverse drug reactions (ADRs)
    - 30-day reporting of other experiences

13. Increased communication with NDA applicants
    Informal meeting 90 days into review cycle

The Pharmaceutical Manufacturers' Association (PMA) has stated that although these proposed revisions to the NDA rules represent a useful first step, a number of serious deficiencies in the drug approval process remain. The PMA contends that any comprehensive reform of the regulatory system must include the following: (a) Regulations governing the conduct of the FDA and the pharmaceutical industry in implementing the IND/NDA process from beginning to end. (b) Regulations establishing the substantive standards by which the safety and effectiveness of drugs will be judged. (c) Regulations governing the content and handling of the IND phase of drug development. (d) Regulations governing the content and handling of the NDA phase of drug development [12]. The current proposals ignore all but the last of these four essential elements and such vital aspects as the IND phase of drug development and postmarketing surveillance. Clearly, a more comprehensive revamping and improvement of the drug approval process is needed if the flow of new drug products is to be encouraged and accelerated.

IV. CURRENT TRENDS AND PERSISTENT CONCERNS

Over the past few years, there has been growing recognition that we need to remove many environmental constraints on R&D in the pharmaceutical industry and provide more incentives for biomedical research. As already noted, some favorable actions are emerging. In addition, the recently enacted federal tax bill provides the business sector with numerous incentives for investment in innovation, including a 25% tax credit for incremental R&D expenditures. Also, the Patent Term Restoration Act of 1981 is designed to improve the climate for innovation in the United States by increasing patent protection for up to 7 years to offset the time required for mandated testing and regulatory review. If enacted into law, this patent restoration will stimulate the application of basic biomedical knowledge to the discovery of new therapeutic modalities. Although there are a number of encouraging developments that promise a more favorable environment for drug innovation and development, the picture is still worrisome, for persistent issues and new concerns continue to bring challenges to the industry.

A. Relative Efficacy

Various consumerist and regulatory groups favor the approval of drugs based on comparisons with existing products. This position of relative efficacy would preclude the introduction of a drug simply because there are existing products which already treat the disease entity "adequately." Such a policy would potentially deny therapy to a patient who might not be able to tolerate one product, but would effectively respond to a similar but alternate product designed to treat the same illness. Furthermore, since we are becoming increasingly aware that patients respond differently to similar drug products, an array of drugs are needed so that the physician can tailor the medication to a particular patient.

B. Postmarketing Surveillance

The adoption of a postmarketing surveillance requirement and an easier NDA withdrawal procedure will probably not alter the basic conservative approach toward drug approval at FDA. The FDA might be severely criticized for allowing distribution of a drug which is subsequently removed. Also, pharmaceutical firms may be reluctant to jeopardize substantial investments in a drug by having it withdrawn when additional pertinent data might be generated toward the end of the development cycle. In other words the "easy-on, easy-off" pathway for new drugs has very real potential negatives if not implemented with some of these factors in mind.

C. Abbreviated/Paper NDAs

The extension of Abbreviated NDAs to certain drugs approved after 1962 could affect unfavorably economic incentives for drug development by reducing the revenues on innovative drug products through increased competition. Consequently, this reduction in revenues might represent a future cost to consumers in the form of the delayed introduction of new and improved therapies.

D. Confidentiality of Research Data

Safety and efficacy data generated by research-intensive pharmaceutical firms represent the greater part of the manufacturers' know-how not protected by patents. Should these data be made available to imitators without the pioneer firms' consent and remuneration, this would result in an unfair competitive advantage and discourage innovation. Postmarketing studies required by the FDA may also represent valuable trade secret data which if released would be economically damaging to manufacturers. The Public Citizen Health Research Group has sought to force release of this research information, which would have profoundly negative impact on research-oriented drug firms in the industry.

E. Look-Alike Drugs

Consumerist groups and many generic drug manufacturers are objecting to the "anti-lookalike" position, indicating that it is against the public interest and will suppress drug competition. They assert that people will resist buying generic drugs because of anxiety stemming from the generic substitutes' failure to resemble the color and shape of the brand-name products to which they have become accustomed. The product lookalike conflict seems to have been addressed temporarily in favor of the research-intensive firm, but the issue may not be completely resolved.

F. Expanding Regulations

Although the trend is toward deregulation, not all recent developments are moving in that direction. Until now, OSHA has not promulgated standards or guidelines specifically for laboratory work as it has for industrial environments. Now, however, it appears that the agency is developing new rules for protecting laboratory workers from toxic substances. The amount of monitoring and documentation that may be demanded could be enormous, further diverting research personnel and resources from more creative pursuits. Another persistent problem has been the mounting concern expressed in the use of animals in drug research. The Animal Welfare Act of 1966 strictly regulates the use of animals in the laboratory. A new proposed Research Modernization Act seems to take the position that current animal protection rules are inadequate, that biomedical researchers are not taking full use of alternative testing methods, and that duplication of tests involving live animals is widespread and unnecessary [13]. Critics of this proposed bill have noted that progress in the biomedical sciences would be severely impeded by the passage of such legislation. In another direction, growing concerns by consumerist groups and others led to a proposed ban on the use of prisoners in penal institutions. Reflecting a more rational approach, the FDA has stayed the effective date of the ban pending a reanalysis of the need for such legislation [14]. It is hoped that any reproposed regulations will permit the use of prisoners in research projects following appropriate guidelines and cautions.

G. Compulsory Licensing

Some groups are calling for compulsory licensing of drugs which benefit from patent extension under patent restoration legislation. Such a move would decrease incentives for research investments and offset the value of the restored protection. It is interesting to note that a Task Force on Biotechnology headed by the Minister of State for Science and Technology in Canada concluded that pharmaceuticals will be markedly influenced by the advent of various biotechnologies, but warns that unless changes are made in legislation regarding compulsory licensing, it is unlikely that firms will invest significantly in R&D in these emerging and exciting areas.

H. Expanding Product Liability Claims

As product liability claims against pharmaceutical firms increase in number and the size of awards, funds that could be directed to R&D are siphoned off. Adverse and often exaggerated negative publicity also frequently follows. However, important secondary impacts from this trend have evolved. Some firms have moved out of research in the vaccine field partly because the costs and risks have become such that they outweigh any possible commercial returns. Some individuals have proposed that the government share the risk of product liability claims in this area and in other situations where the protection of public health involves certain risks.

I. Consumerist Movement

Various consumer representatives now serve on some FDA advisory committees and actively participate in or influence drug regulatory activities through this and other channels. Such consumer representation is valuable, but often their orientation is ultraconservative and anti-industry. A new type of consumer advocate for drugs is needed. Since most consumers of medicines are not hostile to drugs and would favor having better treatment regimens, their voice should also be heard. Perhaps the various disease-oriented foundations and associations should nominate "consumer-representative" candidates for inclusion on specific FDA advisory committees and other groups influencing drug legislation and policy [15].

In any case, the important voice of the consumer is becoming more influential. For example, there is a Center for Medical Consumers in New York City which maintains a library on the side effects of drugs and, according to a radio commercial, "information usually reserved for physicians." Also, in its first foreign drug regulatory involvement, the Health Research Group (HRG) urged the French Ministry of Health and Social Security to remove a drug from the French market, tienilic acid, a diuretic/anti-hypertensive agent which was taken off the U.S. market by Smith Kline. The HRG noted that it will intervene in foreign countries whenever gross, inexplicable differences exist between nations in the regulation of dangerous drugs [16]. Even the Federal Trade Commission (FTC) has become involved and has sought to stimulate awareness of generic drugs with mass mailings to "consumer professionals" explaining the cost-saving benefits of buying drug copies. In the pursuit of economic efficiency and the concern for the immediate improved economic well-being of consumers, the FTC represents another external force depressing drug prices and putting pressure on funds available for biomedical research and new drug development.

J. Orphan Drug Legislation

One positive trend has been the support now being given to aid the development of orphan drugs. President Reagan signed the Orphan Drug Act into law on January 4, 1983. Several major provisions include: (a) that HHS specify in writing all requirements for an NDA approval upon request by a sponsor; (b) that HHS designate a substance as an orphan drug to provide eligibility for special tax and financial considerations; (c) that a 7-year marketing exclusivity be established for nonpatented products; (d) that open protocols be used for Phase III clinical testings; (e) that a 50% tax credit for the cost of human clinical testings be given; and (f) that HHS enter into grants and contracts for qualified clinical studies on designated orphan drugs. The tax credit provision will expire on December 31, 1987 and the grant/contract provision allows $4 million for each of the next 3 fiscal years.

C. CONCLUSIONS

Faced with soaring levels of unpredictability, increasing public criticism, hostile political pressures, and mounting regulation, many of the largest pharmaceutical firms are being forced to redefine objectives and operational strategies. The new demands reaffirm the fact that a corporation can no longer be responsible simply for making a profit or producing goods, but must simultaneously contribute to the solution of complex ecological, moral, political, racial, and social problems. Increasingly, multiple interconnected bottom lines are emerging which influence the research operations not only in terms of funds available for R&D, but also in such areas as the hiring and training of scientific personnel, closer interactions between manufacturing and marketing functions, informational support for various public policy challenges, the generation of data for legal issues and concerns, and the development of new decisional processes and modes within research in order to respond to the generalized speedup of corporate metabolism as the firm seeks to react to powerful environmental changes. As these changes take place, the history of the pharmaceutical industry provides some important general lessons for the current national policy debate on the revitalization of the U.S. economy and the role of the corporation in society.

First, it underscores the importance of industrial innovation for economic prosperity and improved social welfare. In particular, new drug therapies have played a central role in the impressive amount of medical progress that has occurred over the past half-century. In addition, rapid innovation in pharmaceuticals and other high-technology industries has been associated with very favorable economic performance, including high rates of output and productivity growth, rising wage and profit streams, declining price levels in real terms, and positive trade balances.

A second important lesson is that the level of innovation is not independent of basic economic forces. The potential for innovation at any point of time depends of course on the level of scientific opportunities. This varies considerably across industry groups. However, even in industries like pharmaceuticals with rich scientific opportunities, the innovational process involves a long and costly investment of resources and is subject to a high level of uncertainty. When one examines trends in the "bottom line" of expected rewards vs. risks, this has been moving in an unfavorable direction for pharmaceutical innovation and in many other sectors over recent periods.

Third, governmental social and economic policies over the past few decades have been a major factor contributing to these adverse trends. In the case of pharmaceuticals, for example, public policy has been characterized by increasingly stringent regulatory controls, shorter effective patent terms, and increasing encouragement of generic product usage. While all of these policies have been well-intended, in combination they have produced significant unintended adverse side effects on the drug innovation process. They have contributed to the present situation of fewer independent domestic sources of innovation and fewer annual new drug entity introductions. This has occurred despite a steadily expanding base of rich scientific opportunities emerging from basic research endeavors.

It is encouraging to note that the public and policymakers in the United States have gradually recognized these problems and have initiated major regulatory reform programs. Future prosperity depends on today's actions and the ultimate beneficiaries, of course, will be the patient and the public.

REFERENCES

1. E. Caglarcan, R. E. Faust, and J. E. Schnee. In *Impact of Public Policy on Drug Innovation and Pricing*, (S. A. Mitchell and E. A. Link, Eds.). American University, Washington, D.C., 1976, p. 331.
2. R. Hansen, *The Pharmaceutical Development Process: Estimates of Development Costs and Times and the Effects of Proposed Regulatory Changes*, Center for Research in Government Policy

and Business, University of Rochester, Rochester, N.Y., 1981.
3. C. M. Mathias, Senator, United States Senate, *Congressional Record, 127*(14):(Jan. 27, 1981).
4. M. Statman, The effect of patent expiration on the market position of drugs. *Managerial and Decision Economics 2*:61 (1981).
5. J. Virts and J. F. Weston, Returns to research and development in the U.S. pharmaceutical industry. *Managerial and Decision Economics 1*(3):103 (1980).
6. *Prescription Industry Fact Book 1980.* Pharmaceutical Manufacturers Association, Washington, D.C., p. 29 (1980).
7. *IMS Pharmaceutical Newsletter 8*(18): (1981).
8. R. E. Faust, Motivating the scientist in the climate for research today. A paper presented at the Pharm. Tech. Conference, sponsored by Pharmaceutical Technology, Inglewood, California; New York City, Sept. 22-24, 1981.
9. G. McMahon, Pharmaceutical innovation—promises and problems. Testimony before House of Representatives' Subcommittee on Natural Resources, Agricultural Research and Environment and Subcommittee on Investigations and Oversight, April 27, 1981.
10. R. S. Schweiker, Remarks presented at Annual Meeting of Pharmaceutical Manufacturers Association, Boca Raton, Florida, April 6, 1981.
11. J. H. Scheuer, Congressman, Letter to President Reagan, March 6, 1981.
12. *Pharmaceutical Manufacturers Association Newsletter 24*(49):1 (1982).
13. R. J. Griffen, Animals in pharmaceutical research. *Amer. Pharm. 21*(6):14 (1981).
14. *Federal Register 46*(129):35084 (1980).
15. L. Lasagna, Wanted: A new type of consumer advocate for drugs. *N. Engl. J. Med. 298*(16):906 (1978).
16. *SCRIP* No. 591, May 18, 1981.

27
A FUTURISTIC VIEW

Austin Darragh and Ian Brick
Institute of Clinical Pharmacology Ltd.
Dublin, Ireland

Pharmacological clinical research has had an illustrious past, is experiencing a problem-filled present, and—despite prevailing pessimism—has a promising future.

In less than two decades from now we enter the twenty-first century. How different will be the pharmacopoeia of 2080 A.D. from that of the present decade? The comprehensive compendium of materia medica has changed in its content more in the last half century than in the preceding 4,000 years. How many more centuries or millenia will pass before the ultimate edition is finalized cannot even be speculated upon.

There has been an explosion of knowledge since the midpoint of the twentieth century. The deluge of new scientific facts to underpin the science of therapeutics shows little if any sign of diminishing. Already sheer volume has outmoded the conventional textbook. The physical form of the pharmacopoeia will surely alter as dramatically as its content. The volumes on the shelves will be replaced by the computer terminal with visual display and hard copy modules.

In common with structural engineering, space flight, and ballistics, the complexities of present and future therapeutics are such that the capability to compute the consequences of a multidimensional interaction more rapidly than either the brain or a slide rule can do may become a standard necessity. Electronics will handle the masses of intertwined data to give science and safety explicit expression. The personal (computer) desk-top terminal with access to instantly and comprehensively updated global knowledge of drug action and interaction could possibly fulfill the requirements of the physician of the not so distant future.

Although many authorative opinions recently expressed in key scientific journals [1-3] have unanimously predicted a further reduction in new chemical entities entering the area of clinical research, this does not mean a complete halt to innovation. New chemical entities developed not by the serendipity of the past but as a consequence of scientifically predetermined lines of pursuit will be presented for evaluation, especially in the field of biological process modulation. In addition, new delivery systems will be developed utilizing new concepts such as liposomal linked drugs or agents coupled to monoclonal antibodies to achieve, in some cases, precise HLA-DR specific targeting. Pharmacists are dreaming up prescribable prodrug forms, drugs complexed with unique bioavailability enhancers, and drugs to correct the tonic balance of the glomerular membrane to reverse the selective macromolecular leakage of the early stage of nephrotic disease.

As these new modalities of therapy become available, they will endow greater potential to the prescriber's hand. Whether that hand holds an electronic pen or operates a computer keyboard, the resulting prescription will be subjected to computerized registration, checked for correctness of dosage and indications, and cross-checked for contraindications, special precautions, and compliance surveillance requisites before the script is ultimately cleared for dispensing.

The history of therapeutics has followed the evolutionary process—but in the reverse order: at first physicians prescribed for the sick body or its all pervading humors; later medicines were prescribed to restore health to diseased organs; now and in the future the thrust of pharmacological research will be directed at the disordered cell and its component elements.

Alfred, Lord Tennyson, wrote: "The old order changeth, yielding place to new; and God fulfills himself in many ways, lest one good custom should corrupt the world" [4].

Those lines epitomize the essential dynamism of the pharmaceutical industry during the past half century. In the absence of knowledge of the fundamental mechanisms which become disordered by disease, with few exceptions new drugs were developed by researchers on an empirical basis, constantly endeavoring to improve their therapeutic index, the older poisons being discarded in a process of pharmacopaeal purging which persists.

This pattern of evolutional pharmacological serendipity was altered dramatically by the epic work of James Black. He pursued the quest for the "Magic Bullets" dreamed of by Paul Erlich at the opening of this century. But Black had that crucial half a century of scientific progress to enable him to characterize the chemoreceptors postulated by Erlich and to go further by designing the specific antagonistic ligands to block the reception of stimuli at the precisely defined receptors. Erlich would have applauded the remarkable achievement of that one scientist who gave the world β-adrenergic receptor blockers and H_2-receptor blockers within the space of two decades.

Armed now with rapidly increasing knowledge of fundamentals of cell biology and aided by the stereochemical potential of computational quantum analysis, the biologist and the chemist have joined forces to develop precisely designed molecular entities of therapeutic potential which hold promise of high specificity and low toxicity (in conventional terms), and for humans a greater life expectancy than was afforded by the selective poisons of the recent past.

The arbitrary classification of biochemical processes into endocrine, immune, and genetic systems, while convenient for preliminary consideration of their physiology, must be dispensed with if a fuller insight is to be gained. In place of the cumbersome and artificially divisive presentation of three quite distinct systems (dependent for their separate identities on terminological uniqueness only), a unifying concept of a fundamental phenomenon of the living cell, namely, the capability to process information embracing perception, storage, retrieval, transmission, and response, will facilitate an even greater understanding of all biology.

Immunology can be regarded as a system combining features of both the endocrine and genetic systems. Like the process of storing data concerning inherited characteristics, the immune system stores specific information concerning encounters with foreign proteins to which antibodies have been elaborated. In the humoral mechanisms by which T-lymphocytes influence the functions of other leukocytes, the immune system is analogous to the endocrine system with its messenger molecules which T. H. Huxley termed hormones. These in turn can be compared directly to messenger RNA elaborated to convey messages or instructions to ribosomes.

Immunology holds promise of being to therapeutics what microbiology was in the last three decades. Then the market wanted miracle drugs to conquer the infectious diseases that ravaged the young and decimated the old.

The market demanded and the industry accepted the challenge in magnificent fashion. It produced weapon after weapon to wage an unceasing war against disease, pain, disablement. Now the public—or the market, call it by its other name—is raising its voice to demand that the industry provide new drugs, safe drugs, to take up the fight where antibacterials and vaccines have reached the limit of their potential.

Alvin Toffler in his great volume *Third Wave* [5] makes the almost irrefutable case that the Second Wave of human development—the industrial wave—has dashed upon the shore of time and is now a spent force. He lucidly displays the rise and fall of the great industrial complexes of the eighteenth, nineteenth, and twentieth centuries with their shared characteristics of:

Standardization
Specialization
Synchronization
Concentration
Centralization
Maximization

In this industrial revolution, with the inevitable divorce of production from consumption because of the vastly increased output of the production units, the market place assumed a position of paramount importance—politically, economically and socially. The demise of the vast industrial centers in many parts of the world testifies to the change in producer-consumer relations and the diminishing significance of mass production with greater emphasis on catering for more precisely defined market sectors.

To every rule there is an exception. Because of the inescapable legislative controls imposed by public pressures on the manufacture, development, and distribution of drugs, no possibility of complete fragmentation or dismantling of the industry is possible.

Change is nevertheless taking place. In common with other industries that are dependent upon high technological inputs only episodically, companies are now questioning the desirability, in fact the necessity, of maintaining on payroll and inventory expertise and capital items which are utilized only intermittently.

The oil industry has for many years preferred to assign defined functions to contractors who carry out exploration and field development work, the principals harvesting the crude, perhaps sharing refining facilities, but retaining control of the marketing and distribution of the final product. Similar decentralization has been the dominant characteristic of the space program.

The pharmaceutical industry is now exhibiting a similar pattern of retrenchment of investment in high-cost, episodically utilized technology. The advent of the biotechnology explosion coincided with this change in financial policy. As a result, small groups of chemists and biologists have emerged in the role of the exploration companies. Contracted workers in academia and in commercial contract research institutions now regularly supplement (and in some cases supplant) a pharmaceutical company's depleted in-house resources.

This experiment has already proved to be highly successful not only in respect of products of a biotechnical nature but also for other drugs as well. Among the tangible benefits to the pharmaceutical company "buying in" a new product of development expertise is the unexpected major shortening of the drug development cycle by as much as 5 years occasionally and at least 2 years generally.

Theoretically it should be possible to live one's life in good health and then die in good health. That utopian concept comes nearer with each passing decade as new weapons are added to medicine's armamentarium, yet tantalizingly at the same time for each one of us it recedes from our grasp as the tempo of research output by all the world's resources—whether they be in industry, government laboratories, or the universities starved of funding and strangled by excessive legislature controls—slows down more and more.

It is so easy to speculate about more and more wonder drugs, but to bring such fantasy to fruition is yet another day's work. Rudyard Kipling could well have had the innovators of the pharmaceutical world in mind when he wrote [6]:

If you can dream
And let not dreams become your master,
If you can think
And make not thoughts your aim,
If you can meet with triumph and disaster
And treat these two imposters both the same . . .

One could postulate unendingly about the new therapeutic drugs the world would like to have. Undoubtedly some of these would belong in the "orphan" category, while the majority would encompass effective curative treatment of age-related deterioration of the body, cancer, and the progressive diseases of the central nervous system. Pain in its many and distressing forms, expecially the chronic neurological forms of it, has yet to be mastered by nonaddictive medicaments. The expanding horizons presented by recent knowledge of the neuropeptides (and one thinks of Substance P in addition to the endogenous opiates) may provide a rewarding lead to this objective.

The mind, that vast and as yet poorly traversed terrain, is still a frontier land grudgingly yielding a foothold in the battle to understand the delicate mechanisms which hold the balance between sanity and insanity. The endorphins, which are exciting scientists in so many different areas of medicine from neurology to coronary care, may also prove to be the gatekeepers of the deepest recesses of memory; and the opiate antagonists may find among their many new roles a usefulness in pushing aside those gatekeepers to liberate toxic material in the psyche which may be the cause of mental illness.

In the end, when the ultimate pharmacopoeia is written, it is likely that the specific drug for each ailment will be a precisely turned tool, machines to act upon a well-defined cellular component or biochemical process. That day is a long way off, and meanwhile the search now started will proceed, however slowly.

It has been calculated that, despite improvement in therapeutics since the end of World War II, some 20,000 human ailments still await effective remedies. The greater understanding now of the abnormal cellular processes which cause disease opens the way for the development of specific therapeutic agents.

Today, industry with a high technological and scientific content finds itself dependent upon vital components of sometimes a disproportionately small size when compared in purely physical or financial terms to its own mass.

Clinical pharmacology—the science concerned predominantly with the study of drugs in man—is an integral but *essentially independent part* of the cycle of drug development. In a way, the role of clinical pharmacology in the pharmaceutical industry is similar to a microchip in a complex computing machine—in size insignificant, in importance indispensable.

Rigorous controls now monitor the emergence of a new therapeutic drug candidate from the realms of research to acceptance as a clinically proven therapy. The clinical pharmacology unit in which new drugs undergo those critical first test exposures in humans has become in effect "the Pass of Thermopylae"* for each new molecule or drug formulation. Here a drug's prospects will stand or fall on merit alone.

Pharmacological clinical research falls into two major divisions: (1) experimental therapeutics, the importance and future of which is beyond dispute, though the question of who will be responsible for its funding remains to be resolved; and (2) clinical pharmacology, the future of which warrants more detailed consideration.

Although some would place the origins of clinical pharmacology as far back as 1747, when Sir James Lind conducted the first controlled clinical trial and proved the efficacy of citrus fruits in the treatment of scurvy, others would prefer to avoid the persistent confusion between pharmacology and therapeutics engendered by that scientist's epic work and point to three pivotal studies in the field of endocrinology as the beginnings of the discipline of clinical pharmacology.

The first two studies were more pharmacological than clinical. In 1771, John Hunter successfully demonstrated that the male testis contained a factor capable of endrogenizing hens. A century after the study of Lind (which took bureaucracy 40 years to recognize), in 1849 Berthold confirmed the testis as a source of androgens when he successfully reversed the results of castration in roosters by reimplanting testes and restoring male characteristics. The third pivotal study was that of C. E. Brown-Séquard (in 1889), who demonstrated not the androgenic efficacy of the extract he administered to himself but the inescapable necessity to use a placebo control in even the most rudimentary clinical pharmacological protocol.

Clinical pharmacology in recent decades has had great difficulty in establishing its role in an acceptable and uncontroversial way in many medical institutions. Essentially this difficulty has stemmed from

---

*Thermopylae has a long association with therapeutics. The hot spring from which it derives its name has been for centuries renowned for treating rheumatism.

the confusion of clinical pharmacology and experimental therapeutics. Clinical pharmacology is the domain of the clinically orientated researcher whose role is to define the actions and side effects of drugs in physiologically normal humans and in appropriate models of disordered physiology which can be reversibly developed in normal humans. Experimental therapeutics must be seen to be the appropriate domain of the clinician, with research training dedicated to study the therapeutic efficacy of drugs in patients.

The primary interest of the clinical pharmacologist is to evaluate the possibilities of drugs, whereas the primary interest of the experimental therapist is the restoration of health or the alleviation of suffering in the patients being studied. Until and unless this demarcation becomes universally accepted and the clinical pharmacologist is prepared to accept this unarguably artificial limitation of his operational role as distinct from any consultative role he may fulfill, the future of the discipline will be stunted, being regarded as an intrusion by the full-time clinician whose interests may not always be uninvolved with private practice or professional elitism.

The present unsatisfactory situation can be redeemed by redefining the brief of the clinical pharmacologist. Any such definition must ensure that an antagonistic reaction from the practitioners of clinical medicine with an interest in experimental therapeutics will not be evoked.

Clinical pharmacology should regard as its proper domain the evaluation of drug action in normal subjects and in suitably modified physiological models. If by this circumscription of the sphere of activity of the clinical pharmacologist the impression is given that the speciality is to be a very restricted one, then the impression given is erroneous and must be corrected by fuller explanation.

A number of factors will benefitially influence the growth in scope of clinical pharmacology:

1. The ultimate disappearance of the medically qualified teacher from preclinical subjects including pharmacology will accentuate the importance of clinical pharmacology as the ideal transitional subject for introducing the student into the clinical area.
2. The postgraduate physician desirous of pursuing a career within which he may involve himself in experimental therapeutics will have to be fully trained in clinical pharmacology as a sine qua non.
3. The accepted and inevitable limitations of animal and bacterial toxicological screening offer to clinical pharmacology the urgent need to enlarge its role in human toxicology, which is already so much a part of Phase I and Phase II drug evaluation. The amended regulations in the United Kingdom (for example) which allow earlier testing of drugs in humans will call for greater

vigilance of tolerance, acceptability, compatibility, and safety while preliminary clinical pharmacological data are being sought.

4. The emergence of new chemical entities closely related to endogenous chemicals especially peptides will diminish reliance on animal studies. This will be especially apparent as greater numbers of compounds with biological-response-modulating activity become available for evaluation.
5. The spreading controversy engendered by the antivivisectionists is already stimulating greater efforts to find methods for testing drugs other than in animals. Clinical pharmacology will be challenged to provide some of the demanded substitute methodology.
6. The economic necessity for the pharmaceutical industry to be provided with expeditious, expert early evaluation of new compounds in humans now has enhanced the importance of the clinical pharmacologist enormously. Competent assessments can assist management greatly in reaching decisions to conserve diminishing resources.
7. The promised debureaucratization of the regulatory process, with greater reliance on third-party scientific audits, will increase the demand for personnel appropriately trained in the discipline of clinical pharmacology. Third-party evaluation and authentication of the therapeutic potential and safety of new drugs may promote the much-desired lagtime reduction caused by the present regulatory process. Scientific auditing of this kind conducted by suitably qualified and accredited professionals, with all the essential safeguards enshrined in a suitable code of ethics and enforceable discipline, is foreseeable. Such an amendment of the present process utilizing all the available techniques of data handling would make the cavalcades of document-laden vans anachronistic and redundant.

Despite the problems, the disappointments, the frustrations, the financial setbacks, and the unforeseeable mishaps, progress toward that ultimate goal of a cure for every ill will continue unabated. In the creation of the final edition of the pharmacopoeia, clinical pharmacology will have played a critical role in every sense.

REFERENCES

1. L. Lasagna, *Clin. Pharm. Ther. 31*(3):285-289 (1982).
2. M. Weatheral, *Nature 296*:387-390 (1982).
3. J. Cavalia, *Br. Med. J. 1*:1486 (1978).
4. Alfred, Lord Tennyson (1809-1892), excerpt from *Idylls of the King*.
5. Alvin Toffler, *Third Wave*. Collins, Cleveland, 1980.
6. Rudyard Kipling (1865-1936), excerpt from "If."

INDEX

D

J

K

L

M

W